de Gruyter Expositions in Mathematics 7

Editors

O.H. Kegel, Albert-Ludwigs-Universität, Freiburg
V.P. Maslov, Academy of Sciences, Moscow
W.D. Neumann, Ohio State University, Columbus
R.O. Wells, Jr., Rice University, Houston

Infinite Dimensional Lie Superalgebras

by

Yuri A. Bahturin
Alexander A. Mikhalev
Viktor M. Petrogradsky
Mikhail V. Zaicev

Walter de Gruyter · Berlin · New York 1992

Authors

Yu. A. Bahturin, A.A. Mikhalev,
  M.V. Zaicev
Department of Algebra
Faculty of Mathematics and Mechanics
Moscow State University
119899 Moscow, Russia

V.M. Petrogradsky
Department of Mathematics
Branch of Moscow State University in
  Ulianovsk
432700 Ulianovsk, Russia

1991 Mathematics Subject Classification: Primary: 15-02; 16-02; 17-02
Secondary: 15A66, 15A75; 16R10, 16S30, 16W30; 17A70, 17B37, 17B65, 17B70, 17B81; 20E06; 68Q40; 81R10, 81R50

⊚ Printed on acid-free paper which falls within the guidelines of the ANSI to ensure permanence and durability.

Library of Congress Cataloging-in-Publication Data

> Infinite dimensional Lie superalgebras / by Yuri A. Bahturin ...
> [et al.].
>   p.   cm. − (De Gruyter expositions in mathemat-
> ics, ISSN 0938-6572 ; 7)
>   Includes bibliographical references and indexes.
>   ISBN 3-11-012974-4 (acid-free)
>   1. Lie algebras. I. Bakhturin, IU. A. II. Series.
> QA252.3.I54   1992
> 512'.55−dc20   92-29650
>   CIP

Die Deutsche Bibliothek − Cataloging-in-Publication Data

> **Infinite dimensional Lie superalgebras** / by Yuri A. Bahturin
>   ... − Berlin ; New York : de Gruyter, 1992
>   (De Gruyter expositions in mathematics ; 7)
>   ISBN 3-11-012974-4
> NE: Bachturin, Jurij A.; GT

© 1992 by Walter de Gruyter & Co., D-1000 Berlin 30.
All rights reserved, including those of translation into foreign languages. No part of this book may be reproduced or transmitted in any form or by any means, electronic or mechanical, including photocopy, recording, or any information storage and retrieval system, without permission in writing from the publisher. Printed in Germany.
Typesetting: Asco Trade Typesetting Ltd., Hong Kong. Printing: Ratzlow Druck, Berlin.
Binding: Dieter Mikolai, Berlin. Cover design: Thomas Bonnie, Hamburg.

Table of Contents

Preface	vii
List of Symbols	ix

Chapter 1
Basic facts about Lie superalgebras

§0. Some background	1
§1. Graded algebras	4
§2. Identical relations of graded algebras	22
Exercises	35
Comments to Chapter 1	37

Chapter 2
The structure of free Lie superalgebras

§1. The free colour Lie superalgebra, s-regular words and monomials	39
§2. Bases of free colour Lie superalgebras	44
§3. The freeness of subalgebras and its corollaries	53
§4. Bases and subalgebras of free colour Lie p-superalgebras	69
§5. The lattice of finitely generated subalgebras	75
§6. Free colour Lie super-rings	78
Comments to Chapter 2	80

Chapter 3
Composition techniques in the theory of Lie superalgebras

§1. The Diamond Lemma for associative rings	81
§2. Universal enveloping algebras	84
§3. The Composition Lemma	95
§4. Free products with amalgamated subalgebra	105
Comments to Chapter 3	108

Chapter 4
Identities in enveloping algebras

§1. Main results	111
§2. Delta-sets	123
§3. Identities in enveloping algebras of nilpotent Lie superalgebras	129
§4. The case of characteristic zero	136
Comments to Chapter 4	144

Chapter 5
Irreducible representations of Lie superalgebras

§1. The Jacobson radical of universal enveloping algebras	147
§2. Dimensions of irreducible representations	152
§3. More on restricted enveloping algebras	160
§4. Examples	171
Comments to Chapter 5	173

Chapter 6
Finiteness conditions for colour Lie superalgebras with identities

§1. Various types of finiteness conditions. Examples	175
§2. Maximal condition and Hopf property	180
§3. Sufficient conditions for residual finiteness	201
§4. Representability of Lie superalgebras by matrices	210
Comments to Chapter 6	236

Bibliography	237
Author Index	247
Subject Index	249

Preface

In this book we consider some questions of the theory of Lie superalgebras in the spirit of infinite-dimensional Lie algebras. The need for such a book is motivated by the large number of results on this topic scattered through many journals and, of course, by the importance of Lie superalgebras which proved to be useful for quite a few areas of mathematics and physics. There have appeared several books and surveys on Lie superalgebras, but, reflecting the initial stage of the development, they are devoted to special topics of the theory without making an attempt of exposing general features.

In the first chapters of the book we are concerned with general questions of the theory. After introducing the main definitions in Chapter 1 we consider the notion of identity in a (colour) Lie superalgebra, the concept of a variety, and prove some results on varieties of Lie superalgebras. In Chapters 2 and 3 we consider free Lie superalgebras. Chapter 2 is devoted mainly to the question of finding linear bases in free Lie superalgebras i.e., essentially, to the question of the canonical form for the elements of a Lie superalgebra in terms of its generators. The bases obtained are similar to those well-known in the case of ordinary Lie algebras, but they have interesting features of their own. A part of Chapter 2 is devoted to the question of the freeness of subalgebras in free Lie superalgebras. The proof of the corresponding result, which is an analogue of well-known theorems due to A.I. Shirshov and E. Witt, gives an effective procedure for finding free generators of homogeneous subalgebras in Lie superalgebras in a number of important cases. In Chapter 3 we discuss Lie superalgebras given in terms of generators and defining relations. A number of general results and techniques developed are applied to various constructions such as universal enveloping algebras, free products of algebras with amalgamation and some others. The same approach enables us to solve some algorithmic problems in this field.

Chapters 4 and 5 deal with a topic of a more specific nature which is nevertheless of considerable importance, namely the theory of universal enveloping algebras for Lie superalgebras. These are graded associative algebras from which Lie superalgebras can be obtained by introducing the operation of the supercommutator. The importance of universal enveloping algebras lies in the fact that any graded representation of a Lie superalgebra is a representation of this associative algebra; and the converse is also true. Thus, the methods of associative rings used and developed here are of great importance for the representation theory of Lie superalgebras and give a non-standard approach to results in the area previously developed by other methods. In Chapter 4 we study the question of the structure of the universal enveloping algebra. In particular, we consider conditions under which such an algebra satisfies a non-trivial polynomial identity. In fact, we consider these questions in the setting of restricted enveloping algebras, i.e., here the restricted Lie superalgebras are involved. Some of the results proved here provide solutions to some well-known problems. In Chapter 5 we give an application of the results about identities to the question of the boundedness of dimensions of irreducible representations for a Lie superalgebra over fields of any characteristic. Of independent interest are the results on the Jacobson radical as well as those on von Neumann regularity of enveloping algebras.

In the concluding Chapter 6 we consider the questions of residual finiteness, of representability by matrices, and of some other finiteness conditions for Lie superalgebras. Here we use some techniques developed earlier by us; also we suggest some new approaches which are necessary for the solution of the problems arising here.

The book is organized as follows. Each chapter is further divided into sections and these latter into subsections. The three digit numbering k.l.m. means Subsection m in Section §1 of Chapter k. We write §k.l for Section §l in Chapter k. We omit k. or k.l. if we are within Chapter k or Section §k.l.

We do not make any historical or bibliographical remarks in this introduction; at the end of each chapter there is a section of comments on these matters. The bibliography does not strive to embrace all of the literature on Lie superalgebras; though it would be desirable to have a complete list of publications of such an important area of mathematics.

This book owes its existence to the creative atmosphere at the Department of Algebra at Moscow University, a small unrivalled algebraic universe. We would like to thank our teachers and colleagues there for stimulating our research and shaping our view of mathematics. The first author gratefully acknowledges support by DAAD.

Moscow, May 1992

Yu.A. Bahturin
A.A. Mikhalev
V.M. Petrogradsky
M.V. Zaicev

List of Symbols

$[A]$	2	$A(X)$	39
$\operatorname{End}_K M$	2	$A(X)_g$	39
ad	2	$F(X)$	39
$\mathscr{L}^n(M, N)$	2	$F(X)_g$	39
$[S, T]$	3	$L(X)$	39
$L^{(n)}$	3	$\tilde{u}$	40
L^n	3	$[u]$	39
$U(L)$	3, 85	Ψ_n	43
$\mathscr{G}(V)$	6	$\mu(n)$	43
$\operatorname{gr} A$	7	$\Psi(\alpha)$	43
$\mathscr{G}(R)$	8	$\pi: L(X) \to [X]$	47
$\mathfrak{gl}(n, m)$	10	$W(\alpha_1, \ldots, \alpha_k)$	48
$\mathfrak{sl}(n, m)$	11	$SW(\alpha_1, \ldots, \alpha_{t+s})$	49
$\prod_{\lambda \in I} R^\lambda$	12	$l_X(w)$	49, 70
$L \wedge M$	13	$l_x(w)$	49, 70
ε	14	ad'	53
$[A]_\varepsilon$	15	$Z(x)$	54
$x^{[p]}$	18	$W(z)$	54, 71
$\mathfrak{B}$	23	$\bar{Z}$	54
$F(X, \mathfrak{B})$	23	$\bar{W}$	54
var $\mathscr{R}$	24	$a \triangleright b$	54
$\mathfrak{L}(K, G, \varepsilon)$	24	$\omega: S \to L(X)$	57
$\mathfrak{B}(G, \varepsilon)$	25	$PS[X]$	69
S_γ	26	$\Psi: PS[X] \to S(X)$	69
$\operatorname{Sym}(m)$	26	$PZ(x)$	71
$M(d)$	26	$\overline{PZ}(x)$	71
$M(X)$	27	$u(L)$	87
$A \operatorname{wr} L$	28	$Q \hat{\otimes} R$	88
$H(L)$	29	U^n	89
$H(t_1, \ldots, t_m, u_1, \ldots, u_n)$	29	$\operatorname{gr} U(L)$	89
$H(L, t)$	29	$\mathscr{P}(H)$	92
$\mathfrak{A}^2$	34	$\varphi: X \to \mathbb{N}$	96
$\Gamma(X)$	39	$\hat{a}$	96
$\Gamma(X)_g$	39	$\prod_{\alpha \in T} H \circ H_\alpha$	105
$S(X)$	39		
$S(X)_g$	39	Γ_n	112

$[a,b]_\theta$	121	$D_\alpha^m(L)$	157
$f^\theta(X_1,\ldots,X_n,Y_1,\ldots,Y_n)$	122	$D_\alpha(L)$	157
$\Gamma(\alpha,\beta)$	123	$D(L)$	157
$\delta_{\alpha\beta}^m(L)$	124	$l_A(T)$	164
$\Delta_\alpha^m(L)$	124	$r_A(T)$	164
$\Delta_\alpha(L)$	124	$l(L)$	164
$\Delta(L)$	124	$r(L)$	164
$\mathrm{Rad}(V)$	130	$\mathrm{Ass}(B)$	211
$\mathrm{Ker}_l\phi$	132	$\mathfrak{B}_1\mathfrak{B}_2$	233
$\mathrm{Ker}_r\phi$	132	$\mathfrak{N}_k$	233
$\Gamma_n(1,1)$	149	$\mathfrak{A}$	233

Chapter 1

Basic facts about Lie superalgebras

§ 0. Some background

In this chapter we introduce basic notions of the theory of Lie superalgebras. Throughout the book K denotes the ground (or base) ring which is commutative with unity 1. An arbitrary ring R which is a left K-module will be called an *algebra over K* (or a *K-algebra*) provided that for any $\lambda \in K$, $x, y \in R$ the following relations hold:

$$(\lambda x)y = \lambda(xy) = x(\lambda y). \tag{1}$$

In a natural way one can introduce the notions of subalgebra, ideal, quotient-algebra and homomorphism of algebras (see details in, e.g., [Jacobson 1962]). If $\tilde{K}$ is a commutative ring containing K as a subring with the same unity then $\tilde{R} = \tilde{K} \otimes_K R$ is an algebra over $\tilde{K}$; we say that $\tilde{R}$ results from R by extending the ground ring K. An *associative algebra* A is an algebra satisfying the additional axiom $(xy)z = x(yz)$, $x, y, z \in A$. A *Lie algebra* L is an algebra with the identities of *anticommutativity* and *Jacobi*

$$x^2 = 0 \tag{2}$$

$$(xy)z + (yz)x + (zx)y = 0. \tag{3}$$

It is conventional to denote the operation in a Lie algebra by brackets (*commutator brackets*); then the preceding axioms take the form

$$[x, x] = 0 \tag{4}$$

$$[[x, y], z] + [[y, z], x] + [[z, x], y] = 0. \tag{5}$$

A Lie algebra L is called *abelian* if $[x, y] = 0$ for all $x, y \in L$. More generally, if we have $[a, b] = 0$ for some a, b in a Lie algebra L, we say that a and b commute. If, in a Lie algebra L, any two commutators commute, i.e.

$$[[x, y], [z, t]] = 0$$

for any $x, y, z, t \in L$, we say that L is a *metabelian* Lie algebra. If there exists a natural number c such that any $(c+1)$-fold commutator of elements in a Lie algebra L equals zero then L is called a *nilpotent* Lie algebra and if c is the minimal natural number with this property then it is called the *nilpotent index* of L. It is appropriate to introduce the notion of a *left-normed n-fold commutator* $[x_1,\ldots,x_n]$ by setting $[x_1] = x$ for $n = 1$ and defining $[x_1,\ldots,x_n]$ by setting $[x_1,\ldots,x_n] = [[x_1,\ldots,x_{n-1}],x_n]$ for $n > 1$. It is not difficult to verify (see [Bahturin 1987a]) that a nilpotent Lie algebra L of nilpotent index c is a Lie algebra in which one has $[x_1,\ldots,x_{c+1}] = 0$ identically, while $[x_1^0,\ldots,x_c^0] \neq 0$ for a suitable choice of $x_1^0, \ldots, x_c^0 \in L$. Similarly one defines a *right-normed n-fold commutator*.

A *module* M over a Lie algebra L is a left K-module in which for any $x, y \in L$ and $m \in M$ their product is defined to satisfy the following law:

$$[x,y]m = x(ym) - y(xm). \tag{6}$$

The notion of an L-module is equivalent to that of a representation of a Lie algebra. Namely, a *representation* of a K-Lie-algebra L by endomorphisms of a K-module M is a mapping ρ of K-modules $\rho: L \to \operatorname{End}_K M$ such that

$$\rho([x,y]) = [\rho(x),\rho(y)] = \rho(x)\rho(y) - \rho(y)\rho(x) \tag{7}$$

for any $x, y \in L$. Setting $xm = \rho(x)(m)$ we make the module M of the representation ρ into an L-module; reading the same equation from the right to the left we define a representation ρ of a Lie algebra L in an L-module M. Equation (7) generalizes to the commutator in an arbitrary associative algebra A: by setting $[a,b] = ab - ba$ in an associative algebra A we make it into a Lie algebra, denoted by $[A]$. Using this remark one can define a representation of a Lie algebra L by endomorphisms of a K-module M as a Lie algebra homomorphism $\rho: L \to [\operatorname{End}_K M]$.

It should be mentioned that any Lie algebra becomes a module over itself if one sets $xm = [x,m]$ for $x, m \in L$. The corresponding representation is called *adjoint* and it is denoted by $\operatorname{ad}: L \to [\operatorname{End}_K L]$. The endomorphism $\operatorname{ad} x$ is a *derivation* in the sense that for any $y, z \in L$ one has

$$(\operatorname{ad} x)[y,z] = [(\operatorname{ad} x)y,z] + [y,(\operatorname{ad} x)z].$$

If M and N are two L-modules then the module $\mathscr{L}^n(M,N)$ of n-linear mappings on M with values in N becomes an L-module if one defines the mapping xf with $x \in L$ and $f: M \times \cdots \times M \to N$ by setting

$$(xf)(m_1,m_2,\ldots,m_n) = x(f(m_1,m_2,\ldots,m_n)) - f(xm_1,m_2,\ldots,m_n)$$
$$- f(m_1,xm_2,\ldots,m_n) - \cdots - f(m_1,m_2,\ldots,xm_n). \tag{8}$$

An element m of an L-module M is called *invariant* if $xm = 0$ for any $x \in L$ (to shorten the notation we write $Lm = 0$). So, an invariant linear mapping $f: M \to N$ is a mapping satisfying $x(f(m)) = f(xm)$ for any $x \in L$, $m \in M$, i.e. a *homomorphism of L-modules*. An *invariant bilinear mapping* $f: M \times M \to N$ is a mapping with

$$x(f(m_1, m_2)) = f(xm_1, m_2) + f(m, xm_2)$$

for all $x \in L$, $m_1, m_2 \in M$.

Given two subsets S and T in a Lie algebra L we define their *mutual commutator* as the smallest K-submodule containing all commutators $[s, t]$, $s \in S, t \in T$. In particular, if S and T are ideals in L then also $[S, T]$ is an ideal of L. Using this notation, we define L^n as L if $n = 1$ and as $[L^{n-1}, L]$ if $n > 1$. We also set $L^{(0)} = L$ and

$$L^{(n)} = [L^{(n-1)}, L^{(n-1)}]$$

for $n \geq 1$. The series of ideals of the forms

$$L = L^1 \supset L^2 \supset \cdots \supset L^n \supset \cdots \tag{9}$$

$$L = L^{(0)} \supset L^{(1)} \supset \cdots \supset L^{(n)} \supset \cdots \tag{10}$$

are called *lower central* and *derived series* of L, respectively. It is easy to verify that a Lie algebra L is *nilpotent of index c* if, and only if, $L^c \neq \{0\}$ and $L^{c+1} = \{0\}$, while L is *metabelian* if, and only if, $L^{(2)} = \{0\}$. We also say that L is *soluble* if $L^{(l)} = \{0\}$ for some $l \geq 0$. The least l with this property is the *(soluble) length* of L.

In the case where K is an algebraically closed field of characteristic zero, a finite-dimensional L-module M, where L is a soluble Lie algebra, is always *triangulable*, that is, M has a basis $\{e_1, \ldots, e_n\}$, $n = \dim L$, on which the action of any element $x \in L$ can be written in the form of a triangular matrix (Lie's Theorem). We also have Engel's Theorem: in any finite-dimensional L-module M over an arbitrary field K with nilpotent action of L, i.e. with $x^n M = \{0\}$ for any $x \in L$, with n depending on x, there is a nonzero invariant element. (There is a version of this theorem for algebras over an arbitrary ring K, see [Bahturin 1987a]).

An associative K-algebra $U(L)$ with unity is called the *universal enveloping algebra* for L if there exists a homomorphism of Lie algebra $\iota: L \to [U(L)]$ such that for any homomorphism of Lie algebras $\varphi: L \to [A]$ where A is an associative K-algebra with 1 there exists a unique homomorphism of associative algebras with unity $f: U(L) \to A$ such that the diagram of homomorphisms

$$L \xrightarrow{\iota} U(L)$$
$$\varphi \searrow \quad \downarrow f \qquad (11)$$
$$A$$

is commutative, i.e. $f\iota = \varphi$. It is well-known that $U(L)$ always exists; it is unique up to isomorphism and, in a number of important cases, for instance when L is a free K-module (in particular when K is a field) the mapping ι is injective (Poincaré-Birkhoff-Witt Theorem). Thus, in these cases, L is isomorphic to a Lie subalgebra $\iota(L)$ in $[U(L)]$. If $(e_\alpha)_{\alpha \in I}$ is a basis in L and the set I is totally ordered then a basis of $U(L)$ is formed by I and all products of the form

$$\iota(e_{\alpha_1})\iota(e_{\alpha_2})\ldots\iota(e_{\alpha_n}) \qquad (12)$$

where $\alpha_1 \leq \alpha_2 \leq \cdots \leq \alpha_n$. Often, L is identified with $\iota(L)$ and then 'basic monomials' take the form

$$e_{\alpha_1} e_{\alpha_2} \ldots e_{\alpha_n} (\alpha_1 \leq \alpha_2 \leq \cdots \leq \alpha_n). \qquad (13)$$

In all cases it is not difficult to verify that the elements (12) together with 1 generate $U(L)$ as a K-module. The main use of $U(L)$ is based on the fact that any L-module is a $U(L)$-module (and vice versa) which reduces the study of representations and modules over Lie algebras to the study of representations and modules over associative algebras.

§1. Graded algebras

1.1. Definition. Let G be an additively written commutative semigroup, K a commutative ring with unity 1, and let R be an algebra over K admitting a direct decomposition of the form

$$R = \bigoplus_{g \in G} R_g \qquad (1)$$

where each R_g is a K-submodule such that $R_g R_h \subset R_{g+h}$ for all $g, h \in G$. In this case we say that R is a *graded* (G-graded) K-*algebra*. If $K = \mathbb{Z}$ then R is called a *graded ring*. If K is a field then every K-submodule R_g is simply a vector subspace.

An element $s \in R$ is called *homogeneous* if there exists $g \in G$ such that $s \in R_g$. We say that g is the *degree* of the homogeneous element $s \in R_g$ and write $g = d(s)$. A homogeneous subset $S \subset R$ is a subset consisting of homogeneous elements. If $r = s_1 + \cdots + s_n$ where $s_1 \in R_{g_1}, \ldots, s_n \in R_{g_n}$, $g_i \neq g_j$ for $i \neq j$ then we say that s_i is a homogeneous component of r of degree g_i. If all homogeneous components of a K-submodule T are in T themselves then we say that T is a homogeneous (or graded) submodule. This is equivalent to T being representable in the form

$$T = \bigoplus_{g \in G} (T \cap R_g). \tag{2}$$

If this holds we write $T_g = T \cap R_g$. In addition, if T is a subalgebra (or an ideal) of R then we say that T is a graded (= homogeneous) subalgebra (ideal) of R. Such a subalgebra is a G-graded algebra itself. The quotient algebra R/T where T is a graded ideal can be naturally made into a G-graded K-algebra if one sets

$$(R/T)_g = R_g + T/T. \tag{3}$$

Now, by the Homomorphism Theorem one has $(R/T)_g \cong R_g/T_g$ (homomorphism of K-modules). Finally, we say that a homomorphism $\varphi: R \to R'$ of G-graded K-algebras is *graded* if $\varphi(R_g) \subset R'_g$ for any $g \in G$. The class of all G-graded K-algebras as the class of objects and the class of graded homomorphisms as the class of morphisms form a category $\mathfrak{R}_G(K)$. Working within this category, we will often be speaking about algebras, subalgebras, ideals, homomorphisms, etc. bearing in mind graded algebras, graded subalgebras, graded ideals, graded homomorphisms, etc. respectively.

1.2. Examples. (i) Any K-algebra A may be considered as G-graded, G with zero, if one sets $A_0 = A$, $A_g = \{0\}$ for $g \neq 0$.
(ii) The ring $R = K[X_1, \ldots, X_n]$ of polynomials is $\mathbb{N}$-graded where $\mathbb{N}$ is the semigroup of natural numbers (with 0). Here R_k is the K-submodule of R generated by monomials $X_1^{m_1} \ldots X_n^{m_n}$ of degree k, say $m_1 + \cdots + m_n = k$. This same ring can be made $\mathbb{N}^n$-graded if we fix $k_1, \ldots, k_n$ and say that $X_1^{m_1} \ldots X_n^{m_n}$ is of degree $(m_1 k_1, \ldots, m_n k_n) \in \mathbb{N}^n$.
(iii) Any (finite-dimensional) semisimple complex Lie algebra $\mathfrak{g}$ (i.e. with $K = \mathbb{C}$) of rank r is $\mathbb{Z}^r$-graded since it possesses a Cartan decomposition

$$\mathfrak{g} = \mathfrak{g}^0 \oplus \sum_{\alpha \in R} \mathfrak{g}^\alpha$$

where R is the root system of $\mathfrak{g}$ with respect to its Cartan subalgebra $\mathfrak{g}^0$ which is a subset in the free abelian group $\mathbb{Z}^r$ generated by simple roots. It is well-known that $\mathfrak{g}^\alpha \mathfrak{g}^\beta \subset \mathfrak{g}^{\alpha+\beta}$ for any $\alpha, \beta \in R$.

(iv) We consider

$$V = \bigoplus_{g \in G} V_g, \qquad (4)$$

a G-graded vector space over a field K, i.e. simply a vector space with direct decomposition (4). The algebra $E = \operatorname{End}_K V$ of all linear operators on V becomes G-graded as soon as its homogeneous component E_g is defined as the set of all operators of index g, i.e. such ones that

$$\mathscr{A}(V_h) \subset V_{g+h} \qquad (5)$$

for each $\mathscr{A} \in E_g$ and any $h \in G$. It is obvious that

$$E = \bigoplus_{g \in G} E_g \text{ and also } E_g E_h \subset E_{g+h}.$$

(v) The *Grassmann algebra* $\mathscr{G}(V)$ of a vector space V over a field K has a $\mathbb{Z}$-grading

$$\mathscr{G}(V) = \bigoplus_{k=0}^{\infty} \mathscr{G}_k \qquad (6)$$

where $\mathscr{G}_k$ is the linear span of k-vectors of the form

$$v_1 \wedge \cdots \wedge v_k \qquad (v_i \in V) \qquad (7)$$

and also a $\mathbb{Z}_2$-grading

$$\mathscr{G}(V) = \mathscr{G}_{\bar{0}} \oplus \mathscr{G}_{\bar{1}} \qquad (8)$$

where $\mathscr{G}_{\bar{0}}$ is the linear span of k-vectors of the form (7) with k even and $\mathscr{G}_{\bar{1}}$ is the linear span of k-vectors with k odd.

(vi) The Grassmann algebra with the grading (8) may be considered endowed with the operation

$$[x, y] = xy - (-1)^{\alpha \beta} yx \qquad (9)$$

defined on homogeneous elements $x \in \mathscr{G}_{\bar{\alpha}}$ and $y \in \mathscr{G}_{\bar{\beta}}$. It is elementary to verify that the commutator of any two elements in $\mathscr{G}(V)$ introduced by this formula is equal to zero. In this sense $\mathscr{G}(V)$ gives us an example of a commutative (abelian) Lie superalgebra. We will return to Lie superalgebras, which form the core of our exposition, in Subsection 5.

1.3. Filtered algebras. Let A be an algebra over the commutative ring K and G an ordered commutative semigroup with ordering $\leq$ such that, for any $g \in G$ and $h \leq k$ in G we have $g + h \leq g + k$ and also $g + h = g + k$ implies $h = k$. A system of K-modules $\{A^g\}_{g \in G}$ is called an *ascending G-filtration* in A if $g \leq h$ implies $A^g \subset A^h$ and $A^g A^h \subset A^{g+h}$. In this case one can define a G-graded algebra gr A called the *associated graded algebra* defined in the following way. Put

$$\tilde{A}^g = \sum_{h < g} A^h.$$

Then it is obvious that $\tilde{A}^g \tilde{A}^h \subset \tilde{A}^{g+h}$ (use that G is an ordered semigroup with cancellation). In this case there is a well-defined multiplication of classes $A^g/\tilde{A}^g \times A^h/\tilde{A}^h \to A^{g+h}/\tilde{A}^{g+h}$ which makes the direct sum

$$\text{gr } A = \bigoplus_{g \in G} A^g/\tilde{A}^g$$

into a graded algebra over K with grading semigroup G. We write $(\text{gr } A)_g = A^g/\tilde{A}^g$.

Similarly, one can define the notion of a *descending filtration* and its associated graded algebra.

Examples. (i) Let A be a commutative K-algebra generated by some x and y. For any $k, l \in \mathbb{N}$ we define $A^{(k,l)}$ as the K-submodule generated by all $x^s y^r$ with $s \leq k, r \leq l$. Set $G = \mathbb{N} \times \mathbb{N}$ and define a product ordering on G. Then $\{A^{(k,l)}\}_{(k,l) \in G}$ is an ascending filtration on A.
(ii) Let A be an associative or Lie algebra over K, $A^1 = A$, $A^k = \sum_{i+j=k} A^i A^j$, $k = 2, 3, \ldots$. Then $\{A^k\}_{k \in \mathbb{N}}$ is a descending $\mathbb{N}$-filtration where the semigroup $\mathbb{N}$ of natural numbers is ordered in the natural way.

The passage from a filtered algebra to the associated graded algebras leads to a simplified structure, with a number of important properties preserved, and this will be used repeatedly in what follows. In these considerations an important role is played by the following result.

Let $A = \bigoplus_{g \in G} A_g$ be a graded algebra, G a well-ordered semigroup. We set $A^g = \bigoplus_{h \leq g} A_h$. Then $\{A^g\}_{g \in G}$ is an ascending filtration on A. The associated graded algebra gr A is canonically isomorphic to A. If $0 \neq a \in A$ then gr a can be identified with a_g where $a = a_g + \sum_{h < g} a_h$ and $a_g \neq 0$. We often call a_g the *leading term* of a. An assertion needed by us now takes the following form.

Lemma. *Let K be a field, B a subspace in A, gr B the linear span of the set of leading terms of elements in B. If $\dim A/B = m < \infty$ then also $\dim A/\text{gr } B = m$.*

Proof. If $A = B$ there is nothing to prove. If $A \neq B$ then there is an index g such that $(\operatorname{gr} B)_g \neq A_g$. For assume the contrary. Suppose $a \in A$ is a homogeneous element of minimal degree h such that $a \notin B$. Since $(\operatorname{gr} B)_h = A_h$ there exists $b \in B$ such that $a - b \in \bigoplus_{g < h} A_g$. By our hypothesis we have $A_g \subset B$ for all $g < h$. Then $a - b \in B$, i.e. $a \in B$, a contradiction. Now let C_g denote a subspace in A_g such that $A_g = C_g + (\operatorname{gr} B)_g$. If $C = \bigoplus_{g \in G} C_g$ then $A = C + \operatorname{gr} B$. Arguing as above we get $A = B + C$. Now we want to prove that $B \cap C = \{0\}$. Let c be a nonzero element of the least possible degree g which is in $B \cap C$. Then $\operatorname{gr} c \in (\operatorname{gr} B)_g \cap (\operatorname{gr} C)_g = (\operatorname{gr} B)_g \cap C = \{0\}$, a contradiction. Hence $A = B \oplus C$. Now it is obvious that $\dim A/B = \dim A/\operatorname{gr} B = \dim C$, and the proof is complete. $\square$

1.4. Superalgebras. Let G be $\mathbb{Z}_2$, K an arbitrary associative and commutative ring with 1. A G-graded algebra over K will be called a *superalgebra over* K. A homogeneous element of degree $\bar{0}$ will be called *even* while that of degree $\bar{1}$ *odd*. In what follows K is a field. Let $\mathscr{G}(V) = \mathscr{G}_0 \oplus \mathscr{G}_1$ denote the Grassmann superalgebra of an infinite-dimensional vector space V over K. If $R = R_0 + R_1$ is a superalgebra then an algebra of the form $\mathscr{G}(R) = R_0 \otimes \mathscr{G}_0 + R_1 \otimes \mathscr{G}_1$ is called the *Grassmann envelope* of the superalgebra R. $\mathscr{G}(R)$ will be viewed as on ordinary (ungraded) algebra. Now suppose $\mathfrak{K}$ is a class of algebras over K. We define $\mathfrak{K}_S$ as a class of superalgebras such that a superalgebra R is in $\mathfrak{K}_S$ if, and only if, $\mathscr{G}(R)$ is in $\mathfrak{K}$. If we call algebras in $\mathfrak{K}$ $\mathfrak{K}$-algebras then it is natural that the algebras in $\mathfrak{K}_S$ are termed $\mathfrak{K}$-superalgebras. For example, if $\mathfrak{K}$ is the class of associative, or Lie, or Jordan algebras, etc. then $\mathfrak{K}_S$ is the class of superassociative, or Lie superalgebras, or Jordan superalgebras etc., respectively.

It is worth mentioning, however, that a superalgebra R is superassociative if, and only if, it is associative. Indeed, if R is associative then $R \otimes_K \mathscr{G}(V)$ is associative since $\mathscr{G}(V)$ is an associative algebra. Hence the Grassmann envelope $\mathscr{G}(R)$ is also associative. Conversely, if, for instance, $a \in R_1$, $b \in R_0$, $c \in R_1$ and e_1, e_2 are distinct elements in a basis of V then $a \otimes e_1$, $b \otimes 1$, $c \otimes e_2$ are some elements in $\mathscr{G}(R)$. Since $\mathscr{G}(R)$ is associative we have

$$a(bc) \otimes (e_1 \wedge e_2) = (a \otimes e_1)((b \otimes 1)(c \otimes e_2)) = ((a \otimes e_1)(b \otimes 1))(c \otimes e_2)$$

$$= (ab)c \otimes (e_1 \wedge e_2).$$

Using properties of the basis of a tensor product one gets $a(bc) = (ab)c$. The verification of the associative law in the other cases of a, b, c lying in various components of R can be dealt with similarly.

The pattern is different when we consider $\mathfrak{K}$ as the class of Lie algebras. For example, it follows from the anticommutativity $[x, y] = -[y, x]$ in the Grassmann envelope $\mathscr{G}(R)$ of a superalgebra $R = R_0 \oplus R_1$ that for any

$a, b \in R_1$ and e_1, e_2 in the basis of V the following holds:

$$[a,b] \otimes (e_1 \wedge e_2) = [a \otimes e_1, b \otimes e_2] = -[b \otimes e_2, a \otimes e_1]$$

$$= -[b,a] \otimes (e_2 \wedge e_1) = [b,a](e_1 \wedge e_2).$$

Thus, for $a, b \in R_1$ we have

$$[a,b] = [b,a]. \tag{10}$$

Now if $a \in R_0$ and $b \in R_1$, or $a \in R_1$, $b \in R_0$, or $a, b \in R_0$ then

$$[a,b] = -[b,a].$$

It is easy to observe that the validity of (9) and (10) for the homogeneous elements of R guarantees the anticommutativity of the Grassmann envelope $\mathscr{G}(R)$. Now starting with the validity of the Jacobi identity in $\mathscr{G}(R)$ we easily arrive at the validity of the Jacobi identity

$$[[a,b],c] + [[b,c],a] + [[c,a],b] = 0 \tag{11}$$

in all cases where either all a, b and c are even or odd or one of them odd. But if $a \in R_0$, $b, c \in R_1$ then choosing $e_1, e_2 \in V$ yields $0 = [[a,b],c] \otimes (e_1 \wedge e_2) + [[b,c],a] \otimes (e_1 \wedge e_2) + [[c,a],b] \otimes (e_2 \wedge e_1)$. Obviously, this implies

$$[[a,b],c] + [[b,c],a] - [[c,a],b] = 0. \tag{12}$$

It is easy to verify that, conversely, the validity of (9)–(12) in R (with appropriate restrictions on homogeneous $a, b, c \in R$) guarantees that $\mathscr{G}(R)$ is a Lie algebra.

1.5. Lie superalgebras. Abstracting from the procedure of the last subsection we give the definition of a Lie superalgebra over a commutative and associative ring K with unity. It is usually assumed that 2 and 3 are invertible in K. (However, see Subsection 1.10 for a discussion of the remaining cases.) Now $L = L_0 \oplus L_1$ is a *Lie superalgebra* if for any $x \in L_\alpha$, $y \in L_\beta$ and $z \in L$ the following identities hold:

$$[x,y] = -(-1)^{\alpha\beta}[y,x] \tag{13}$$

$$[[x,y],z] = [x,[y,z]] - (-1)^{\alpha\beta}[y,[x,z]]. \tag{14}$$

It is easy to verify that (13) and (14) coincide with (9)–(12).

The *even subspace*, i.e. the set of all even elements of a Lie superalgebra $L = L_0 \oplus L_1$ is a Lie algebra. Since $[L_0, L_1] \subset L_1$ and by (14), which with $\alpha = 0$, $\beta = 1$ and $z \in L_1$ takes the form

$$[[x, y], z] = [x, [y, z]] - [y, [x, z]],$$

we observe that the commutator of L makes L_1 into an L_0-module. Furthermore, the restriction of the commutator to L_1 defines a bilinear symmetric mapping $\Phi: L_1 \times L_1 \to L_0$. Since L_0 is the adjoint L_0-module one may speak about the action of L_0 on the bilinear mappings from L_1 into L_0 (see formula (8) in the preceding section). We observe by (12) that the action of L_0 on the bilinear mapping Φ given by the commutator in L_1 is trivial, that is, Φ is an L_0-invariant. Summing up all the remarks above, one may conclude that a superalgebra $L = L_0 \oplus L_1$ is a Lie superalgebra provided that
(1) L_0 is a Lie algebra;
(2) L_1 is an L_0-module;
(3) the bilinear mapping $[\,,\,]: L_1 \times L_1 \to L_0$ is symmetric and L_0-invariant;
(4) $[x, y] = -[y, x]$ for $x \in L_0$, $y \in L_1$;
(5) for $y_1, y_2, y_3 \in L_1$ one has $J(y_1, y_2, y_3) = [[y_1, y_2], y_3] + [[y_2, y_3], y_1] + [[y_3, y_1], y_2] = 0$.

1.6. Examples. (i) Any Lie algebra L is a Lie superalgebra with $L_0 = L$, $L_1 = \{0\}$.
(ii) If $A = A_0 \oplus A_1$ is an associative superalgebra (= simply a $\mathbb{Z}_2$-graded associative algebra) then, introducing a (super) commutator on A by the formula

$$[x, y] = xy - (-1)^{\alpha\beta} yx \qquad (15)$$

with $x \in A_\alpha$, $y \in A_\beta$, one turns A into a Lie superalgebra usually denoted by $[A]$.
(iii) A specific example of the construction of (ii) is supplied by the matrix superalgebra $L = \mathfrak{gl}(n, m)$ the elements of which are all the square matrices of order $n + m$. The even elements of $\mathfrak{gl}(n, m)$ are all matrices with zeros beyond two square diagonal blocks of order n upstairs and m downstairs while the odd elements are the matrices with zeros in the blocks just mentioned:

$$L_0 = \begin{pmatrix} * & 0 \\ \hline 0 & * \end{pmatrix} \begin{matrix} n \\ m \end{matrix}, \quad L_1 = \begin{pmatrix} 0 & * \\ \hline * & 0 \end{pmatrix} \begin{matrix} n \\ m \end{matrix}. \qquad (16)$$

§1. Graded algebras

Any matrix in $\mathfrak{gl}(n,m)$ can be written in the form

$$X = \left(\begin{array}{c|c} A & B \\ \hline C & D \end{array}\right) \qquad (17)$$

where A and D are square matrices of order n and m, respectively, while B and C are rectangular matrices of order $n \times m$ and $m \times n$ respectively. The *supertrace* str X is defined by the formula

$$\operatorname{str} X = \operatorname{tr} A - \operatorname{tr} D \qquad (18)$$

where tr is the usual trace function. The set $\mathfrak{sl}(n,m)$ of matrices with zero supertrace forms a subalgebra called the special Lie superalgebra.

(iv) In the case where K is a field of characteristic zero the special Lie superalgebra $\mathfrak{sl}(n,m)$ is an example of a simple Lie superalgebra, i.e. it has no nonzero proper graded ideals. The classification of the simple finite-dimensional Lie superalgebras over an algebraically closed field of characteristic zero has been obtained by V. G. Kac (see [Kac 1977a] and also [Scheunert 1979]).

(v) If $\mathfrak{g}$ is a Lie algebra, V a $\mathfrak{g}$-module, then setting $L_0 = \mathfrak{g}$, $L_1 = V$ gives rise to a Lie superalgebra $L = L_0 \oplus L_1$ if we define zero multiplication in L_1 and if the multiplication of the other elements is defined by that in $\mathfrak{g}$ and by the action of $\mathfrak{g}$ on V.

(vi) An abelian Lie superalgebra is one with zero commutator. A *soluble Lie superalgebra* is a superalgebra L with a series of ideals

$$L = H_0 \supset H_1 \supset \cdots \supset H_{l-1} \supset H_l = \{0\}$$

such that for any $i = 0, 1, \ldots, l-1$, the quotient algebra H_i/H_{i+1} is abelian. Of course, one can define the *lower central* and the *derived series* in the same way as in the preceding section ((9) and (10)); then a Lie superalgebra L is called *nilpotent* or *soluble* if we have $L^{c+1} = \{0\}$ or $L^{(l)} = \{0\}$ for suitable c and l, respectively.

(vii) In the case of finite-dimensional Lie superalgebras over a field of characteristic zero the following holds. (For the proof see [Scheunert 1979].)

Theorem. *A finite-dimensional Lie superalgebra $L = L_0 \oplus L_1$ over a field K of characteristic zero is soluble if, and only if, L_0 is a soluble Lie algebra.*

(viii) If $R = R_0 \oplus R_1$ is a superalgebra then a linear mapping $\delta \colon R \to R$ is called a *superderivation* of degree $\alpha = 0, 1$, if, for any $y \in R_\beta$ and $z \in R$, $\beta = 0, 1$, one has

$$\delta(yz) = \delta(y)z + (-1)^{\alpha\beta} y\delta(z). \qquad (19)$$

It is obvious that the linear span D_α of superderivations of degree α consists of superderivations of the same kind and if $\delta_1 \in D_\alpha$, $\delta_2 \in D_\beta$ then the supercommutator (15) $[\delta_1, \delta_2]$ is an element in $D_{\alpha+\beta}$. Indeed, if $y \in R_\gamma$ then

$$[\delta_1, \delta_2](yz) = (\delta_1\delta_2 - (-1)^{\alpha\beta}\delta_2\delta_1)(yz)$$

$$= \delta_1(\delta_2(y)z + (-1)^{\beta\gamma}y\delta_2(z))$$

$$- (-1)^{\alpha\beta}\delta_2(\delta_1(y)z + (-1)^{\alpha\gamma}y\delta_1(z))$$

$$= \delta_1\delta_2(y)z + (-1)^{\alpha(\beta+\gamma)}\delta_2(y)\delta_1(z) + (-1)^{\beta\gamma}\delta_1(y)\delta_2(z)$$

$$+ (-1)^{\beta\gamma}(-1)^{\alpha\gamma}y(\delta_1\delta_2)z - (-1)^{\alpha\beta}\delta_2\delta_1(y)z$$

$$- (-1)^{\alpha\beta}(-1)^{\beta(\alpha+\gamma)}\delta_1(y)\delta_2(z) - (-1)^{\alpha\beta}(-1)^{\alpha\gamma}\delta_2(y)\delta_1(z)$$

$$- (-1)^{\alpha\beta}(-1)^{\alpha\gamma}(-1)^{\beta\gamma}y\delta_2\delta_1(z)$$

$$= (\delta_1\delta_2 - (-1)^{\alpha\beta}\delta_2\delta_1)(y)z + (-1)^{(\alpha+\beta)\gamma}y(\delta_1\delta_2 - (-1)^{\alpha\beta}\delta_2\delta_1)(z)$$

$$= [\delta_1,\delta_2](y)z + (-1)^{(\alpha+\beta)\gamma}y[\delta_1,\delta_2](z).$$

The underlined terms cancel. Hence $D = D_0 \oplus D_1$ is a Lie superalgebra called the superalgebra of (super) derivations of R and denoted by Der R.

1.7. Semidirect product. Given a family $(R^\lambda)_{\lambda \in I}$ of superalgebras over K it is natural to define their *Cartesian product* $R = \prod_{\lambda \in I} R^\lambda$ as the set of all functions $f: I \to \bigcup_{\lambda \in I} R^\lambda$ such that $f(\lambda) \in R^\lambda$. A function f is even if $f(\lambda) \in R_0^\lambda$ for all $\lambda \in I$ and odd if $f(\lambda) \in R_1^\lambda$ for all $\lambda \in I$. The subset of functions which take almost all (up to a finite number) values zero forms the *direct product* of the family which coincides with the Cartesian product in the case of finite index set I. If $I = \{1,2\}$ then we arrive at the direct product $R = R^1 \times R^2$ where R^1 and R^2 are ideals.

A more general construction, the *semidirect product*, can be defined in the case where $R^1 = L$ and $R^2 = M$ are Lie superalgebras. In this case we consider the derivation algebra $D = \text{Der } M$ and an arbitrary homomorphism $\varphi: L \to D$ (it can be zero, then we get back to the direct product); we write $\varphi(a) = a^*$, $a^*(b) = a^*b$ for $a \in L$, $b \in M$. Then the direct sum $P = L \oplus M$ of K-modules becomes a Lie superalgebra if one sets $P_0 = L_0 \oplus M_0$, $P_1 = L_1 \oplus M_1$ and defines the commutator in P by taking the commutator of elements in L or M using the operations of these superalgebras while for $a \in L$, $b \in M$ we set $[a, b] = a^*b$. As a result we get a Lie superalgebra called the *semidirect product* of L by M with respect to φ; it is denoted by $P =$

$L \wedge_\varphi M$, or simply by $P = L \wedge M$ if there is no confusion concerning the action of L on M. In this superalgebra, L is a subalgebra and M an ideal. For instance, a Lie superalgebra $L = L_0 \oplus L_1$ with zero multiplication in L_1 is a semidirect product of $L = L_0 \oplus \{0\}$ and $M = \{0\} \oplus L_1$.

A number of results on finite-dimensional Lie algebras (e.g. Engel's Theorem, the Poincaré-Birkhoff-Witt Theorem and some others) hold valid or can be easily adapted to Lie superalgebras, but many others persist. For instance Lie's Theorem is no more true in this new setting. To be specific we consider the Grassmann algebra $\mathcal{G}(V)$ of an n-dimensional vector space $V = \langle x_1, \ldots, x_n \rangle$ over a field K of characteristic zero. Let the operators ξ_i and η_j, $i, j = 1, \ldots, n$, be defined as superderivations $\partial/\partial x_i$ of degree one and multiplication by x_j on the left which is also a linear transformation of degree one on V. We have

$$[\xi_i, \eta_j] = \xi_i \eta_j + \eta_j \xi_i$$

and for any x_k

$$[\xi_i, \eta_j](x_k) = \xi_i \eta_j x_k + \eta_j \xi_i x_k = \xi_i x_j x_k + \delta_{ik} \eta_j 1_V$$

$$= \delta_{ij} x_k - \delta_{ik} x_j + \delta_{ik} x_j = \delta_{ij} x_k = \delta_{ij} 1_V(x_k),$$

i.e. $[\xi_i, \eta_j] = \delta_{ij} 1_V$ and $L = \langle 1_V \rangle_0 \oplus \langle \xi_1, \eta_1, \ldots, \xi_n, \eta_n \rangle_1$ is a class 2 nilpotent Lie superalgebra of linear transformations in V. It is easy to verify that V is an irreducible L-module. Since its dimension is 2^n, Lie's Theorem fails. If we take the semidirect product $P = L \wedge V$ then we get an example of a soluble Lie superalgebra of soluble length 3 whose derived algebra is not nilpotent (since $1_V \in P^2$ and its action on V is not nilpotent), which is true for soluble algebras over an arbitrary field of characteristic zero. Bearing this fact in mind, one might say that Lie superalgebras in characteristic zero have much resemblance to modular Lie algebras, and this view will be further supported in what follows.

1.8. Colour Lie superalgebras. The ordinary Lie superalgebras considered above arise from the natural desire to unify algebra and analysis with both commuting and anticommuting variables [Berezin 1983]. If we wish to have more than two different features combined in the same structure then we are on the way leading to colour Lie superalgebras. There are a number of papers (see Comments) where the authors show that the axioms below are not artificial.

Consider a commutative ring K with unity 1 and with 2, 3 invertible, i.e. 2, $3 \in K^*$, the multiplicative group of invertible elements in K. Let G be an abelian additive group which will serve as the grading group of certain graded algebras of the form $L = \bigoplus_{g \in G} L_g$. To construct an algebra with some properties which are close to Lie algebras and Lie superalgebras we consider a *bilinear alternating form* $\varepsilon: G \times G \to K^*$, i.e. a function of two arguments in G taking values in K^* and satisfying the following properties:

$$\varepsilon(g+h, k) = \varepsilon(g, k)\varepsilon(h, k), \qquad \varepsilon(g, h+k) = \varepsilon(g, h)\varepsilon(g, k) \qquad (20)$$

for $g, h, k \in G$, and

$$\varepsilon(g, h) = \varepsilon(h, g)^{-1}. \qquad (21)$$

If K is a ring without zero divisors then we always have $\varepsilon(g, g) = \pm 1$ which induces a decomposition

$$G = G_+ \cup G_- \text{ where } G_\pm = \{g \in G | \varepsilon(g, g) = \pm 1\}.$$

(It is obvious that G_+ is a subgroup in G of index ≤ 2.) Now a G-graded algebra $L = \bigoplus_{g \in G} L_g$ is called an (ε-) *colour Lie superalgebras* if for any $a \in L, b \in L$ and $c \in L$ we have

$$[a, b] = -\varepsilon(g, h)[b, a], \qquad (22)$$

$$[[a, b], c] = [a, [b, c]] - \varepsilon(g, h)[b, [a, c]]. \qquad (23)$$

If $G = \{0\}$ then we get an ordinary Lie algebra; if $G = \mathbb{Z}_2$, $\varepsilon(0, 0) = \varepsilon(0, 1) = \varepsilon(1, 0) = 1$, $\varepsilon(1, 1) = -1$ then L becomes an ordinary Lie superalgebra. Natural examples of Lie superalgebras arise, e.g., in physics. For instance, if $G = \mathbb{Z}_2 \oplus \mathbb{Z}_2$, $K = \mathbb{C}$ then the following forms $\varepsilon_1, \varepsilon_2$ have been considered in [Rittenberg-Wyler 1978a]. If $f = (f_1, f_2)$, $g = (g_1, g_2) \in \mathbb{Z}_2 \oplus \mathbb{Z}_2$ then one sets

$$\varepsilon_1(f, g) = (-1)^{(f_1+f_2)(g_1+g_2)}$$

$$\varepsilon_2(f, g) = e^{2\pi i(1/2(f_1 g_2 - f_2 g_1) + (f_1 g_1 + f_2 g_2))}.$$

In the case of ε_1 we have $G_+ = \{(0, 0), (1, 1)\}$, $G_- = \{(0, 1), (1, 0)\}$ while in the case of ε_2 it follows that $G = G_+$, $G_- = \emptyset$.

It should be stressed that properties (20) and (21) are indeed natural. For instance, if we wish to have $[x, y] = -\varepsilon(g, h)[y, x]$ for $x \in L_g$ and $y \in L_h$ then, obviously, $[x, y] = \varepsilon(g, h)\varepsilon(h, g)[x, y]$ and, in a "non-degenerate" case, we are doomed to have $\varepsilon(g, h)\varepsilon(h, g) = 1$, i.e. (21). Now if we wish to have an identity expressing the property that the commutator $[a, b]$ acts on c as a generalization of the supercommutator (see formula (15) in Subsection 1.4) then an easy verification of the above kind shows that having (23) is inevitable.

A general way of obtaining colour Lie superalgebras is to introduce a "colour" commutator on a G-graded algebra $A = \bigoplus_{g \in G} A_g$ by setting

$$[x, y] = xy - \varepsilon(x, y)yx \qquad (24)$$

where we write $\varepsilon(x, y)$ in place of $\varepsilon(g, h)$ if x is homogeneous of degree g and y is homogeneous of degree h.

The following lemma shows why it is useful to have $2, 3 \in K^*$.

Lemma. *Let L be a colour Lie superalgebra, x, y, z homogeneous in L, $d(x) \in G_+$, $d(y) \in G_-$, $2, 3 \in K^*$. Then*

$$[x, x] = 0, \qquad [[y, y], y] = 0, \qquad [[y, y], z] = 2[y, [y, z]].$$

Proof. (i) Since $[x, x] = -\varepsilon(x, x)[x, x] = -[x, x]$ we have $2[x, x] = 0$ and by the hypothesis $[x, x] = 0$.
(ii) If $y \in L_h$ then

$$[[y, y], y] = [y, [y, y]] - \varepsilon(h, h)[y, [y, y]] = 2[y, [y, y]]$$

$$= -2\varepsilon(h, 2h)[[y, y], y] = -2\varepsilon(h, h)^2[[y, y], y]$$

$$= -2[[y, y], y]$$

hence $3[[y, y], y] = 0$ and by the hypothesis $[[y, y], y] = 0$.
(iii) As previously, with $y \in L_h$, $z \in L$ we easily derive

$$[[y, y], z] = [y, [y, z]] - \varepsilon(h, h)[y, [y, z]] = 2[y, [y, z]].$$

The proof is now complete. □

1.9. Universal enveloping algebra. Given a colour Lie superalgebra $L = \bigoplus_{g \in G} L_g$ one easily adapts the notion of the universal enveloping algebra of an ordinary Lie algebra to obtain the notion of the *universal enveloping algebra* in this general setting. Namely, a G-graded associative algebra $U = U(L)$ is such an algebra if there exists a homomorphism ι of colour Lie superalgebras $\iota: L \to [U]_\varepsilon$, i.e. such that

$$\iota([x, y]) = \iota(x)\iota(y) - \varepsilon(x, y)\iota(y)\iota(x)$$

and such that for any homomorphism $\mu: L \to [A]_\varepsilon$ where $A = \bigoplus_{g \in G} A_g$ is an associative G-graded algebra there exists a unique (graded) homomorphism of associative algebras $f\colon U \to A$ which makes the diagram

commutative, that is, $f \circ \iota = \mu$. Now the uniqueness of $U = U(L)$ is a trivial matter. To prove its existence it is useful to consider the tensor algebra $T = T(L)$ of the K-module L. To obtain $U(L)$ it is sufficient to consider the ideal I generated by all tensors of the form $u \otimes v - \varepsilon(g,h)(v \otimes u)$ where $u \in L_g, v \in L_h, g, h \in G$. Any graded linear mapping $\mu\colon L \to A$ into an associative algebra obviously extends to a unique graded homomorphism φ of $T(L)$ into A. But if $\mu\colon L \to [A]_\varepsilon$ is a homomorphism then $I \subset \operatorname{Ker} \varphi$. Thus φ factors through a homomorphism f of $T(L)/I$ so that this algebra can be taken for $U(L)$. Also, the natural embedding of L into $T(L)$ factors through a homomorphism $\iota\colon L \to [U(L)]$ and, obviously, the diagram (25) is commutative. It will be shown in Chapter 3 that under certain restrictions on K and on the K-module structure of L the mapping ι is injective. If $E = (e_\alpha)_{\alpha \in I}$ is a homogeneous basis of L_+, $F = (f_\beta)_{\beta \in I}$ a homogeneous basis of L_- then, under the same restrictions, e.g. when K is a field, a basis of $U(L)$ is formed by 1 and by all monomials of the form

$$e_{\alpha_1} \ldots e_{\alpha_k} f_{\beta_1} \ldots f_{\beta_l} \quad \text{with} \quad \alpha_1 \leq \cdots \leq \alpha_k, \beta_1 < \cdots < \beta_l. \tag{26}$$

An important particular case is that of L being abelian. Were L an ordinary Lie algebra then $U(L)$ would be the polynomial ring $K[E]$, E being a basis of L. In the case of abelian Lie superalgebras (with $G = \mathbb{Z}_2$), $L = L_0 \oplus L_1$ where if $L_0 = \{0\}$ we would arrive at the Grassmann algebra $U(L) = \Lambda(L_1)$. The richest structure will be attained in the case where L is an arbitrary colour Lie superalgebra $L = \bigoplus_{g \in G} L_g = L_+ \oplus L_-$ defined by a form $\varepsilon\colon G \times G \to K^*$. It is obvious from (26) that, as a vector space, $U(L) = K[E] \otimes \Lambda(L_-)$. If L is abelian, then, with $L_- = \{0\}$ we arrive at a generalization of the polynomial ring $K[E]$, the so called (ε-) *colour polynomial ring* $K^\varepsilon[L_+]$ which is an associative algebra generated by E subject to defining relations of the form $e_\alpha e_\beta - \varepsilon(g,h) e_\beta e_\alpha$ where $e_\alpha \in L_g$, $e_\beta \in L_h$ and $\varepsilon(g,g) = \varepsilon(h,h) = 1$. Its basis has the same form as in the case of the polynomial ring in E. Now if $L_+ = \{0\}$ then we get a generalization of the Grassmann algebra $\Lambda^\varepsilon(L_-)$ which is an associative algebra generated by F subject to defining relations of the form $f_\alpha f_\beta - \varepsilon(g,h) f_\beta f_\alpha$ but now $f_\alpha \in L_g$, $f_\beta \in L_h$ and $\varepsilon(g,g) = \varepsilon(h,h) = -1$. Its basis is the same as in the case of the Grassmann algebra of L. We will also denote $\Lambda^\varepsilon(L_-)$ by $\Lambda(F)$ where F is a homogeneous basis of L_-.

§1. Graded algebras

Now we give a construction of the *Grassmann envelope* $\mathscr{G}(L)$ of a colour Lie superalgebra $L = \bigoplus_{g \in G} L_g$ defined by a form $\varepsilon \colon G \times G \to K^*$. To this end we consider a G-graded space $V = \bigoplus_{g \in G} V_g$ as an abelian ε^{-1}-colour Lie superalgebra. Let $P = U(V)$ be its universal enveloping algebra. Consider $L \otimes P$ and, in this algebra, its zero component $\mathscr{G}(L) = \bigoplus_{g \in G} L_g \otimes P_{-g}$. We want to verify that $\mathscr{G}(L)$ is a Lie algebra. To shorten the notation we write $a \otimes \xi = a\xi$, $a \in L$, $\xi \in P$. Now if $a \in L_g$, $\xi \in P_{-g}$, $b \in L_h$, $\eta \in P_{-h}$ then

$$[a\xi, b\eta] = [a,b]\xi\eta = -\varepsilon(g,h)[b,a]\varepsilon^{-1}(g,h)\eta\xi$$

$$= -\varepsilon(g,h)\varepsilon(h,g)[b,a]\eta\xi = -[b\eta, a\xi].$$

If also $c \in L_k$, $\zeta \in P_{-k}$ then

$$[[a\xi, b\eta], c\zeta] = [[a,b],c]\xi\eta\zeta,$$

$$[a\xi, [b\eta, c\zeta]] = [a,[b,c]]\xi\eta\zeta,$$

$$[b\eta, [a\xi, c\zeta]] = [b,[a,c]]\eta\xi\zeta = \varepsilon^{-1}(h,g)[b,[a,c]]\xi\eta\zeta$$

$$= \varepsilon(g,h)[b,[a,c]]\xi\eta\zeta.$$

Since we have $[[a,b],c] = [a,[b,c]] - \varepsilon(g,h)[b,[a,c]]$ in L it follows that

$$[[a\xi, b\eta], c\zeta] = [a\xi, [b\eta, c\zeta]] - (b\eta, [a\xi, c\zeta]]$$

i.e. the Jacobi identity holds in $\mathscr{G}(L)$. Therefore $\mathscr{G}(L)$ is an ordinary Lie algebra.

This construction enables us to define ε-superalgebras for arbitrary classes of algebras. Again, as in the case of associative algebras, ε-associative algebras are simply G-graded algebras.

1.10. Lie superalgebras with positive characteristic. The definition of a colour Lie superalgebra over a field K of characteristic different from 2 and 3 has been given by us earlier. There are real difficulties showing that we cannot take the same definition in the remaining cases. One of them is that we cannot imbed a Lie superalgebra of characteristic 2 or 3 into an associative algebra if we do not demand $[x,x] = 0$ for even x if the characteristic is 2 and $[[y,y],y] = 0$ for odd y if the characteristic is 3. Thus, to obtain a correct definition of a colour Lie superalgebra in characteristic 2 we have to require, in addition to the usual axioms (20) and (21), the existence of a quadratic operator $q: L_g \to L_{2g}$ for $g \in G_-$ such that

$$q(\alpha a) = \alpha^2 q(a)$$

$$[a,b] = q(a+b) - q(a) - q(b)$$

$$[a,[a,b]] = [q(a),b]$$

where $\alpha \in K$, $a, b \in L_g$, $g \in G_-$. (If char $K \neq 2$ then one can set $q(a) = (1/2)[a,a]$.) In the case of char $K = 3$ one has to require additionally that $[[y,y],y] = 0$ for all homogeneous $y \in L_-$. (If L is a Lie superalgebra, $G = \mathbb{Z}_2$, char $K = 3$, generated by one odd y and $[[y,y],y] \neq 0$ then it is 3-dimensional: $L = \langle [y,y] \rangle_0 \oplus \langle y, [[y,y],y] \rangle_1$, but its image in $U(L)$ is 2-dimensional since, obviously, $[[y,y],y] = [2y^2, y] = 0$ in any associative algebra.)

An important feature of Lie algebras in characteristic $p > 0$ is the possibility of introducing an operation of raising to p-th powers in a number of important cases. In this way we get restricted Lie algebras or, which is the same, Lie p-algebras. Their definition extends to Lie superalgebras in the following way.

Let K be a commutative and associative ring with 1 of characteristic $p > 0$ (i.e. $\underbrace{1 + \cdots + 1}_{p} = 0$). A colour Lie superalgebra $L = \bigoplus_{g \in G} L_g$ is called a *restricted* or a *p-superalgebra* if, for any even homogeneous element $x \in L_g$, $g \in G_+$ there exists an element $x^{[p]}$ such that this (partial) operation of "raising to p-th power" satisfies the following conditions:

$$(\alpha x)^{[p]} = \alpha^p x^{[p]}$$

$$(\text{ad } x^{[p]})(z) = [x^{[p]}, z] = (\text{ad } x)^p (z)$$

$$(x+y)^p = x^p + y^p + \sum_i s_i(x,y)$$

where y is homogeneous of the same degree as x, $z \in L$, is $s_i(x,y)$ is the coefficient of t^{i-1} in the polynomial $(\text{ad}(tx+y))^{p-1}(x)$. (If $G = \{0\}$ then the above is the standard definition of a Lie p-algebra.)

The motivation of this definition is contained in the following.

Proposition. *If K is a ring with 1, char $K = p > 0$, $p \neq 2, 3$, A a G-graded associative algebra over K, $[A]_\varepsilon$ a colour Lie superalgebra associated with A, $x^{[p]} = x^p$ for homogeneous $x \in A_+$ then $[A]_\varepsilon$ with $x \to x^{[p]}$ is a colour Lie superalgebra (denoted by $[A]_\varepsilon^p$).*

§1. Graded algebras

Proof. The first property is obvious. If z is homogeneous in A then $(\operatorname{ad} x)z = xz - \varepsilon(x,z)zx = (l_x - \varepsilon(x,y)r_x)(z)$ where $l_x(z) = xz$, $r_x(z) = zx$. Now

$$(\operatorname{ad} x)^2(z) = (\operatorname{ad} x)([x,z]) = (l_x - \varepsilon(x,[x,z])r_x)([x,z])$$

$$= (l_x - \varepsilon(x,z)r_x)([x,z])$$

since $\varepsilon(x,[x,z]) = \varepsilon(x,x)\varepsilon(x,z) = \varepsilon(x,z)$ by the choice of $x \in A_+$. Finally,

$$(\operatorname{ad} x)^p(z) = (l_x - \varepsilon(x,z))^p(z) = (l_x^p - \varepsilon(x,z)^p r_x^p)(z)$$

$$= (l_{x^p} - \varepsilon(x^p,z)r_{x^p})(z) = \varepsilon(\operatorname{ad} x^{[p]})(z).$$

We have made use of char $K = p > 0$. Since any $z \in L$ is the sum of its homogeneous components, the second axiom has been verified.

To verify the third axiom we consider the following identity in the polynomial ring $A[t]$:

$$(tx + y)^p = t^p x^p + y^p + \sum_{i=1}^{p-1} t^i s_i(x,y).$$

After formally differentiating both sides by t we obtain the following:

$$\sum_{i=0}^{p-1} (tx + y)^i x (tx + y)^{p-1-i} = \sum_{i=1}^{p-1} i t^{i-1} s_i(x,y).$$

Now let us consider $(\operatorname{ad}(tx + y))^{p-1}(x)$. Since x and y are of the same degree and $\varepsilon(x,x) = 1$ we can write

$$\operatorname{ad}(tx + y)(x) = (l_{tx+y} - \varepsilon(x,x)r_{tx+y})(x) = (l_{tx+y} - r_{tx+y})(x),$$

$$(\operatorname{ad}(tx + y))^2(x) = \operatorname{ad}(tx + y)([tx + y, x])$$

$$= (l_{tx+y} - \varepsilon(x,x^2)r_{tx+y})[tx + y, x]$$

$$= (l_{tx+y} - r_{tx+y})[tx + y, x]$$

since $\varepsilon(x,x^2) = \varepsilon(x,x)^2 = 1$ and, finally,

$$\operatorname{ad}(tx+y)^{p-1}(x) = (l_{tx+y} - r_{tx+y})^{p-1}(x)$$

$$= \sum_{i=0}^{p-1} (-1)^i \binom{p-1}{i} l_{tx+y}^i r_{tx+y}^{p-1-i}(x)$$

$$= \sum_{i=0}^{p-1} l_{tx+y}^i r_{tx+y}^{p-1-i}(x). \tag{27}$$

Here we have used the equation $\binom{p-1}{i} \equiv (-1)^i \pmod{p}$ which can be easily verified by induction on i. Now the third equation follows by comparison of equations (26) and (27). □

1.11. Schur's Lemma and Burnside's Theorem. We conclude this section with several remarks related to graded finite-dimensional irreducible representations of colour Lie superalgebras which, in fact, are representations of their universal enveloping algebras.

Suppose we are given an irreducible representation $\rho: L \to [\text{End } V]$, $L = \bigoplus_{g \in G} L_g$, $V = \bigoplus_{g \in G} V_g$. We want first to describe the structure of the *centralizer* C of ρ, i.e. the subalgebra of $\text{End}_K V$ such that $\phi a = a\phi$ for all $\phi \in C$ and $a \in A$ where A is the associative subalgebra generated by $\phi(L)$. Immediate observation considering $\text{Ker } \phi$ and $\text{Im } \phi$ for ϕ homogeneous shows that any such ϕ is either trivial or it is an automorphism of V. Moreover, is $\phi, \psi \in C_g$ for the same $g \in G$ then $\phi\psi^{-1} \in C_0$ and we can say that if $C_g \neq 0$ then $C_g = C_0 \phi$. It is also obvious that the set of all g with $C_g \neq 0$ is a subgroup of G, say H. Thus, C, as a vector space over C_0, which is a skew field, can be identified with the group algebra $C_0[H]$. Examples will follow to show that, even in the case of an algebraically closed field, C need not be commutative although the product is almost the same as in the group algebra since, for $\phi_g \in C_g$, $\phi_h \in C_h$ we have $\phi_g \phi_h = \lambda \phi_{g+h}$, $\lambda \in C_0$. Now if we call a G-graded algebra $S = \bigoplus_{g \in G} S_g$ a *graded (skew) field*, provided that $0 \neq 1$ and every homogeneous element of S invertible, then we have shown

Schur's Lemma. *The centralizer of a graded irreducible module is a graded field whose underlying space is a group algebra of a subgroup of the grading group with coefficients in the zero component which is a usual field. If the dimension of the module over the ground field is finite and the field is algebraically closed then the coefficients of the group algebra can be taken from the ground field.* □

To get some examples of graded fields we consider two subsets S' and S'' in the set of real matrices of order two. S' consists of all matrices of the form $\begin{pmatrix} a & b \\ -b & a \end{pmatrix}$ while S'' contains those of the form $\begin{pmatrix} a & b \\ b & a \end{pmatrix}$. We give S' and S'' a $\mathbb{Z}_2$-grading by calling $\begin{pmatrix} a & 0 \\ 0 & a \end{pmatrix}$ even in both cases and $\begin{pmatrix} 0 & b \\ -b & 0 \end{pmatrix}, \begin{pmatrix} 0 & b \\ b & 0 \end{pmatrix}$ odd in S', S'', respectively. Clearly, each homogeneous element is invertible. Therefore S' and S'' are graded fields. But $S' \cong \mathbb{C}$ is a field, while every degenerate matrix in S'' provides an example of a non-invertible element.

There is no problem to develop some linear algebra over graded fields. So, a left S-module V, where S, V are as above, is called a (left) vector space over S if it is graded, i.e. $S_g V_h \subset V_{g+h}$ for all $g, h \in G$, and unital, i.e. $1v = v$ for all $v \in V$.

The usual concepts of linear dependence, elementary transformations of matrices, basis and dimension work in this setting, but one must be careful in the sense that all this should be done for homogeneous elements only. This is important since it may well happen that $\lambda v = 0$ for $0 \neq \lambda \in S$ and $0 \neq v \in V$. For instance, we have

$$\begin{pmatrix} 1 & -1 \\ -1 & 1 \end{pmatrix} \begin{pmatrix} 1 \\ 1 \end{pmatrix} = \begin{pmatrix} 0 \\ 0 \end{pmatrix}$$

with $\begin{pmatrix} 1 & -1 \\ -1 & 1 \end{pmatrix} \in S''$, $\begin{pmatrix} 1 \\ 1 \end{pmatrix} \in \mathbb{R}^2$ where $\mathbb{R}^2$ is a graded S''-vector space. A basis in this case consists of any non-zero homogeneous vector, say $e_1 = \begin{pmatrix} 1 \\ 0 \end{pmatrix}$ or $e_2 = \begin{pmatrix} 0 \\ 1 \end{pmatrix}$.

Having in mind this understanding of linear dependence and dimension, one has

Burnside's Theorem. *Suppose that V is finite-dimensional over the centralizer C of the action of a graded associative subalgebra $\mathscr{A} \subset \operatorname{End} V$. Then $\mathscr{A}$ is the centralizer of the action of C on V.*

There is no real need to repeat the well-known proofs of this result (see, e.g., [Jacobson 1956]). We only have to make sure that every argument should be applied to homogeneous elements.

If the base field is algebraically closed we cannot deduce that an irreducible algebra of linear operators should coincide with $\operatorname{End} V$ as in the non-graded case. Now the conclusion is that $\mathscr{A}$ is the set of operators commuting with those in H if $C = K[H]$ where H is the subgroup in Schur's Lemma.

We now give some examples of irreducible algebras. Let $E = E_n$ be the identity matrix of order n. Then the set $\mathscr{C}$ of all matrices of the form $\begin{pmatrix} E_n & E_n \\ E_n & E_n \end{pmatrix}$ is a graded field isomorphic to S''. The set of matrices centralizing $\mathscr{C}$ is the set of all matrices $\begin{pmatrix} A & B \\ C & D \end{pmatrix}$ such that $\begin{pmatrix} A & B \\ C & D \end{pmatrix} \begin{pmatrix} 0 & E \\ E & 0 \end{pmatrix} = \begin{pmatrix} 0 & E \\ E & 0 \end{pmatrix} \begin{pmatrix} A & B \\ C & D \end{pmatrix}$. Performing the calculations we find that $\mathscr{A}$ consists of all matrices of the form $\begin{pmatrix} A & B \\ B & A \end{pmatrix}$ where A, B are arbitrary matrices of order n.

More complicated examples arise in the case $G = \mathbb{Z}_2 \oplus \mathbb{Z}_2$. We restrict ourselves to the case of a 4-dimensional space V labelled by the elements 0, a, b, $a + b$ of G. We give two examples of G-graded fields and indicate the form of an irreducible algebra whose centralizer is the second field. So, any of the two sets of matrices forms a G-graded subfield in the algebra of all 4 by 4 square matrices:

$$\begin{bmatrix} \alpha & -\beta & -\gamma & -\delta \\ \beta & \alpha & -\delta & \gamma \\ \gamma & \delta & \alpha & -\beta \\ \delta & -\gamma & \beta & \alpha \end{bmatrix}, \begin{bmatrix} \alpha & \beta & \gamma & \delta \\ \beta & \alpha & \delta & \gamma \\ \gamma & \delta & \alpha & \beta \\ \delta & \gamma & \beta & \alpha \end{bmatrix}, \quad \alpha, \beta, \gamma, \delta \in K.$$

If $K = \mathbb{R}$, the first is, obviously the quaternion field $\mathbb{H}$. If $K = \mathbb{C}$, then this is not a field, but still a graded field and it is not commutative. The second is commutative but has zero divisors (it contains S'' as above). The algebra centralized by this takes the form of all matrices which commute both with $\begin{pmatrix} I & 0 \\ 0 & I \end{pmatrix}$ and $\begin{pmatrix} 0 & I \\ I & 0 \end{pmatrix}$ where $I = \begin{pmatrix} 0 & 1 \\ 1 & 0 \end{pmatrix}$. Immediate calculations show that $\mathscr{C}$ is self-centralizing. "Blowing-up" the pattern from the 4-dimensional to the $4n$-dimensional case we arrive at an irreducible algebra of the form

$$\begin{bmatrix} A & A' & B & B' \\ A' & A & B' & B \\ B & B' & A & A' \\ B' & B & A' & A \end{bmatrix}$$

where A, A', B, B' are arbitrary square matrices of order n.

In conclusion we mention that if all components V_g of V have pairwise distinct dimensions then we still must have $\mathscr{A} = \operatorname{End} V$ as in the usual Burnside Theorem.

§ 2. Identical relations of graded algebras

2.1. Definitions. Let G be an additive semigroup, K an associative and commutative ring with 1, $X = \bigcup_{g \in G} X_g$ a G-graded set, $F(X)$ the free nonassociative algebra with free generating set X and with coefficients in K. If $f = f(x_1, \ldots, x_n)$ is an element in $F(X)$ then an expression of the form $f(x_1, x_2, \ldots, x_n) = 0$ is called an *identical relation* (or an *identity*). Suppose also that $R = \bigoplus_{g \in G} R_g$ is a graded algebra over K. We say that $f(x_1, x_2, \ldots, x_n) = 0$ is satisfied in R if $f(a_1, a_2, \ldots, a_n) = 0$ for any choice of $a_1, a_2, \ldots, a_n \in R$ such that all a_i are G-homogeneous and $d(a_i) = d(x_i)$, $i = 1, 2, \ldots, n$.

If $f_1(x_1, \ldots, x_n) = 0$ and $f_2(x_1, \ldots, x_m) = 0$ are two identities then we say that the latter identity is a *consequence* of the former one if $f_2(x_1, \ldots, x_m) = 0$ is satisfied in any G-graded algebra satisfying $f_1(x_1, \ldots, x_n) = 0$. If also the former identity is a consequence of the latter then the two identities are said to be *equivalent*.

The class of all G-graded K-algebras satisfying a fixed system of identical relations is called a *variety*. Any variety of ordinary non-graded algebras is a variety of G-graded algebras for any semigroup G with 0. In this case the identities defining such a variety should be written in the elements of X_0 and one has to set $x_g \equiv 0$ for all $g \neq 0$.

The set of all elements $f = f(x_1, \ldots, x_n) \in F(X)$ such that $f(x_1, \ldots, x_n) = 0$ is an identity of a fixed variety $\mathfrak{V}$ forms a homogeneous ideal V of the algebra $F(X)$. The ideal V is closed relative to substitutions of any homogeneous $h_1, \ldots, h_n$ of $F(X)$ in place of $x_1, \ldots, x_n$ in such a way that $d(h_i) = d(x_i)$, $i = 1, \ldots, n$. Any ideal with this property is called *verbal*. The quotient algebra $F(X, \mathfrak{V}) = F(X)/V \in \mathfrak{V}$ is called the free algebra of $\mathfrak{V}$ with free generating set X. The cardinality of X is called the *rank* of $F(X, \mathfrak{V})$. In all cases to follow in this book the rank of a free algebra is an invariant and X is identified with its image in $F(X, \mathfrak{V})$. The following result is obvious.

Proposition. *Let* $R = \bigoplus_{g \in G} R_g$ *be an arbitrary algebra of the variety* $\mathfrak{V}$ *of G-graded K-algebras and $F(X, \mathfrak{V})$ the free algebra of* $\mathfrak{V}$ *with free generating set X. Then any graded mapping* $\varphi \colon X \to R$ *(i.e. with $\varphi(X_g) \subset R_g$) uniquely extends to a homomorphism* $\overline{\varphi} \colon F(X, \mathfrak{V}) \to R$ *such that* $\overline{\varphi}|_X = \varphi$. □

Similarly to the non-graded case due to G. Birkhoff one can prove the following result (see [Bahturin 1987a]).

2.2. Theorem. *Let K be an associative and commutative ring with 1, G an additive and commutative semigroup. A non-empty class of G-graded K-algebras is a variety if, and only if, it is closed under forming subalgebras, quotient algebras and Cartesian products.* □

The construction in the proof of Birkhoff's Theorem enables us to construct the free algebra of a variety from "below" starting with the generating algebras of a variety. To be precise, we say that a family $\mathscr{R} = (R^\alpha)$ of G-graded algebras generates a variety $\mathfrak{V}$ if $\mathfrak{V}$ is the least variety containing all algebras in $\mathscr{R}$. In this case we usually write $\mathfrak{V} = \mathrm{var}\,\mathscr{R}$. It turns out in the proof of Birkhoff's Theorem that any $F(X, \mathfrak{V})$ is isomorphic to a subcartesian product of algebras in $\mathscr{R}$, i.e. it is isomorphic to a subalgebra in the Cartesian product such that its projections to the components are surjective.

2.3. Types of identities. A polynomial $f(x_1, \ldots, x_n) \in F(X)$ is called *normal* if it is equal to the sum of monomials each of which depends on a fixed set of variables. It is obvious that any system of identities is equivalent to a system of normal identities, i.e. such identities with normal polynomials as left hand sides. (This follows easily by substituting zeros in place of some variables.)

An identity is called *homogeneous* if it is the sum of monomials which have the same degree in X. A well-known procedure in which a fixed x is replaced by λx, $\lambda \in K$, with the use of Vandermonde's determinant enables us to claim that any system of identities over an infinite field K is equivalent to a system of *multihomogeneous* identities, i.e. such that each monomial in the expression of any of these identities contains each x to the same degree, depending only on the identity under consideration.

Finally, we say that an identity is *multilinear* if it is multihomogeneous and its degree with respect to any variable occurring is equal to 1. Any identity has a multilinear consequence which can be obtained by using the linearization (or the polarization) procedure replacing $f(x_1, x_2, \ldots, x_n)$, where x_1 occurs to a degree greater than one, by

$$f(y_1 + z_1, x_2, \ldots, x_n) - f(y_1, x_2, \ldots, x_n) - f(z_1, x_2, \ldots, x_n).$$

If K is of characteristic zero then an inverse procedure is possible, i.e. each identity is equivalent to a system of multilinear identities.

2.4. Varieties of colour Lie superalgebras. If K, G and ε are fixed then the identities (22) and (23) of §1 define a variety called the variety $\mathfrak{L} = \mathfrak{L}(K, G, \varepsilon)$ of colour Lie superalgebras. Its free algebra is denoted simply by $L(X)$ and is called the free colour Lie superalgebra with free generating set X. (We also write $L(X) = [X]$ which is a general notation for a Lie superalgebra generated by a set X.) In what follows we are dealing with the study of identities and varieties of colour Lie superalgebras. In doing so it will be assumed that K, G and ε are fixed, that all varieties to be considered are in $\mathfrak{L} = \mathfrak{L}(K, G, \varepsilon)$ and that the left hand side of an arbitrary identity $f(x_1, \ldots, x_n)$ is an element of an appropriate free colour Lie superalgebra $L(X)$. Accordingly, the ideal V of identities of a variety $\mathfrak{V}$ will be considered as an ideal in $L(X)$. The Lie superalgebra $L(X)$, in addition to the G-grading, possesses also a grading relative to the length of monomials with respect to X. This grading is induced by the corresponding grading of the free algebra $F(X)$. Thus, the notions of the normal, homogeneous, multihomogeneous and multilinear identities transfer without any confusion to this new understanding. For instance, if K is an infinite field then any ideal of identities V can be written in any of the forms

$$V = \bigoplus_{\substack{g \in G \\ n \in \mathbb{N}}} V_{g,n}, \quad V = \bigoplus_{\substack{g \in G \\ \alpha}} V_{g,\alpha}. \tag{1}$$

In the former decomposition n stands for the total degree (= length) of elements in $V_{g,n}$ with respect to X; in the latter one α is the multidegree, i.e. the function $\alpha: X \to \mathbb{N}$ such that $w \in V_{g,\alpha}$ if, and only if, $d(w) = g$ and the length of any monomial in w with respect to any $x \in X$ is equal to $\alpha(x)$.

In the case where K is a field of characteristic zero any identity of length n is equivalent to a multilinear identity of length n. If $\gamma: G \to \mathbb{N}$ is a function which is zero almost everywhere then, as usual, we may set $|\gamma| = \sum_{g \in G} \gamma(g)$. We denote by P_γ the multilinear component of $L(X)$, $X = \bigcup_{g \in G} X_g$ of length $\gamma(g)$ in the variables X_g with these variables fixed (it is assumed that each X_g is countable). By P_n we denote the sum

$$P_n = \sum_{|\gamma|=n} P_\gamma. \tag{2}$$

This is exactly the space (for suitable n) where we find the left hand sides of identities when we consider superalgebras over a field of characteristic zero. In Subsection 2.8 we will see that the dimension of each of the spaces P_γ, $|\gamma| = n$, equals $(n-1)!$ as in the case of ordinary Lie algebras (see [Bahturin 1987a, Subsection 4.8.1]).

2.5. Examples. a) In many cases we will be using usual non-graded identities to define varieties of colour Lie superalgebras. Of course, any such identity is equivalent to a number of graded identities. In this way we obtain such varieties as the variety $\mathfrak{A}$ of abelian superalgebras given by $[x, y] \equiv 0$, $\mathfrak{N}_c$, $c = 1, 2, \ldots$, of class at most c nilpotent superalgebras given by $[x_1, x_2, \ldots, x_{c+1}] \equiv 0$, $\mathfrak{S}_l$, $l = 0, 1, 2, \ldots$, of length $\leq l$ soluble superalgebras given by $\delta_l(x_1, \ldots, x_{2^l-1}) \equiv 0$ where $\delta_1(x_1) = x_1$ and

$$\delta_{l+1}(x_1, \ldots, x_{2^l}) = [\delta_l(x_1, \ldots, x_{2^{l-1}}), \delta_l(x_{2^{l-1}+1}, \ldots, x_{2^l})]. \tag{3}$$

b) Given a variety $\mathfrak{B}$ of Lie algebras one can define the variety $\mathfrak{B}(G, \varepsilon)$ of colour Lie superalgebras by saying that a Lie superalgebra $L = \bigoplus_{g \in G} L_g$ is in $\mathfrak{B}(G, \varepsilon)$ if, and only if, any of its Grassmann envelopes $\mathscr{G}(L)$ is in $\mathfrak{B}$. It is obvious that for any variety $\mathfrak{B}$ given by a multilinear monomial (or a set of these) its corresponding variety $\mathfrak{B}(G, \varepsilon)$ is given by the same set of "global" identities. But in many other cases the identities differ drastically. For example, if we consider the variety of ordinary superalgebras $\mathfrak{B}(\mathbb{Z}_2, \varepsilon)$ corresponding to the variety of Lie algebras given by the n-th Engel identity $[x, y^{(n)}] = 0$ then, x being of arbitrary parity, $y_1, \ldots, y_n$ being odd, we find the following, so called "standard" identity satisfied in $\mathfrak{B}(\mathbb{Z}_2, \varepsilon)$:

$$\sum_{\sigma \in \text{Sym}(n)} (\text{sign } \sigma)[x, y_{\sigma(1)}, \ldots, y_{\sigma(n)}] = 0. \tag{4}$$

An abridged form of (4) is $[x, \bar{y}_1, \ldots, \bar{y}_n] = 0$*⁾. If we have two or more groups of "alternating" variables then we will be using also tildes " ~ ", hats " ^ ", etc. in addition to bars " ¯ " used just before. This notation is especially useful when we apply the representation theory of the symmetric group $\text{Sym}(n)$ to the study of identities (where the base field is of characteristic zero).

c) If we consider the multilinear component P then it is natural to consider the natural action of the group $S_\gamma = \prod_{g \in G} \text{Sym} \gamma(g)$ on it by interchanging variables of the same G-degree. It is obvious that any irreducible S_γ-module is isomorphic to the tensor product of $\text{Sym} \gamma(g)$-modules and if V is an ideal of identities in $L(X)$ then its γ-th multilinear component is an S_γ-submodule in P_γ. Now it is known from the representation theory of the symmetric group $\text{Sym}(m)$ (cf. [Bahturin 1987a, Chapter 3]) that the irreducible modules of this group are in one-one correspondence with the so-called *Young diagrams*, or, which is the same, with partitions $m = m_1 + m_2 + \cdots + m_k$ where $m_1 \geq m_2 \geq \cdots \geq m_k$ (for some $k \geq 1$). Moreover, if we fill a Young diagram d with numbers $1, 2, \ldots, m$ then we obtain a Young tableaux σd (σ stands for the corresponding permutation of $1, \ldots, m$) and we can write a polynomial $f_{\sigma d}$ which, if non-zero, generates an irreducible $\text{Sym}(m)$-submodule associated with the Young diagram d. Every submodule in the module under consideration is equal to the sum of the ones just described. (All this can be found in [Bahturin 1987a], Chapters 3 and 4.) Now if we denote by $M(d)$ the irreducible $\text{Sym}(m)$-module associated with the Young diagram d then any irreducible S_γ-submodule takes the form $M(d_1) \otimes M(d_2) \otimes \cdots \otimes M(d_s)$ where $d_1, d_2, \ldots, d_s$ are partitions of numbers $\gamma(g)$ which are different from zero.

For example, let $K = \mathbb{C}, G = \mathbb{Z}_2, \varepsilon(1, 1) = -1, n = 3$. Then

$$P_3 = P_{3,0} \oplus P_{2,1} \oplus P_{1,2} \oplus P_{0,3}.$$

The acting groups here are $\text{Sym}(3)$, $\text{Sym}(2) \times \text{Sym}(1)$, $\text{Sym}(1) \times \text{Sym}(2)$ and $\text{Sym}(3)$. Since the second and the third are commutative groups and $\dim P_{2,1} = \dim P_{1,2} = 2$, the modules $P_{2,1}$ and $P_{1,2}$ are not irreducible and they are isomorphic to the sum of two pairwise non-isomorphic irreducible modules. It is easy to show that $P_{3,0}$ and $P_{0,3}$ are irreducible. Thus P_3 splits

---

*⁾ Here and everywhere in §2 of Chapter 1 the commutators are left-normed; everywhere else in this book the commutators are *right*-normed.

into the direct sum of six pairwise non-isomorphic submodules and, accordingly, there are $2^6 - 1 = 63$ different submodules in P_3. Hence we have 63 pairwise distinct varieties given by identities of degree 3 in X.

To introduce some techniques which can be used in the study of varieties of colour Lie superalgebras we devote some of the following sections to a study of the varieties of metabelian, i.e. soluble of length 2, Lie superalgebras over a field.

2.6. Free metabelian Lie superalgebras. Bases. A superalgebra of the form $M(X) = L(X)/L^{(2)}(X)$ is called *free metabelian* with free generating set X. We first want to establish a theorem on the basis of $M(X)$ as a vector space over the field K. We recall that $X = \bigcup_{g \in G} X_g = X_+ \cup X_-$ where $X_\pm = \bigcup_{g \in G_\pm} X_g$. The elements in X_+, even variables, will be denoted by x's while odd variables will be denoted by y's. We will also need the so-called *abelian wreath product* $W = A \text{ wr } L$ of two abelian Lie superalgebras A and L with graded bases $R = \bigcup_{g \in G} R_g$ and $U = \bigcup_{g \in G} U_g$ which are in bijective correspondence with one another. By this we mean a semidirect product $L \wedge T$ of L by the free L-module T freely generated by R. A basis for W, according to the Poincaré-Birkhoff-Witt Theorem (cf. p. 85) is formed by U and by all elements of the form

$$[r, e_1, \ldots, e_m, f_1, \ldots, f_n]$$

with $r \in R$, $e_1 \leq \cdots \leq e_m$, $f_1 < \cdots < f_n$, where $e_i \in E_+$, $f_j \in E_-$ and $E_+ \cup E_-$ is a homogeneous basis in L.

Theorem. *Let $X = \bigcup_{g \in G} X_g = X_+ \cup X_-$ be a G-graded set, $M(X)$ the free metabelian algebra in X over a field K, char $K \neq 2, 3$. Then a basis in $M(X)$ is formed by all monomials of the form*

$$x_i \in X_+, \qquad y_j \in X_-$$

$$[x_{i_1}, x_{i_2}, \ldots, x_{i_p}, y_{j_1}, y_{j_2}, \ldots, y_{j_p}], \quad i_1 > i_2 \leq \cdots \leq i_p, j_1 < j_2 < \cdots < j_q, p > 1, \tag{5}$$

$$[y_{j_1}, y_{j_2}, \ldots, y_{j_q}], \quad j_1 \geq j_2 < \cdots < j_q, q > 1, \tag{6}$$

$$[y_{j_1}, x_{i_1}, \ldots, x_{i_p}, y_{j_2}, \ldots, y_{j_q}], \quad i_1 \leq \cdots \leq i_p, j_2 < \cdots < j_q, p, q \geq 1. \tag{7}$$

Proof. Let X be a totally ordered set such that all even elements, i.e. elements in X_+, precede the odd ones. Then all elements in (5) to (7) have the form

$$[x_1, x_2, x_3, \ldots, x_n] \quad \text{with } x_1 \geq x_2 \leq x_3 \leq \cdots \leq x_n. \tag{8}$$

According to the metabelian identity $[[x,y],[u,v]] = 0$ and to the defining identities (22), (23) in §1 of the class of colour Lie superalgebras, any element in $M(X)$ can be written in the form of a linear combination of monomials of the form (8). Indeed, it follows from the metabelian identity that

$$[[x,y],u,v] = \varepsilon(u,v)[[x,y],v,u],$$

which enables us to permute the variables starting from the third one. Now the places of the first three variables can be changed using the defining identities.

The monomials of the form (8) which are not contained among those in (5) split into three groups:
(i) If $i_1 = i_2$ then using "anticommutativity" we obtain zero.
(ii) Strict inequalities for $y_{j_1}, y_{j_2}, \ldots, y_{j_q}$ in (5), $y_{j_3}, \ldots, y_{j_q}$ in (6) and $y_{j_2}, y_{j_3}, \ldots, y_{j_q}$ in (7) hold since for $a \in L^2$ and $y \in X_-$ we have $2[[a,y],y] = [a,[y,y]] = 0$.
(iii) If we have $j_2 = j_3$ in $u = [y_{j_1}, y_{j_2}, y_{j_3}, \ldots, y_{j_q}]$ then $j_1 > j_2$ following from $3[[y,y],y] = 0$ which holds for y odd. Then, by the Jacobi identity we have

$$u = -\tfrac{1}{2}[y_{j_2}, y_{j_3}, y_{j_1}, \ldots, y_{j_q}]$$

and moving y_{j_1} to its place to the right (if necessary) we arrive at a monomial of the form (6).

Now we want to verify that the system of all monomials of the form (5) to (7) is linearly independent. To this end we use the abelian wreath product $W = A \operatorname{wr} L$ of two abelian superalgebras defined above in the case of ordinary superalgebras, i.e. with $G = \mathbb{Z}_2$ (see 1.7). Again W is a semidirect product of an abelian Lie superalgebra L with G-graded basis $U = \bigcup_{g \in G} U_g$ and an ideal T which is a free $U(L)$-module with free basis $R = \bigcup_{g \in G} R_g$. It follows from the Poincaré-Birkhoff-Witt Theorem for colour Lie superalgebras (see §2.2) that a basis of $W = L \wedge T$ is formed by U and by the set of elements of the form

$$r_{i_1} u_{i_2} \ldots u_{i_p} v_{j_1} \ldots v_{j_q} \qquad i_2 \leq \cdots \leq i_p, j_1 < \cdots < j_q, u_i \in U_+, v_j \in U_-.$$

By a property of free algebras of varieties we can extend the mapping φ: $x \mapsto r + u$ (x, r and u correspond to each other under natural bijections of X, R and U) to a homomorphism $\bar\varphi: M(X) \to W$. Then $\bar\varphi$ brings (5) to (7) to the form:

$$[r_{i_1}, u_{i_2}, \ldots, u_{i_p}, v_{j_1}, \ldots, v_{j_q}] - [r_{i_2}, u_{i_1}, \ldots, u_{i_p}, v_{j_1}, \ldots, v_{j_q}] \qquad (9)$$

$$(i_1 > i_2 \leq \cdots \leq i_p, j_1 < \cdots < j_q, p > 1)$$

§ 2. Identical relations of graded algebras

$$[s_{j_1}, v_{j_2}, \ldots, v_{j_q}] + [s_{j_2}, v_{j_1}, \ldots, v_{j_q}] \quad (j_1 \geq j_2 < \cdots < j_q, q > 1) \qquad (10)$$

$$[s_{j_1}, u_{i_1}, \ldots, u_{i_p}, v_{j_1}, \ldots, v_{j_q}] - [r_{i_1}, v_{j_1}, u_{i_2}, \ldots, u_{i_p}, v_{j_2}, \ldots, v_{j_q}] \qquad (11)$$

$$(i_1 \leq \cdots \leq i_p, j_2 < \cdots < j_q, p, q \geq 1).$$

Here $r_i \in R_+$, $s_j \in R_-$. Now let $w_1, w_2, \ldots, w_n$ be linearly dependent elements in (5) to (7). Then there exists a non-trivial linear combination

$$\alpha_1 \overline{\varphi}(w_1) + \alpha_2 \overline{\varphi}(w_2) + \cdots + \alpha_n \overline{\varphi}(w_n) = 0.$$

Now it is obvious that if an expression of the form (10) linearly depends on other elements in (9) to (11) then we can restrict to a dependence on other elements in (10). But considering such elements with j_1 greatest possible shows that a dependence of this kind is impossible. In the same way we can handle other types of elements in (9) and (11). Thus (9) to (11) is a linearly independent system and then (5) to (7) is indeed a basis of $M(X)$, completing the proof of the theorem. $\square$

Corollary 1. *$M(X)$ is isomorphic to a subalgebra in $W = A \operatorname{wr} L$ generated by all elements of the form $r + u$ where r and u correspond to the same element of X in R and U.* $\square$

Let P_n denote the multilinear component of degree n in the free metabelian superalgebra $L(X)$. As in the case of the absolutely free superalgebra $L(X)$ we have $P_n = \sum_{|\gamma|=n} P_\gamma$.

Corollary 2. *For any $\gamma: G \to \mathbb{N} \cup \{0\}$ a basis in P_γ is formed by all monomials of the form*

$$[x_1, x_2, \ldots, x_n] \quad \text{where } x_1 > x_2 < x_3 < \cdots < x_n$$

such that the number of variables in X_g is equal to $\gamma(g)$. In particular, $\dim P_\gamma = |\gamma| - 1$. $\square$

2.7. Free metabelian superalgebras. Hilbert series. In this subsection X is a finite set. We set $X_+ = \{x_1, \ldots, x_n\}$, $X_- = \{y_1, \ldots, y_m\}$. Given a free algebra $L = L(X, \mathfrak{B})$ of a variety $\mathfrak{B}$ generated by X, a formal series of the form

$$H(L) = H(t_1, \ldots, t_m, u_1, \ldots, u_n) = \sum (\dim L_{\gamma, \delta}) t_1^{\gamma_1} \ldots t_m^{\gamma_m} u_1^{\delta_1} \ldots u^{\delta_n}$$

is called the *Hilbert series* of L; this is an important invariant of the variety $\mathfrak{B}$. Similarly, one defines the Hilbert series $H(L, t)$ in one variable t as

$$H(L, t) = \sum (\dim L_s) t^s \quad \text{where } L_s = \sum L_{\gamma, \delta}, |\gamma| + |\delta| = s.$$

In this subsection we are going to compute the Hilbert series in the case of a free metabelian algebra.

Theorem. *Let $X_+ = \{x_1,\ldots,x_m\}$, $X_- = \{y_1,\ldots,y_n\}$, $M(X)$ a free metabelian algebra on X. Then*

$$H(M(X)) = 1 + (t_1 + \cdots + t_m + u_1 + \cdots + u_n)$$

$$+ \frac{(t_1 + \cdots + t_m + u_1 + \cdots + u_n - 1)(1 + u_1)\ldots(1 + u_n)}{(1 - t_1)(1 - t_2)\ldots(1 - t_m)}.$$

Proof. By the theorem in Subsection 2.6 a basis of $M(X)$ is formed by X and monomials of the form (5) to (7). Therefore,

$$H(M(X)) = \sum_{i=1}^{m} t_i + \sum_{j=1}^{n} u_j + H(A_1) + H(A_2) + H(A_3)$$

where A_1, A_2 and A_3 stand for the graded linear spaces with bases formed by (5), (6) and (7), respectively. Now A_1 is the quotient space of a space B_1 with basis

$$x_{i_1} \otimes x_{i_2} \ldots x_{i_p} \otimes y_{j_1} \ldots y_{j_q}, \quad i_2 \leq \cdots \leq i_p, \quad j_1 < \cdots < j_q, \; p > 1,$$

by the span C_1 of all tensors of the form

$$x_{i_1} \otimes x_{i_2} \ldots x_{i_p} \otimes y_{j_1} \ldots y_{j_q}, \quad i_1 \leq i_2 \leq \cdots \leq i_p, \, j_1 < \cdots < j_q, \, p > 1.$$

Hence $H(A_1) = H(B_1) - H(C_1)$. Now

$$H(B_1) = (\textstyle\sum t_i)(\sum t_{i_2}\ldots t_{i_p})(\sum u_{j_1}\ldots u_{j_q}), \quad i_2 \leq \cdots \leq i_p, j_1 < \cdots < j_q.$$

It is obvious that

$$\textstyle\sum t_{i_2}\ldots t_{i_p} = \prod (1 - t_i)^{-1} - 1 \quad \text{where } i_2 \leq \cdots \leq i_p, p > 1$$

and

$$\textstyle\sum u_{j_1}\ldots u_{j_q} = \prod (1 + u_j) \quad \text{where } j_1 < \cdots < j_q.$$

Thus,

$$H(B_1) = (\textstyle\sum t_i)(\prod (1 - t_i)^{-1} - 1)\prod (1 + u_j).$$

§2. Identical relations of graded algebras

Similarly,
$$H(C_1) = (\prod(1-t_i)^{-1} - 1 - \sum t_i)\prod(1+u_j)$$

Finally,
$$H(A_1) = (\sum t_i - 1)\prod(1+u_j)\prod(1-t_i)^{-1} + \prod(1+u_j).$$

To determine $H(A_2)$ we consider the linear space B_2 with basis $y_{j_1} \otimes y_{j_2} \ldots y_{j_q}$, $j_2 < \cdots < j_q$, $q > 1$, and C_2 spanned by the tensors with the additional requirement $j_1 < j_2 < \cdots < j_q$. Then

$$A_2 \cong B_2/C_2, \qquad H(B_2) = (\sum u_j)\sum u_{j_2}\ldots u_{j_q},$$

where $j_2 < \cdots < j_q$, i.e.

$$H(B_2) = (\sum u_j)(\prod(1+u_j) - 1)$$

and, similarly $H(C_2) = \prod(1+u_j) - 1 - \sum u_j$ so that

$$H(A_2) = (\sum u_j - 1)\prod(1+u_j) + 1.$$

Now let B_3 be the space with basis $y_{j_1} \otimes x_{i_1}\ldots x_{i_p} \otimes y_{j_1}\ldots y_{j_q}$, $i_1 \leq \cdots \leq i_p$, $j_2 < \cdots < j_q$, $p, q > 0$. Then also $A_3 \cong B_3$ and

$$H(A_3) = \sum u_j \sum t_{i_1}\ldots t_{i_p} \sum u_{j_2}\ldots u_{j_q} = (\sum u_j)(\prod(1-t_i)^{-1} - 1)\prod(1+u_j).$$

It follows from the expression for $H(A_1)$, $H(A_2)$, $H(A_3)$ that

$$H(M(X)) = \sum t_i + \sum u_j + (\sum t_i - 1)\prod(1+u_j)\prod(1-t_i)^{-1} + \prod(1+u_j)$$
$$+ (\sum u_j - 1)\prod(1+u_j) + 1$$
$$+ (\sum u_j)(\prod(1-t_i)^{-1} - 1)\prod(1+u_j)$$
$$= 1 + \sum t_i + \sum u_j + (\sum t_i + \sum u_j - 1)\prod(1+u_j)\prod(1-t_i)^{-1}. \quad \square$$

Corollary. *Let $|X_+| = m$, $|X_-| = n$. Then*

$$H(M(X), t) = 1 + (m+n)t + ((m+n)t - 1)\frac{(1+t)^n}{(1-t)^m}.$$

In the case of ordinary Lie algebras we have

$$H_m(\mathfrak{A}^2, t) = 1 + mt + \frac{mt - 1}{(1-t)^m}.$$ □

It would be interesting to know for what varieties the Hilbert series of the free algebra of the variety is rational, i.e. it can be written as a quotient of two polynomials (just as above).

2.8. Free metabelian algebras. Multilinear elements. The action of the group $S = \prod_{g \in G} \mathrm{Sym}(\gamma(n))$ on the set of L_γ of multilinear elements in a free colour Lie superalgebra induces a similar action on the set P_γ of multilinear elements in $M(X)$. If $n = |\gamma| > 1$ then it follows from the description of $M(X)$ (Subsection 2.6) that dim $P_\gamma = n - 1$ and that P_γ has a basis of monomials of the form $[x_{i_1}, x_{i_2}, \ldots, x_{i_n}]$, $i_1 > i_2 < \cdots < i_n$. We want to describe P_γ as an S-module. Let g, g_i denote some elements in G_+, h, h_j in G_-. We will assume also that the set of elements in G is totally ordered.

Theorem. *Let γ have the form*

$$\gamma = (\gamma(g_1), \ldots, \gamma(g_r), 0, 0, \ldots, \gamma(h_1), \ldots, \gamma(h_s), 0, 0, \ldots),$$

where $g_1 < \cdots < g_r$, $h_1 < \cdots < h_s$, $g_i \in G_+$, $h_j \in G_-$ and $\gamma(g_i) \neq 0$, $\gamma(h_j) \neq 0$. Then P_γ is the direct sum of the following irreducible S-submodules.
(i) $M_1(f) = M(\gamma(f) - 1, 1) \otimes (\bigotimes_{g \neq f} M(\gamma(g)) \otimes (\bigotimes_h M(1^{\gamma(h)}))$ *where $f \in G_+$, $\gamma(f) > 1$ and the first factor in the tensor product corresponds to $f \in G$.*
(ii) $M_2(f) = M(2, 1^{\gamma(f)-2}) \otimes (\bigotimes_g M(\gamma(g)) \otimes (\bigotimes_{h \neq f} M(1^{\gamma(h)}))$ *where $f \in G_-$, $\gamma(f) > 1$ and the first factor in the tensor product corresponds to $f \in G$.*
(iii) $r + s - 1$ *isomorphic copies of the module*

$$M_3 = \left(\bigotimes_g M(\gamma(g)) \right) \otimes \left(\bigotimes_h M(1^{\gamma(h)}) \right).$$

Proof. Bearing in mind that $M(p)$ and $M(1^p)$ are 1-dimensional while dim $M(p-1, 1) = $ dim $M(2, 1^{p-2}) = p - 1$, $p > 1$, (see "Hook Formula" in [Bahturin, 1987a, Chapter 3]) we have

$$\sum_{f \in G_+} \dim M_1(f) + \sum_{f \in G_-} \dim M_2(f) + (r + s - 1) \dim M_3$$

$$= \sum (\gamma(g) - 1) + \sum (\gamma(h) - 1) + (r + s - 1)$$

$$= \sum \gamma(g) + \sum (\gamma(h) - 1) = |\gamma| - 1.$$

§2. Identical relations of graded algebras

Thus, it is sufficient to verify that, indeed, there are submodules of the form $M_1(f)$, $M_2(f)$ and a direct sum of $(r + s - 1)$ copies of M_3 in P.

We set $p_i = \gamma(g_i)$, $q_j = \gamma(h_j)$, $X_{g_i} = \{x_{i1}, x_{i2}, \ldots\}$, $X_{h_j} = \{y_{j1}, y_{j2}, \ldots\}$. Moreover, if we set $X_f = \{x_1, x_2, \ldots\}$, $q = \gamma(f)$ for $f \in G_-$. We start with $f \in G_+$ and $p > 1$. Consider

$$u = \left[[x_1, x_2] - x_2, x_1, x_1^{p-2}, \prod_{g_i \neq f} x_{i1}^{p_i}, \bar{y}_{11}, \ldots, \bar{y}_{1q_1}, \ldots, \tilde{y}_{s1}, \ldots, \tilde{y}_{sq_s}\right]. \quad (12)$$

It is obvious that $u \neq 0$ in $M(X)$. A standard argument (see, e.g., [Bahturin, 1987a, §4.8]) shows that the linearization of u generates an irreducible S-submodule in P isomorphic to M_1.

Similarly, if $q > 1$ and $f \in G_-$ then the linearization of

$$v = \left[y_1, \bar{y}_1, \ldots, \bar{y}_{q-1}, \prod_{g_i} x_{i1}^{p_i}, y_{11}, \ldots, y_{1q_1}, \ldots, \bar{\bar{y}}_{s1}, \ldots, \bar{\bar{y}}_{sq_s}\right] \quad (13)$$

generates an irreducible S-submodule isomorphic to M_2 (where there are no occurrences of $y_1, \ldots, y_{q-1}$ among $y_{11}, \ldots, \bar{\bar{y}}_{sq_s}$).

Finally, for each $f = g_{i_0} \in G_+$ with $f > g_1$ and $\gamma(f) \neq 0$ one can define

$$u_f = \sum \left[x_1, \left(\prod_{i \neq i_0} x_{i1}^{p_i}\right), x_1^{p-1}, \bar{y}_{11}, \ldots, \bar{y}_{1q_1}, \ldots, \tilde{y}_{s1}, \ldots, \tilde{y}_{sq_s}\right] \quad (14)$$

while for $f \in G_-$, $\gamma(f) \neq 0$ one can write

$$v_f = [y_1, \prod x_{i1}^{p}, \bar{y}_{11}, \ldots, \bar{y}_{1q_1}, \ldots, \tilde{y}_1, \ldots, \tilde{y}_{q-1}, \ldots, \bar{\bar{y}}_{s1}, \ldots, \bar{\bar{y}}_{sq_s}]. \quad (15)$$

Now the set of u_f and v_f is linearly independent in $M(X)$ and their linearizations "simultaneously" generate S-modules which are isomorphic to M_3. Thus P_γ contains $(r + s - 1)$ copies of M_3 and any submodule isomorphic to M_3 is generated by a linearization of a non-zero polynomial of the form $\sum \alpha_f u_f + \sum \beta_f v_f$, $\alpha_f, \beta_f \in K$. Hence the proof of the theorem is complete. □

Remark. Each of $M_1(f)$ or $M_2(f)$ in the preceding theorem can be generated by a single left-normed commutator of the form

$$w = [x_1, \ldots, x_p, x_{11}, \ldots, x_{1p_1}, \ldots, x_{r+s,1}, \ldots, x_{r+s,q_s}], \quad (16)$$

where $x_1, \ldots, x_p \in X_f$ while $x_{11}, \ldots, x_{r+s,q}$ are in $X_{g_1}, \ldots, X_{h_s}$ with $g_i, h_j \neq f$. Indeed, if, e.g., $f \in G_+$ then $u = 0$ (u as in (12)) follows easily from $w = 0$.

Thus it is sufficient to show that (16) generates an irreducible S-module. But if Q is an irreducible S-module generated by w then, x_1 and x_2 being skew-symmetric in w, it follows that the partition of p corresponding to the f-component of Q is $(p-1, 1)$, i.e.

$$Q \subset M(p-1, 1) \otimes \left(\bigotimes_{g_i \neq f} M(p_i) \right) \otimes \left(\bigotimes_{h_j} M(1^{q_j}) \right)$$

and then Q is indeed irreducible and equal to $M_1(f)$ or $M_2(f)$.

2.9. Subvarieties in $\mathfrak{A}^2$. A variety of algebras is called *Specht* if every of its subvarieties (including itself) has a finite basis for its laws, i.e. any of its identities follows from a finite subset of the laws. We are going to show that $\mathfrak{A}^2$ is Specht whenever $|G| < \infty$. This will be a consequence of the following (we assume that G is finite).

Theorem. *Any proper subvariety $\mathfrak{B}$ in $\mathfrak{A}^2$ is multinilpotent, i.e. there exists $\gamma = (\gamma_1, \ldots, \gamma_r)$ such that $P_\delta(\mathfrak{B}) = 0$ if the multidegree $\delta = (\delta_1, \ldots, \delta_r)$ satisfies $\delta_i \geq \gamma_i$, $i = 1, \ldots, r$, where r is the order of G.*

Proof. Without any loss of generality we may assume that $\mathfrak{B}$ satisfies $w = 0$ where w is a multilinear polynomial generating an irreducible S-submodule in P_0. If w generates $M_1(f)$ (or $M_2(f)$) in Theorem 2.8 then it is clear from the remark following Theorem 2.8 that $w = 0$ is equivalent to $[x_1, \ldots, x_p, \prod_{g_i \neq f} x_{i1}, \ldots, x_{ip}] = 0$ (see (13)) in P_{γ_0} where $\gamma_0 = (p_1, p_2, \ldots, p_r)$, $x_1, \ldots, x_p \in X_f$, $x_{ij} \in X_{g_j}$. Let g', g'' be arbitrary in G and $g''' = f - g' - g''$. Taking $y_1 \in X_{g'}$, $y_2 \in X_{g''}$, $y_3 \in X_{g'''}$ we get a consequence of $w = 0$ of the form

$$\left[y_1, y_2, y_3, x_2, \ldots, x_p, \prod_{g_i \neq f} x_{i1}, \ldots, x_{ip_i} \right] = 0.$$

Thus, if $u \in P_\delta$ and $\delta_1 \geq p + 2$, $\delta_i \geq p_i + 2$, $i = 2, \ldots, r$, then $u = 0$ is a consequence of $w = 0$.

Now we assume that w generates an S-module of the form

$$\prod_{g \in G_+} M(p_i) \otimes \prod_{h \in G_-} M(1^{q_j}).$$

Again we assume that $|G_+| = r$, $|G_-| = s$, $x_i \in X_{g_i} \subset X_+$, $y_{j_t} \in X_{h_j} \subset X_-$, $\gamma(g_i) = p_i$, $\gamma(h_j) = q_j$. Then $w = 0$ is equivalent to

$$w' = \sum \alpha_i [x_i, x_i^{p_i-1}, \ldots, x_r^{p_r}, \bar{y}_{11}, \ldots, \bar{y}_{1q_1}, \ldots, \tilde{y}_{s1}, \ldots, \tilde{y}_{sq_s}]$$

$$+ \sum \beta_j [y_j, x_1^{p_1}, \ldots, x_r^{p_r}, \bar{y}_{11}, \ldots, \bar{y}_{1q_1}, \ldots, \tilde{y}_j, \ldots, \tilde{y}_{jq_j-1}, \ldots, \bar{\bar{y}}_{s1}, \ldots, \bar{\bar{y}}_{sq_s}]$$

$$= 0$$

where at least one of the coefficients $\alpha_i, \beta_j \in K$ is different from zero. Here for any i with $\alpha_i \neq 0$ there exists $t < i$ with $p_t \neq 0$. Similarly, if $\beta_j \neq 0$ then either there exists t for which $p_t \neq 0$ or $s < j$, with $q_s = 0$. To specify, let $\alpha_i \neq 0$, $f = g_i$ and $g', g'' \in G$. We set $g''' = f - g' - g''$. Consider $z_1 \in X_{g'}, z_2 \in X_{g''}$ and $z_3 \in X_{g'''}$ and replace x_i by $x_i + [z_1, z_2, z_3]$. Then we get a consequence of the form

$$\alpha_i[z_1, z_2, z_3, x_1^{p_1}, \ldots, x_i^{p_i-1}, \ldots, x_r^{p_r}, \bar{y}_{11}, \ldots, \bar{y}_{1q_1}, \ldots, \tilde{y}_{s1}, \ldots, \tilde{y}_{sq_s}] = 0. \quad (17)$$

It is easy to observe that, for $g \in G_+$, the linearization of $[z_1, z_2, x^p] = 0$, $x \in X_g$, is equivalent to $[z_1, z_2, x_1, \ldots, x_p] = 0$, $x_i \in X_g$, is equivalent to $[z_1, z_2, \bar{y}_1, \ldots, \bar{y}_q] = q![z_1, z_2, y_1, \ldots, y_q]$, $y_j \in X_h \subset X_-$. Hence, with g', g'' running through the whole of G we find that (17) implies the multi-homogeneous property of $\mathfrak{B}$ under consideration. The case $\beta_j \neq 0$ is quite similar. □

Corollary. *Any variety of metabelian colour Lie superalgebras has a finite basis for its identities (in other words, $\mathfrak{A}^2$ is Specht).*

Proof. It has been established in [Bahturin 1987a, Chapter 5] that any variety is Specht if, and only if, it is finitely based and if all descending chains of subvarieties have finite length. Since $\mathfrak{A}^2$ is obviously finitely based it is the latter property that is of importance. But a multinilpotency identity with $\gamma = (\gamma_1, \ldots, \gamma_r)$ implies that with $\gamma' = (\gamma'_1, \ldots, \gamma'_r)$ if, and only if, $\gamma_1 \leq \gamma'_1, \ldots, \gamma_r \leq \gamma'_r$. Now a descending chain of varieties in $\mathfrak{A}^2$ means a descending chain of multinilpotencies, i.e. of r-tuples under the ordering just described. It is well-known (see, e.g. [Bahturin 1987a, Chapter 5]) that this ordering is a partial well-ordering and so the chain must terminate, as required. □

Exercises

We begin this section with an example of an algebra which is close to a colour superalgebra but which cannot be made into such one. Then we suggest a number of exercises leading to the result on the number of various forms ε on an abelian group with values in an algebraically closed field.

Let H be a three-dimensional Lie algebra over the real number field with basis a, b, c and the multiplication table $[a, b] = c$, $[a, c] = [b, c] = 0$. We denote by M the space of functions $f(t): \mathbb{R} \to \mathbb{C}$ which are exponentially decreasing with $t \to \pm\infty$. One can make M into an H-module by setting $af(t) = (d/dt)f(t)$, $bf(t) = itf(t)$, $cf(t) = if(t)$ where $i = \sqrt{-1}$. Giving M zero multiplication, we consider the semidirect product $H \oplus M$. A more complicated structure will be given to the space $L = \mathbb{C} \oplus H \oplus M$. We define the commutator on M by setting

$$[f,g] = \int_{-\infty}^{+\infty} f(t)\overline{g(t)}\,dt$$

where the "bar" above $g(t)$ means complex conjugation. If T is the subspace of real valued functions in M then $M = T \oplus iT$ and $[f,g] = [g,f]$ for $f, g \in T$ or $f, g \in iT$. Also, $[f,g] = -[g,f]$ if $f \in T$, $g \in iT$. If we set $[H,\mathbb{C}] = [M,\mathbb{C}] = [\mathbb{C},\mathbb{C}] = 0$ then L becomes a G-graded algebra where $G = \mathbb{Z}_2 \oplus \mathbb{Z}_2$ and the grading is defined by $L_{(0,0)} = \langle a \rangle \oplus \mathbb{C}$, $L_{(1,0)} = T$, $L_{(0,1)} = iT$, $L_{(1,1)} = \langle b,c \rangle$.

We remark that L is a graded algebra of Lie type, that is, for all homogeneous elements x, y, z we have equations of the form

$$[x,y] = -\alpha[y,x]$$

$$[x,[y,z]] = \beta[[x,y],z] + \gamma[y,[x,z]]$$

in which $\alpha, \beta, \gamma \in \mathbb{R}$. In our case $\beta = \gamma = 1$ for all x, y, z. Also α equals 1 in all cases except $x, y \in T$ or $x, y \in iT$ when $\alpha = -1$. For instance, if $f, g \in M$ then we have $[a,[f,g]] = [b,[f,g]] = 0$. On the other hand,

$$[[a,f],g] + [f,[a,g]] = \int_{-\infty}^{+\infty} (f'\overline{g} + f\overline{g}')\,dt = f\overline{g}\Big|_{-\infty}^{+\infty} = 0,$$

$$[[b,f],g] + [f,[b,g]] = \int_{-\infty}^{+\infty} (itf\overline{g} + f\overline{itg})\,dt = i\int_{-\infty}^{+\infty} (tf\overline{g} - tf\overline{g})\,dt = 0.$$

Similarly one can verify the other relations. But:

Exercise 1. Prove that no bilinear alternating form ε on G can make L into a colour Lie superalgebra.

Now let G be an additive abelian group and K an algebraically closed field, $\varepsilon: G \times G \to K^*$ an alternating bilinear form. The aim of exercises 2) to 8) is to compute the number of these forms.

Exercise 2. Let α and β be of orders n and m, respectively. Then $(\varepsilon(\alpha,\beta))^t = 1$ where t is the greatest common divisor of n, m.

Exercise 3. Let $G = \bigoplus_p \mathrm{Syl}_p(G)$ be the decomposition of G, a finite group, into its primary components. Then distinct components are orthogonal with respect to ε, i.e., $\varepsilon(\alpha,\beta) = 1$ for $\alpha \in \mathrm{Syl}_p(G)$, $\beta \in \mathrm{Syl}_q(G)$ with $p \neq q$.

Exercise 4. Let K be a field with char $K = p > 0$ and $|G| = p^a b$ where $(p, b) = 1$. Then $(\varepsilon(\alpha, \beta))^b = 1$ for all $\alpha, \beta \in G$.

Exercise 5. $G_+ \subset G$ is a subgroup such that either $G = G_+$ or G_+ is of index 2 in G.

Exercise 6. The set $\hat{G}$ of all forms ε for given G (with K fixed) forms a group with respect to the operation

$$(\varepsilon, \varepsilon')(x, y) = \varepsilon(x, y)\varepsilon'(x, y), \qquad \varepsilon, \varepsilon' \in G.$$

Exercise 7. Let $G = n\mathbb{Z}$. Then

$$G \cong \frac{n(n-1)}{2} K^* \oplus n\mathbb{Z}_2.$$

Exercise 8. Let $G = \bigoplus_p (\bigoplus_{m \geq 1} (n_p^{(m)} \mathbb{Z}_{p^m}))$ be a primary decomposition of G. Then, if K is of characteristic zero, one must have

$$\hat{G} = \bigoplus_p \left(\bigoplus_{m \geq 1} n_p^{(m)} \left(\frac{n_p^{(m)} - 1}{2} + n_p^{(m+1)} + n_p^{(m+2)} + \cdots \right) \mathbb{Z}_{p^m} \right)$$

$$\oplus (n_2^{(1)} + n_2^{(2)} + n_2^{(3)} + \cdots) \mathbb{Z}_2.$$

If K has positive characteristic p then the p-component should be dropped from the above expression.

Comments to Chapter 1

There are only few monographs and surveys on Lie superalgebras. Among them we list [Berezin 1983], [Kac 1977a], [Leites 1984], [Scheunert 1979b]. In fact, the theory under consideration in this book is young and so some concepts presented here are due to the authors. For instance, the notion of a restricted colour Lie superalgebra first appeared in [Mikhalev 1988]. However, the notion of colour Lie superalgebras is much older and, without getting into further details, we refer to [Scheunert 1979a]. In this substantial paper the author shows that in many important cases the introduction of a new operation on a colour Lie superalgebras by setting $[a, b]^\sigma = \sigma(d(a), d(b))[a, b]$, where $\sigma: G \times G \to K^*$ is a 2-cocycle, makes it into an ordinary superalgebra with G-grading. This can be used to prove Ado's Theorem for colour Lie superalgebras, showing that each finite-dimensional Lie superalgebra has a faithful finite-dimensional representation; see also [Hannabus 1987],

[Mosolova 1981]. In connection with graded versions of Schur's Lemma and Burnside's Theorem we mention the paper [Van Geel–van Oystayen 1981] where the reader may find further references to graded rings and modules. The last section consists mainly of the results in [Bahturin–Drensky 1987].

Chapter 2

The structure of free Lie superalgebras

In this chapter we construct a linear basis of the free colour Lie superalgebras and of the free colour Lie p-superalgebras (Theorem 2.6, Lemma 4.3) and prove that any subalgebra of the free colour Lie (p-)superalgebra is also free (Theorems 3.15, 4.15, 6.4). This latter theorem has some corollaries (3.17–3.33). We give a proof of the fact that the intersection of finitely many finite rank subalgebras in free colour Lie (p-)superalgebras in the case char $K = p > 3$ is also a finite rank subalgebra (Theorem 5.6).

§ 1. The free colour Lie superalgebra, s-regular words and monomials

1.1. The free colour Lie superalgebra. Let $X = \bigcup_{g \in G} X_g$ be a G-graded set, i.e. $X_g \cap X_f = \emptyset$ for $g \neq f$, $d(x) = g$ for $x \in X_g$; let also $\Gamma(X)$ be the groupoid of *nonassociative monomials* in the alphabet X, $u \circ v = (u)(v)$ for $u, v \in \Gamma(X)$, and $S(X)$ be the free semigroup of *associative words* with the bracket removing homomorphism $^-: \Gamma(X) \to S(X)$; $[u]$ being an arrangement of brackets on the word $u \in S(X)$. For $u = x_1 \ldots x_n \in S(X)$, $x_i \in X$ we consider the word length $l_X(u) = n$, the *multidegree* (structure) $m(u)$, $d(u) = \sum_{i=1}^n d(x_i) \in G$. For $z \in \Gamma(X)$ we set $m(z) = m(\bar{z})$, $l_x(z) = l_x(\bar{z})$ and $d(z) = d(\bar{z})$. By analogy, we define the length relative to a subset $Y \subset X$.

Let K be a commutative associative ring with 1, $A(X)$ and $F(X)$ the *free associative* and *nonassociative* K-algebras, respectively, $S(X)_g = \{u \in S(X), d(u) = g\}$, $\Gamma(X)_g = \{v \in \Gamma(X), d(v) = g\}$. Let $A(X)_g$ and $F(X)_g$ be the K-linear spans of the subsets $S(X)_g$ and $\Gamma(X)_g$ respectively, $A(X) = \bigoplus_{g \in G} A(X)_g$ and $F(X) = \bigoplus_{g \in G} F(X)_g$ being the free G-graded associative and nonassociative algebras respectively. Let now I be the G-graded ideal of $F(X)$ generated by the homogeneous elements of the form

$$a \circ b + \varepsilon(a,b) b \circ a \quad \text{and} \quad (a \circ b) \circ c - a \circ (b \circ c) + \varepsilon(a,b) b \circ (a \circ c),$$

where $a, b, c \in \Gamma(X)$; then $L(X)$ is the *free colour Lie K-superalgebra* (i.e. every G-map φ of degree zero from X into any colour Lie K-superalgebra R with the same group G and form ε ($d(\varphi(x)) = d(x)$, $x \in X$) can be uniquely extended to a colour Lie superalgebra homomorphism $\bar{\varphi}: L(X) \to R$. It is clear that this universal property defines the free colour Lie superalgebra $L(X)$ uniquely up to a colour Lie superalgebra isomorphism. For $u \in F(X)$ we set $\tilde{u} = u + I \in L(X)$.

Suppose that the set $X = \bigcup_{g \in G} X_g$ is totally ordered and the set $S(X)$ is ordered *lexicographically*, i.e. for $u = x_1 \ldots x_t$ and $v = y_1 \ldots y_m$ where x_i, $y_j \in X$ we have $u < v$ if either $x_i = y_i$ for $i = 1, \ldots, t - 1$ and $x_t < y_t$ or $x_i = y_i$ for $i = 1, 2, \ldots, m$ and $r > m$. Then $S(X)$ is linearly ordered. Consider the order relation on $\Gamma(X)$ defined by setting $a < b$ for $a, b \in \Gamma(X)$ if either $\bar{a} < \bar{b}$ in $S(X)$ or $\bar{a} = \bar{b}$, $a = (u)(v)$ and $b = (c)(d)$ with $l(u) > l(c)$ or $l(u) = l(c)$, but $u < c$, or $u = c$ and $v < d$. It is clear that this ordering is total.

If in the product $u \circ v$ we have $u > v$ then we say that v is the least factor (analogously, if $u \leq v$ then we say that u is the least factor).

1.2. Definition. The word $u \in S(X)$ is said to be *regular* if for any decomposition $u = ab$, where $a, b \in S(X)$, we have $u > ba$. The word $w \in S(X)$ is said to be *s-regular* if either w is a regular word or $w = vv$ with v a regular word and $d(v) \in G_-$.

A monomial $u \in \Gamma(X)$ is said to be *regular* if either $u \in X$ or:
a) from $u = u_1 \circ u_2$ it follows that u_1, u_2 are regular monomials with $\bar{u}_1 > \bar{u}_2$;
b) from $u = (u_1 \circ u_2) \circ u_3$ it follows that $\bar{u}_2 \leq \bar{u}_3$.

Observe that if $x \in X$, v is a regular monomial and $x > \bar{v}$ then $x(v)$ is a regular monomial.

A monomial $u \in \Gamma(X)$ is said to be *s-regular* if either u is a regular monomial or $u = (v)(v)$ with v a regular monomial and $d(v) \in G_-$.

For $u \in S(X)$ we use the notation $B(u)$ for the linear span in $A(X)$ of all s-regular words of multidegree $m(u)$.

1.3. Lemma. *Let $u, v \in S(X)$, u be a regular word and $u < v$. Then $uv < vu$.*

Proof. If $u < v$ and $uv \geq vu$, then $u = vz$ and therefore $vzv \geq vvz$. Thus $zv \geq vz$ and we have a contradiction to the regularity of u. $\square$

1.4. Lemma. *If $u = vv_1$ is a regular word then $u > v_1$.*

Proof. Since u is a regular word, $u = vv_1 > v_1v$. If now $u = vv_1 < v_1$ then $u = v_1w$ and therefore from $u = v_1w > v_1v$ we have $w > v$. As u is a regular word, $u = v_1w > wv_1$; from $w > v$ it follows that $wv_1 > vv_1 = u$, and we come to a contradiction. $\square$

1.5. Remark. The condition (if $u = u_1u_2$ for $u_1, u_2 \in S(X)$, then $u > u_2$) is equivalent to the regularity of u. Indeed, in one way this follows from Lemma 1.4, in the other direction we have $u > u_1u_2 > u_2 > u_2u_1$.

1.6. Lemma. *Let u be a regular word, $u > v$. Suppose that $u > v_0$ for all endings v_0 of v. Then $uv > vu$.*

Proof. The proof will be given by induction on the length $l(v)$ of v. The basis of induction: $v \in X$. Then from $u > x$ it follows that $ux > xu$. Suppose that our assertion is true for all v with $l(v) < n$. Let now $l(v) = n$. If $v \neq uv_0$ then $uv > vu$ (comparing first letters). If $v = uv_0$, then by our assumption $u > v_0$ (since v_0 is an ending of v). By the induction hypothesis $uv_0 > v_0 u$ (observe that any ending of v_0 is also an ending of v). Therefore $uv = uuv_0 > uv_0 u = vu$. □

1.7. Lemma. *Let a and b be regular words and $a > b$. Then ab is a regular word.*

Proof. Consider possible decompositions of ab:
a) $ab > ba$ by Lemma 1.3;

b) $a_1 \underline{a_2 b} > a_2 a_1 b > a_2 b a_1$ by regularity of a and Lemma 1.3;

c) $\underline{ab_1} b_2 > ab_2 b_1$ by regularity of b. Since $a > b$, by Lemma 1.4 we have $a > b_0$ for any ending b_0 of b, therefore $a > b_2$ and $a > b_0'$ for any ending b_0' of b_2. By Lemma 1.6 $ab_2 > b_2 a$ and therefore $ab_2 b_1 > b_2 ab_1$. Thus we have $ab_1 b_2 > b_2 ab_1$. This completes the proof of the lemma. □

1.8. Lemma. 1) *If u is a regular monomial then $\bar{u}$ is a regular word.*
2) *Conversely, for any regular word u there is a unique arrangement of brackets $[u]$ on u such that $[u]$ is a regular monomial.*

Proof. 1) We use induction on $l(u)$. If $l(u) = 1, 2$ then our assertion is evident. If $l(u) > 2$ then $u = (u_1)(u_2)$ where u_1 and u_2 are regular monomials with $\bar{u}_1 > \bar{u}_2$. By the induction hypothesis and Lemma 1.7 we see that $\bar{u} = \bar{u}_1 \bar{u}_2$ is a regular word.
2) Again we use induction on $l(u)$. The starting point $l(u) = 1$ is obvious. Let $X = \{x_1, \ldots, x_n\}$ and x_1 be the least letter from X. We may suppose that x_1 occurs in u. Now we show that the letter x_1 does not occur as the first letter in any regular word distinct from x_1. Indeed, if $w = x_1 v$ is a regular word, then $x_1 v > v x_1$ and therefore $v = x_1 v_1$, i.e. $w = x_1^l$ and w is not a regular word. Thus we may rewrite our word using the following as new letters: $x_i x_1^k = x_i \underbrace{x_1 \ldots x_1}_{k \text{ times}}$, $i \neq 1$, $k \in \mathbb{N} \cup \{0\}$. Consider the following order relation on the new letters: $x_i x_1^k > x_j x_1^l$ for $i > j$; $x_i x_1^k > x_i x_1^l$ for $k < l$. For the words which can be rewritten using the new letters we have the same ordering as in the alphabet X. The length of u in the new letters $x_i x_1^k$ is less than $l(u)$ in X because x_1 occurs in u. Thus by the induction hypothesis and the uniqueness

of the regular arrangement of brackets on $x_i x_1^k = (\ldots \underbrace{(x_i x_1) x_1 \ldots)x_1}_{k \text{ times}}$ we have our assertion. □

1.9. Remark. If we consider the regular arrangement of brackets of a regular word u (see Claim 2 of Lemma 1.8) then the external brackets $[u] = (a)(b)$ are such that $\bar{b}$ is the longest proper regular subword in u from the right hand side. Indeed, we will show this by induction on $l(u)$. The starting point of induction: $u = yx$, $y, x \in X$. If $l(u) = n > 2$ then, as in the proof of Lemma 1.8, we consider new letters $x_i x^k$. If $\bar{a} = a_1 x$ and $\bar{b} = xb_1$ where $a_1, b_1 \in S(X)$ then $a_1 \neq 1$ and $xb_1 \geq x$ (by the definition of a regular monomial, Remark 1.5 and Claim 1 of Lemma 1.8). Therefore, $b_1 = 1$, $\bar{b} = x$. In the other case a and b can be written in letters $x_i x^k$ and we may apply our induction hypothesis to complete the proof.

Now we compute the number Ψ_n of regular words of fixed length n in the alphabet $X = \{x_1, \ldots, x_q\}$. Consider a word u with $l(u) = n$ and all its cyclic conjugates, i.e. the words of the form $u_2 u_1$ where $u = u_1 u_2$. Consider two possible cases.

Case 1. All cyclic conjugates of u are distinct and therefore in the conjugate class of u we have exactly n words. We may take as a representative of this class the lexicographically greatest word which is, obviously, regular.

Case 2. Some cyclic conjugates may coincide. To consider this situation we need the following.

1.10. Lemma. *Let $a, b \in S(X)$ and $ab = ba$. Then $a = v^k$, $b = v^l$, $v \in S(X)$ and all cyclic conjugates of v are distinct.*

Proof. We give the proof using induction on $l(a) + l(b)$. The basis of induction: $l(a) = 1$, $l(b) = 1$, $a, b \in X$. Then from $ab = ba$ it follows that $a = b$. Suppose that our assertion holds for all $a, b \in S(X)$ with $l(a) + l(b) < n$ and let $l(a) + l(b) = n$.

A: $a = b$. If all cyclic conjugates of a are distinct then our claim is obvious. If $u = u_1 u_2 = u_2 u_1$ for some cyclic conjugate u of a then $l(u_1) + l(u_2) = l(a) < 2l(a)$ and by the induction hypothesis $u_1 = v_1^k$, $u_2 = v_1^l$, and all cyclic conjugates of v_1 are distinct. Therefore $a = v^{k+l}$ where v is a cyclic conjugate of v_1, and the proof in the case (A) is now complete.

B: $a \neq b$. Let $l(a) < l(b)$. Then from $ab = ba$ it follows that $b = aw$. Therefore the equalities $aaw = awa$ and $aw = wa$ are equivalent. But $l(a) + l(w) < l(a) + l(b)$. By induction we have $a = v^k$, $w = v^l$, $b = aw = v^{k+l}$ and all cyclic conjugates of v are distinct. A similar argument works in the case $l(a) \geq l(b)$. This completes the proof of the lemma. □

§1. The free colour Lie superalgebra, s-regular words and monomials

Now by Lemma 1.10 in the Case 2 we have that some cyclic conjugate of u has the form v_1^k where all cyclic conjugates of v_1 are distinct. Therefore, $u = v^k$ where v is a cyclic conjugate of v_1 and all cyclic conjugates of v are distinct.

From Cases 1 and 2 we have

$$q^n = \sum_{l|n} l\Psi_l, \tag{1}$$

where $q = |X|$ and Ψ_l is the number of all regular words of length l.

We would like to resolve this equation relative to Ψ_l.

1.11. Definition. The *Möbius function* is given by $\mu(1) = 1$ and

$$\mu(n) = \begin{cases} (-1)^r & \text{if } n \text{ is the product of } r \text{ distinct primes,} \\ 0 & \text{if } n \text{ is divisible by the square of a prime.} \end{cases}$$

1.12. Lemma (Möbius inversion formula, see for example [Bahturin, 1987a, 3.1.2] or [Cohn, 1989, §2.4]). *Given any functions f, g on $\mathbb{N}$ such that $f(n) = \sum_{m|n} g(m)$, we have $g(n) = \sum_{m|n} \mu(m) f(n/m)$.*

1.13. Theorem (Witt's formula). *If Ψ_n is the number of regular words of length n in the alphabet $X = \{x_1, \ldots, x_q\}$ then*

$$\Psi_n = \frac{1}{n} \sum_{m|n} \mu(m) q^{n/m}.$$

Proof. The assertion of our theorem follows immediately from (1) and the Möbius Inversion Formula (1.12). □

1.14. Theorem (Witt's formula). *If $\Psi(\alpha)$ is the number of regular words in the alphabet $X = \{x_1, \ldots, x_q\}$ of multidegree $\alpha = (\alpha_1, \ldots, \alpha_q)$, $\alpha_i \in \mathbb{N} \cup \{0\}$, then*

$$\Psi(\alpha) = \frac{1}{|\alpha|} \sum_{l|\alpha} \mu(l) \frac{\left(\frac{|\alpha|}{l}\right)!}{\left(\frac{\alpha}{l}\right)!},$$

where $0! = 1! = 1$, $l|\alpha$ means $l|\alpha_i$ for all i, $\frac{\alpha}{l} = \left(\frac{\alpha_1}{l}, \ldots, \frac{\alpha_q}{l}\right)$, $\alpha! = \alpha_1! \ldots \alpha_q!$, $|\alpha| = \sum_{i=1}^q \alpha_i$. In particular, $\sum_{|\alpha|=n} \Psi(\alpha) = \Psi_n$.

Proof. By analogy with (1) we observe that for any multidegree $\beta = (\beta_1, \ldots, \beta_q)$ one has

$$\sum_{l|\beta} l\Psi\left(\frac{\beta}{l}\right) = \frac{|\beta|!}{\beta!}.$$

(It is necessary to replace the number q^n of all words of length n by the number $\frac{|\beta|!}{\beta!}$ of all words of multidegree β.) Applying Lemma 1.12 completes the proof. □

1.15. Lemma. *Let u be an s-regular monomial. Then $\bar{u}$ is an s-regular word. Conversely, if u is an s-regular word, then there is a unique arrangement of brackets $[u]$ on u such that $[u]$ is an s-regular monomial.*

Proof. Let u be a regular monomial. Then by Lemma 1.8 $\bar{u}$ is a regular word and, in particular, an s-regular word. Let now $u = (v)(v)$ where v is a regular monomial. Then by Lemma 1.8 $\bar{v}$ is a regular word. Since $d(v) \in G_-$, $\bar{u} = \bar{v}\bar{v}$ is an s-regular word.

Let u be a regular word. By Lemma 1.8 there is a unique arrangement of brackets $[u]$ on u such that $[u]$ is a regular monomial. But u is a regular word, therefore, there is no arrangement of brackets on u such that $[u] = ([v])([v])$. Let now $u = vv$ where v is a regular word and $d(v) \in G_-$. By Lemma 1.8 there is a unique arrangement of brackets $[v]$ on v such that $[v]$ is a regular monomial. Since $u = vv$ is not a regular word, there exists a unique arrangement of brackets $[u] = ([v])([v])$ on u such that $[u]$ is an s-regular monomial. The proof is now complete. □

§ 2. Bases of free colour Lie superalgebras

In this section we construct a basis in the free colour Lie superalgebra (Theorem 2.6). As a corollary we have the dimension formula for the subspace of elements of given multidegree in the free colour Lie superalgebra (Corollary 2.8 is an analogue of Witt's Formula). We produce a superanalogue of the Specht-Wever Criterion (Theorem 2.12) and describe the algorithm for "extracting square roots" of even elements in the free colour Lie superalgebra (Theorem 2.14).

2.1. Lemma. *Let $2, 3 \in K^*$ and u, v be s-regular monomials. Then:*
1) $u \circ v = \sum_k \alpha_k w_k$ *in $L(X)$ where w_k is an s-regular monomial for all k;*
2) $m(w_k) = m(uv)$;
3) *if $\bar{u} \neq \bar{v}$ then $\bar{w}_k > \min(\bar{u}, \bar{v})$.*

Proof. We use induction on $n = l(u) + l(v)$. The starting point is $n = 2$. Then $u, v \in X$. If $u = v$ and $d(v) \in G_-$ then $u \circ v$ is an s-regular monomial by definition. If $u = v$ and $d(v) \in G_+$ then $u \circ v = 0$. If $u > v$ then $u \circ v$ is a regular monomial and $\overline{uv} > \overline{v} = \min(\overline{u}, \overline{v})$. Finally, if $u < v$, then $u \circ v = -\varepsilon(u,v)v \circ u$, where $v \circ u$ is a regular monomial and $\overline{vu} > \overline{u} = \min(\overline{u}, \overline{v})$. Suppose that our assertion is true for all s-regular monomials w and z with $l(w) + l(z) < n$ and $l(u) + l(v) = n$. We need the following

2.2. Remark. Suppose that our inductive assumption takes place and $u = p \circ p$ where p is a regular monomial, $d(p) \in G_-$, $v \neq p$ and $v \neq u$. Then $u \circ v = \sum_i \delta_i u_i \circ v_i$, where $\delta_i \in K$ and u_i, v_i are regular monomials, $m(u_i v_i) = m(uv)$, $\overline{u}_i > \min(\overline{u}, \overline{v})$, $\overline{v}_i > \min(\overline{u}, \overline{v})$.

Proof. It is clear that $u \circ v = (p \circ p) \circ v = 2p \circ (p \circ v)$. By induction hypothesis we have $p \circ v = \sum_k \alpha_k v_k + \sum_l \beta_l p_l \circ p_l$ where $\alpha_k, \beta_l \in K$, v_k, p_l are regular monomials, $d(p_l) \in G_-$, $m(v_k) = m(p_l p_l) = m(pv)$, $\overline{v}_k > \min(\overline{p}, \overline{v})$, $\overline{p}_l \overline{p}_l > \min(\overline{p}, \overline{v})$. Therefore

$$u \circ v = 2 \sum_k \alpha_k p \circ v_k - 2 \sum_l \varepsilon(p, p_l p_l)\beta_l (p_l \circ p_l) \circ p. \tag{1}$$

Moreover, $m(pv_k) = m(p_l p_l p) = m(uv)$ for all k, l, and since $\overline{p} > \overline{pp}$ we have $\overline{v}_k > \min(\overline{pp}, \overline{v})$ and $\overline{p}_l \overline{p}_l > \min(\overline{pp}, \overline{v})$ for all k, l (thus the least factor in any summand of the right hand side in (1) is greater than the least factor in $u \circ v$). We set $T = \{t, \beta_t \neq 0\}$. If $T = \emptyset$ then the proof is complete. Otherwise, if $T \neq \emptyset$ then for any product $(p_l \circ p_l) \circ p$, $l \in T$ we apply the above procedure setting $u = p_l \circ p_l$, $v = p$ (in the case $p_l = p$ we have $(p_l \circ p_l) \circ p = 0$).

Since the set of all nonassociative monomials of multidegree $m(uv)$ is finite, after a finite number of steps we complete the proof. □

Now we continue the proof of Lemma 2.1.

Case 1. u and v are regular monomials.

a) $u = v$. If $d(u) \in G_-$, then $u \circ v$ is a regular monomial. If $d(u) \in G_+$, then $u \circ v = 0$.

b) $u \neq v$. We may assume that $\overline{u} > \overline{v}$ (otherwise, $u \circ v = -\varepsilon(u,v)v \circ u$).

b.1) $l(u) = 1$. Then $u \circ v$ is a regular monomial and by Lemma 1.4 we have $\overline{uv} > \overline{v} = \min(\overline{u}, \overline{v})$, i.e. our statement holds.

b.2) $l(u) > 1$. Then by the definition of regular monomials we have $u = u_1 \circ u_2$ where u_1, u_2 are regular monomials and $\overline{u}_1 > \overline{u}_2$.

b.2.1) $\overline{v} \geq \overline{u}_2$. Then, by definition, $u \circ v$ is a regular monomial and by Lemma 1.4 $\overline{uv} > \overline{v} = \min(\overline{u}, \overline{v})$.

b.2.2) $\bar{v} < \bar{u}_2$. Now we consider "the procedure of increasing the least factor":

$$(u_1 \circ u_2) \circ v = -\varepsilon(u_1, u_2 v)(u_2 \circ v) \circ u_1 + \varepsilon(u_2, v)(u_1 \circ v) \circ u_2. \quad (2)$$

(A) We consider $(u_2 \circ v) \circ u_1$. By induction hypothesis we have

$$u_2 \circ v = \sum_j \alpha_j z_j + \sum_l \beta_l p_l \circ p_l \quad \text{where} \quad \alpha_j, \beta_l \in K,$$

where z_j, p_l are regular monomials, $d(p_l) \in G_-$, $m(z_j) = m(p_l p_l) = m(u_2 v)$, $\bar{z}_k > \min(\bar{u}_2, \bar{v}) = \bar{v}$, $\bar{p}_l \bar{p}_l > \min(\bar{u}_2, \bar{v}) = \bar{v}$ for all j, l. By Remark 2.2, if $p_l \neq u_1$ then $(p_l \circ p_l) \circ u_1 = \sum_i \gamma_{l,i} w_{i1}^l \circ w_{i2}^l$ where $\gamma_{l,i} \in K$, w_{i1}^l, w_{i2}^l are regular monomials, $m(w_{i1}^l w_{i2}^l) = m(u_2 v u_1)$, $\bar{w}_{i1}^l > \min(\bar{p}_l \bar{p}_l, \bar{u}_1)$, $\bar{w}_{i2}^l > \min(\bar{p}_l \bar{p}_l, \bar{u}_1)$ (in the case $p_l = u_1$ we have $(p_l \circ p_l) \circ u_1 = 0$). Thus,

$$(u_2 \circ v) \circ u_1 = \sum_j \alpha_j z_j \circ u_1 + \sum_l \sum_i \beta_l \gamma_{l,i} w_{i1}^l \circ w_{i2}^l, \quad m(z_j u_1) = m(u_2 v u_1),$$

and since $\bar{p}_l \bar{p}_l > \bar{v}$, $\bar{u}_1 > \bar{v}$ we find that $\bar{w}_{i1}^l > \min(\bar{u}, \bar{v})$, $\bar{w}_{i2}^l > \min(\bar{u}, \bar{v})$. We use the following notation

$$J = \{j, z_j = u_1\}; \quad M = \{(l, i), w_{i1}^l = w_{i2}^l\}.$$

$1°$. If $j \in J$ or $(l, i) \in M$, then $z_j \circ u_1$ or $w_{i1}^l \circ w_{i2}^l$ are either some s-regular monomials or zero in $L(X)$. If w, v are regular words and $w > v$ then $ww > v$. Indeed, let $ww \leq v$. Then $v = wa$, $ww \leq wa$, $w \leq a$. Since v is a regular word we derive, by Lemma 1.4, $v > a$. Therefore, $w > wa = v > a$, which gives us a contradiction. Thus, in this case, $\bar{z}_j \bar{u}_1 > \bar{v}$ or $\bar{w}_{i1}^l \bar{w}_{i2}^l > \bar{v}$, respectively.

$2°$. If $j \notin J$ or $(l, i) \notin M$, then $z_j \circ u_1$ or $w_{i1}^l \circ w_{i2}^l$, respectively, is a product of two distinct regular monomials each of which has the least factor greater than v (we may assume that the least factor is the second one in the product).

(B) Analogously, we consider the second summand in (2) (changing places of u_1 and u_2 in (A)).

(C) From (A) and (B) it follows that $u \circ v = a + b$ where $a \in B(uv)$, $b = \sum_k \xi_k u_k \circ v_k$, $\xi_k \in K$, u_k, v_k are regular monomials, $u_k > v_k > v$, $m(u_k v_k) = m(uv)$. Therefore, in $u_k \circ v_k$, the least factor is greater than in $u \circ v$. Moreover, all summands of a have the least factor greater than in $u \circ v$. If $b = 0$ then the assertion of our lemma in the Case b.2.2 is proved. If now $b \neq 0$ then we apply the "increasing the least factor" procedure (see (2)) to the product $u_k \circ v_k$, setting $u = u_k$ and $v = v_k$. As the set of all monomials of multidegree $m(uv)$ is finite, after a finite number of steps of our procedure we get $u \circ v = c \in B(uv)$. Moreover, all summands in c have the least factor greater than in $u \circ v$. Thus we have proved the statement of our lemma in the Case b.2.2. This completes our consideration in the Subcase b) and in the whole of Case 1.

Case 2. One of u or v has the form $p \circ p$ where p is a regular monomial with $d(p) \in G_-$ (we may consider $u = p \circ p$, otherwise $u \circ v = -\varepsilon(u,v) v \circ u$). If $v = p$ or $v = u$ then $u \circ v = 0$. For $v \neq p$ and $v \neq u$ by Remark 2.2 we have that $u \circ v = \sum_i \delta_i u_i \circ v_i$ where $\delta_i \in K$, u_i, v_i are regular monomials, $m(u_i v_i) = m(uv)$, $\bar{u}_i > \min(\bar{u}, \bar{v})$, $\bar{v}_i > \min(\bar{u}, \bar{v})$. It follows from Case 1 that $u_i \circ v_i = \sum_k \alpha_{i,k} w_{i,k}$ for any i, where $\alpha_{i,k} \in K$, $w_{i,k}$ is an s-regular monomial, $m(w_{i,k}) = m(u_i v_i)$, $\bar{w}_{i,k} > \min(\bar{u}_i, \bar{v}_i)$ for any k. Thus $u \circ v = \sum_i \sum_k \delta_i \alpha_{i,k} w_{i,k}$ where $w_{i,k}$ is an s-regular monomial for all i, k, $m(w_{i,k}) = m(uv)$, $\bar{w}_{i,k} > \min(\bar{u}_i, \bar{v}_i) > \min(\bar{u}, \bar{v})$, completing the proof of our lemma. □

2.3. Lemma. *For any $w \in \Gamma(X)$ the coset with representative w is a linear combination of cosets whose representatives are s-regular monomials.*

Proof. The proof will be given by induction on the length $l(w)$. The basis of induction is $l(w) = 1$. Here, obviously, w is a regular monomial. Let $l(w) > 1$ and suppose that our statement holds for all $w' \in \Gamma(X)$ with $l(w') < l(w)$. Then $w = u \circ v$, $u, v \in \Gamma(X)$, $l(u) < l(w)$, $l(v) < l(w)$. By induction hypothesis, the cosets with representatives u and v are linear combinations of cosets such that their representatives are s-regular monomials. Applying distributivity and Lemma 2.1, we have proved our assertion. □

2.4. Lemma. *If $[X]$ is a subalgebra of the colour Lie superalgebra $[A(X)]$ generated by X and $\pi: L(X) \to [X]$ is the homomorphism induced by the identity mapping of X, then $\pi(\tilde{u}) = \bar{u} + u^*$ for any regular monomial u and $\pi(\tilde{u}) = 2\bar{u} + u^*$ for $u = v \circ v$, where v is a regular monomial with $d(u) \in G_-$. Moreover, in both cases u^* is a linear combination of some words which are less than $\bar{u}$ and have multidegree equal to $m(u)$. In particular, π is an isomorphism of colour Lie superalgebras.*

In what follows we shall often call $\pi(\tilde{u})$ the *associative form* of $u \in L(X)$.

Proof. For a regular monomial we argue by induction on its length. The basis of induction: $l(u) = 1$, $u \in X$; here $\pi(\tilde{u}) = u$. Suppose that $l(u) \geq 2$ and that our assertion is valid for all regular monomials v such that $l(v) < l(u)$. Then $u = u_1 \circ u_2$, where u_1, u_2 are regular monomials. By the induction assumption, $\pi(\tilde{u}_1) = \bar{u}_1 + u_1^*$, $\pi(\tilde{u}_2) = \bar{u}_2 + u_2^*$ and therefore

$$\pi(\tilde{u}) = [\pi(\tilde{u}_1), \pi(\tilde{u}_2)] = \bar{u}_1 \bar{u}_2 + u_1^* \bar{u}_2 + \bar{u}_1 u_2^* + u_1^* u_2^*$$

$$- \varepsilon(u_1, u_2)(\bar{u}_2 \bar{u}_1 + \bar{u}_2 u_1^* + u_2^* \bar{u}_1 + u_2^* u_1^*).$$

It is obvious that $\bar{u}_1 \bar{u}_2$ is the greatest word among the words occuring in the first four summands and $\bar{u}_2 \bar{u}_1$ is the greatest word among the words occuring in the last four summands. Since $\bar{u} = \bar{u}_1 \bar{u}_2$ is a regular word, $\bar{u}_1 \bar{u}_2 > \bar{u}_2 \bar{u}_1$.

Let now $u = v \circ v$ where v is a regular monomial, $d(v) \in G_-$. As we saw above $\pi(\tilde{v}) = \bar{v} + v^*$. Since $\varepsilon(v,v) = -1$, we have $\pi(\tilde{u}) = [\pi(\tilde{v}), \pi(\tilde{v})] = 2\bar{v}\bar{v} + 2(\bar{v}v^* + v^*\bar{v} + v^*v^*)$ and therefore $\pi(\tilde{u}) = 2\bar{u} + u^*$. □

2.5. Corollary. *The cosets in $L(X) = F(X)/I$ whose representatives are s-regular monomials are linearly independent.*

Proof. Let $\sum_i \alpha_i u_i = 0$ where the u_i are pairwise distinct s-regular monomials, $0 \neq \alpha_i \in K$. We choose the greatest s-regular monomial u_k. Since $\pi(\sum \alpha_i \tilde{u}_i) = 0$, by Lemmas 1.8 and 2.4 we have that either $\alpha_k \bar{u}_k = 0$ or $2\alpha_k \bar{u}_k = 0$ (and then since $2 \in K^*$ we have $\alpha_k = 0$). Thus we have a contradiction. □

2.6. Theorem. *Let K be a commutative ring with 1. Let $2, 3 \in K^*$, and let $L(X) = F(X)/I$ be the free colour Lie superalgebra. Then the cosets in $F(X)/I$ whose representatives are s-regular monomials form a basis of $L(X)$ as a free G-graded K-module.*

Proof. From Lemma 2.3 and the fact that $\{z + I, z \in \Gamma(X)\}$ is a set of generators of the K-module $L(X)$ it follows that the cosets with s-regular monomials as representatives also generate the K-module $L(X)$. By Corollary 2.5 they are linearly independent. □

2.7. Remark. The statement of our theorem is false if we do not assume that $2, 3 \in K^*$. Indeed, the basis constructed above does not contain the squares of even elements and the cubes of odd elements. It is possible to overcome this difficulty if we add to the defining identities of the class of colour Lie superalgebras the identities of the form $[x, x] = 0$ and $[y, [y, y]] = 0$ with x homogeneous even and y homogeneous odd.

Observe that, in Theorem 2.6, instead of regular monomials we can use the arrangement of brackets [ ] on each regular word u such that the greatest associative word in the expression of $\pi(\widetilde{[u]})$ is equal to u. Examples of such arrangements of brackets which are different from the arrangement of brackets on a regular monomial will be given in Lemma 3.2 of Chapter 3.

2.8. Corollary (an analogue of Witt's formula). *Let $2, 3 \in K^*$, $X = \bigcup_{g \in G} X_g$, $X_+ = \{x_1, \ldots, x_t\}$, $X_- = \{x_{t+1}, \ldots, x_{t+s}\}$, and let $L(X)$ be the free colour Lie K-superalgebra, $\mu(l)$ the Möbius function, and $W(\alpha_1, \ldots, \alpha_k)$ the rank of the free module of elements of multidegree $\alpha = (\alpha_1, \ldots, \alpha_k)$ in the free Lie algebra of rank k,*

$$W(\alpha_1, \ldots, \alpha_k) = \frac{1}{|\alpha|} \sum_{e|\alpha} \mu(e) \frac{(|\alpha|/e)!}{(\alpha/e)!},$$

where $|\alpha| = \sum_{i=1}^{k} \alpha_i$ (Witt's Formula, cf. Theorems 1.14 and 2.6). Let $SW(\alpha_1, \ldots, \alpha_{t+s})$ be the rank of the free module of elements of multidegree $\alpha = (\alpha_1, \ldots, \alpha_{t+s})$ in the free colour Lie superalgebra $L(X)$ of rank $t + s$. Then

$$SW(\alpha_1, \ldots, \alpha_{t+s}) = W(\alpha_1, \ldots, \alpha_{t+s}) + \beta W\left(\frac{\alpha_1}{2}, \ldots, \frac{\alpha_{t+s}}{2}\right),$$

where

$$\beta = \begin{cases} 0 & \text{if there exists an } i \text{ such that} \\ & \alpha_i \text{ is odd, or if } \frac{1}{2}\sum_{i=t+1}^{t+s} \alpha_i d(\alpha_i) \in G_-. \\ 1 & \text{otherwise.} \end{cases}$$

2.9. Exercise (See [Molev, Tsalenko, 1986]). Let $L(X)$ be the free Lie superalgebra, $X = X_- = X_1 = \{x_1, \ldots, x_s\}$, K a field, char $K = 0$. Then

$$SW(\alpha_1, \ldots, \alpha_s) = \frac{(-1)^{|\alpha|}}{|\alpha|} \sum_{e \mid \alpha} \mu(e) \frac{(-1)^{|\alpha|/e}(|\alpha|/e)!}{\alpha!}.$$

2.10. Definition. For $w \in L(X)$ we define the *length* $l_X(w)$ as the greatest length of s-regular monomials in the presentation of w as a linear combination of s-regular monomials. If $\pi(\tilde{w}) = \sum_{i=1}^{k} \alpha_i u_i$ where $0 \neq \alpha_i \in K$, $u_i \in S(X)$, $u_i \neq u_j$ for $i \neq j$, then $l_X(w) = \max_{1 \leq i \leq k}\{l_X(u_i)\}$. Therefore $l_X(w)$ does not depend on the ordering of X.

Let now w_0 denote the leading part of $w \in L(X)$ (i.e. the sum of summands whose length is equal to $l_X(w)$ in the s-regular presentation of w). If $w = w_0$ then we say that w is *l-homogeneous*. By analogy, we define the length $l_x(w)$, the leading part and the property of being homogeneous relative to $x \in X$. We say that an element $w \in L(X)$ is *multihomogeneous* if w is homogeneous relative to all $x \in X$. The greatest monomial of the leading part of w will be called the *leading term* of w.

The cardinality $|X|$ is called the *rank* of the free colour Lie superalgebra $L(X)$. It follows from Theorem 2.6 that

$$\text{rank } L(X) = \dim L(X)/[L(X), L(X)]; \qquad |X_g| = \dim(L(X)/[L(X), L(X)])_g$$

and therefore, rank $L(X)$ and $|X_g|$ do not depend on the choice of a set of free generators in $L(X)$.

2.11. Remark. Any associative derivation δ with $d(\delta) \in G$ of the G-graded associative algebra Q gives us the derivation of the colour Lie superalgebra

[Q]:

$$\delta([a,b]) = \delta(ab - \varepsilon(a,b)ba)$$

$$= \delta(a)b + \varepsilon(d(\delta),d(a))a\delta(b) - \varepsilon(a,b)(\delta(b)a + \varepsilon(d(\delta),d(b))b\delta(a))$$

$$= (\delta(a)b - \varepsilon(d(\delta(a)),d(b))b\delta(a))$$

$$+ \varepsilon(d(\delta),d(a))(a\delta(b) - \varepsilon(d(a),d(\delta(b)))\delta(b)a)$$

$$= [\delta(a),b] + \varepsilon(d(\delta),d(a))[a,\delta(b)].$$

All ε-derivations δ with $d(\delta) \in G$ of the free G-graded associative algebra $A(X)$ have the following form:

$$\delta(x_1 \ldots x_n) = \delta(x_1)x_2 \ldots x_n + \varepsilon(d(\delta),d(x_1))x_1\delta(x_2)x_3 \ldots x_n + \cdots$$

$$+ \varepsilon\left(d(\delta), \sum_{i=1}^{n-1} d(x_i)\right) x_1 \ldots x_{n-1}\delta(x_n)$$

where $i \in \mathbb{N}$, $x_i \in X$.

The proof is easy by induction on n.

Since $[X] \subset A(X)$ is the free colour Lie superalgebra generated by X, any mapping $\delta: X \to [X]$ with $d(\delta) \in G$ (i.e. $d(\delta(x)) = d(x) + d(\delta)$) has a unique extension to a derivation of the free colour Lie superalgebra $[X]$.

Let $x_i \in X$ and $[x_i] = x_1, \ldots, [x_1 x_2 \ldots x_n] = [x_1, [x_2 \ldots x_n]]$ be right-normed monomials. Consider a linear mapping $\delta: A(X) \to [X]$ such that $\delta(x_1 \ldots x_n) = [x_1 \ldots x_n]$ for all $x_i \in X$, $n \in \mathbb{N}$. Observe that $d(\delta(u)) = d(u)$ for all $u \in S(X)$.

Since $A(X)$ is the universal enveloping algebra for $[X]$ there exists a homomorphism of the G-graded associative algebra $\theta: A(X) \to \text{End}[X]$ such that the following diagram is commutative.

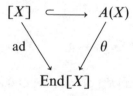

We may write

$$\delta(x_{i_1} \ldots x_{i_n} x_{j_1} \ldots x_{j_m}) = \theta(x_{i_1} \ldots x_{i_n})\delta(x_{j_1} \ldots x_{j_m}).$$

For homogeneous elements $a, b \in [X]$ we have

$$\delta([a,b]) = \delta(ab - \varepsilon(a,b)ba) = \theta(a)\delta(b) - \varepsilon(a,b)\theta(b)\delta(a)$$

$$= [a, \delta(b)] - \varepsilon(a,b)[b, \delta(a)]$$

$$= [a, \delta(b)] + \varepsilon(a,b)\varepsilon(d(b), d(\delta(a)))[\delta(a), b] = [\delta(a), b] + [a, \delta(b)]$$

because of $d(\delta) = 0$, i.e. δ is a derivation with $d(\delta) = 0$ of the colour Lie superalgebra $[X]$.

2.12. Theorem (Specht-Wever criterion). *Let K be a field, char $K = 0$. Then an element $a \in A(X)$ with $l(a) = m$ belongs to $[X]$ if, and only if, $\delta(a) = ma$.*

Proof. If $\delta(a) = ma$, then $a = (1/m)\delta(a) \in [X]$. We show by induction on the length that any element $a \in [X]$ is a linear combination of monomials of the form $[x_{i_1} \ldots x_{i_k}]$ where $x_{i_j} \in X$. Clearly, it is necessary to prove this only for l-homogeneous elements. The basis of induction: $l(a) = 1$. Then a has the required form. Let now $a = \sum_i \gamma_i [u_i, v_i]$ where $\gamma_i \in K$ and $[u_i, v_i]$ are s-regular monomials, $l(u_i v_i) = l(a)$. It is sufficient to consider only one summand $[u_j, v_j]$. By the induction hypothesis,

$$u_j = \sum_i \alpha_i \omega_i, \quad l(\omega_i) = l(u_j), \quad v_j = \sum_k \beta_k w_k, \quad l(w_k) = l(v_j), \quad \alpha_i, \beta_k \in K$$

and ω_i, w_k have the desired form for all i, k. Then $[u_j, v_j] = \sum_{i,k} \alpha_i \beta_k [\omega_i, w_k]$ and it is sufficient to consider the elements

$$[[x_{i_1} \ldots x_{i_n}], [x_{j_1} \ldots x_{j_m}]] = \alpha[[x_{i_2} \ldots x_{i_n}], [x_{i_1}, [x_{j_1} \ldots x_{j_m}]]]$$

$$+ \beta[x_{i_1}, [[x_{i_2} \ldots x_{i_n}], [x_{j_1} \ldots x_{j_m}]]]$$

where $\alpha, \beta \in K$, $x_{i_j} \in X$. The first summand has the same form, but with first factor of shorter length. The second factor in the second summand is, by induction, a linear combination of monomials having the form wanted. It is clear that the second summand also is such a linear combination. The proof is completed by induction on the length of the first factor.

Thus, it is sufficient to prove the theorem only for right-normed monomials $[x_{i_1} \ldots x_{i_n}]$. We use induction on n. If $n = 1$ then $\delta(x_i) = x_i$. Furthermore

$$\delta([x_{i_1}, [x_{i_2} \ldots x_{i_n}]]) = [\delta(x_{i_1}), [x_{i_2} \ldots x_{i_n}]] + [x_{i_1}, \delta([x_{i_2} \ldots x_{i_n}])]$$

$$= (1 + (n-1))[x_{i_1} \ldots x_{i_n}] = n[x_{i_1} \ldots x_{i_n}]$$

since δ is a derivation with $d(\delta) = 0$. □

Observe that in the free colour Lie superalgebras we can have nonzero squares of odd elements which are certainly even. It is a natural question to decide whether an even element is a square of some odd element, and if so to give an algorithm for "finding a square root".

2.13. Lemma. *Let K be a field, char $K \neq 2, 3$, $L(X) = L_+ \oplus L_-$ be the free colour Lie superalgebra, $v \in L_-$, $w = [v, v]$. If $[w, z] = 0$ with $z \in L_-$ then $z = \alpha v$, $\alpha \in K$.*

Proof. If $z = \alpha v$, $\alpha \in K$ then $[w, z] = 0$ (see Lemma 1.8 in Chapter 1). If now $z \neq \alpha v$, $\alpha \in K$, then $z = \beta v + u$, $\beta \in K$, and the leading term $u°$ of u is different from $v°$. Since the associative leading term of $[w, z]$ is equal to either $\gamma \bar{v} \bar{v} u°$ or $\gamma' \bar{u}° \bar{v} \bar{v}$, $0 \neq \gamma, \gamma' \in K$, we have $[w, z] = [w, u] \neq 0$. This completes the proof. $\square$

2.14. Theorem. *Let K be a field, char $K \neq 2, 3$. Suppose that, in K, we have an algorithm for square root extraction (for example, in $\mathbb{Q}$, $\mathbb{R}$, $\mathbb{C}$, p-adic numbers $\mathbb{Q}_p$ and in finite fields). Let $L(X)$ be the free colour Lie K-superalgebra. Then there exists an algorithm which decides whether an even element w is a square of some odd element and, if so, to find a square root.*

Proof. We suppose that our element w is a sum of monomials. First of all we express it as a sum of s-regular monomials. It is possible to write any monomial as a linear combination of s-regular monomials in the following way. If $x \in X$ then x is a regular monomial. In the monomials $[x_i, x_j]$, x_i, $x_j \in X$, it is sufficient just to transpose the letters. If we consider a monomial $[u, v]$ with $l(u) + l(v) \geq 2$ then, by induction, we may assume that u and v have the form wanted. Furthermore, the summands of $[u, v]$ which have the form of products of s-regular monomials can be reduced to the s-regular form by the procedure of increasing the least factor (see Lemma 2.1, (2)). Hence, we may assume that the element $w \in L(X)_+$ is already in an s-regular form. If its leading term (according to its length and, for equal lengths, to its lexicographical order) does not have the form $\alpha[u, u]$, where $0 \neq \alpha \in K$, $\sqrt{\alpha} \in K$ and u is a regular monomial with $d(u) \in G_-$, then w is not a proper square. If the leading term of w has the above form, then we write down inductively, by definition of regular monomials, consecutively the set $\mathscr{P}$ of all even regular monomials which are smaller than u (at first by length and then lexicographically). By Lemma 2.13 the "square root" sought z, $d(z) \in G_-$, satisfies the equation $[w, z] = 0$. We shall resolve the equation $[w, z] = 0$ where z is a linear combination of monomials from $\mathscr{P} \cup u$ (the coefficient of u is equal to 1). We resolve this equation in the following way.

1. We write down $[w, z]$ in a regular form (as it has been described above).

2. We reduce the equation $[w, z] = 0$ to a system of linear equations on coefficients of z in the regular form equating to zero the coefficients of regular monomials in the regular expression of $[w, z]$.

3. We resolve the system of linear equations using one of the standard methods (for example, the Gauss method). If our system is incompatible then w is not a square. Otherwise, by Lemma 2.13 we have the unique solution z with the leading term u. Multiplying it by $\pm\sqrt{\alpha}$ we have all the desired square roots of w. □

2.15. Remark. Using our constructions of bases of the free colour Lie superalgebra L over a commutative ring K it is possible to prove that the image of the inner derivation ad x of an element x of L is a direct summand of a free K-module L in the following cases: a) when the leading associative word in x is regular and its coefficient is invertible in K; b) in the case when $x = [v, v]$ where v is an odd element of L as in a). For the details (also for the case of colour Lie p-superalgebra) see [Mikhalev, 1992a, 1992b]. Remark that for the free Lie algebra in the case when x is an element of a free generating set of L this result was proved by D. Doković in [Doković, 1988].

§ 3. The freeness of subalgebras and its corollaries

In this section we prove a theorem about the freeness of G-homogeneous subalgebras in the free colour Lie superalgebras (Theorem 3.15). As a corollary we have: an analogue of Schreier's Formula for the rank of a subalgebra which has finite odd codimension (Theorem 3.19); an analogue of P.M. Cohn's Theorem on generators of the automorphism group of a free Lie algebra of finite rank (Theorem 3.25); Theorem 3.28 about the infiniteness of the rank of the even component; a theorem on the possibility to recover a subalgebra of finite rank by its odd component (if it is non-trivial) (Theorem 3.30), etc.

Throughout this section K stands for a field, char $K \neq 2, 3$ and $L(X) = L_+ \oplus L_-$ is a free colour Lie superalgebra.

Recall that, in a colour Lie superalgebra R, for $x, y \in R$ we write ad $x(y) = [x, y]$. Sometimes we use the notation y ad′ $x = [y, x]$. We introduce it since the construction of a basis in free colour Lie superalgebras uses regular words in the sense of A.I. Shirshov (see Definition 1.2 and also [Shirshov, 1958]). Remark that it is possible to consider the symmetric version of regular words (i.e. from $a = uv$ it follows that $uv < vu$).

3.1. Definition. Let $X = \bigcup_{g \in G} X_g$, $x \in X_+$, $z \in X_-$. The following notation is used systematically:

$$Z(x) = \{y, y(\text{ad}' x)^n | y \in X \setminus x, n \in \mathbb{N}\} \subset L(X),$$

$$W(z) = \{y, [y, z], [z, z] | y \in X \setminus z\} \subset L(X),$$

$$\bar{Z} = \{\bar{u}, u \in Z(x)\}, \qquad \bar{W} = \{\bar{w}, w \in W(z)\}.$$

We consider a well-ordering on X. Let x be the least element. We extend this ordering to a partial ordering $\rhd$ on $S(X)$ in the following way: for $a, b \in S(X)$ we have $b \rhd a$ if, and only if, $m(a) = m(b)$ and a has a smaller number of inversions. It is clear that this is a semigroup ordering, i.e. $a \rhd b$, $c \unrhd d$ or $a \unrhd b$, $c \rhd d$ implies that $ab \rhd cd$.

3.2. Lemma. *Let $u \in Z(x)$. Then $u = \bar{u} + u^*$ where $u^* = \sum_{j=1}^{t} \alpha_j u_j$, $\alpha_j \in K$, $u_j \in S(X)$, $m(u_j) = m(u)$, $\bar{u} \rhd u_j$ for $1 \leq j \leq t$.*

Proof. Since $x < y$, u is a regular monomial. By Lemma 2.4, $u = \bar{u} + u^*$ where

$$u^* = \sum_{j=1}^{t} \alpha_j u_j, \alpha_j \in K, u_j \in S(X), m(u_j) = m(u),$$

and $\bar{u} > u_j$ for $1 \leq j \leq t$. But, for our u and u_j, the inequalities $\bar{u} > \bar{u}_j$ and $\bar{u} \rhd \bar{u}_j$ are equivalent. □

3.3. Lemma. *Let x be the least element in X and let w be a regular monomial, $w \neq x$. Then w is a monomial in the elements of $Z(x)$. Moreover, if $x \in X_-$ then $w = 2^{-k} u$, where u is a monomial in the elements of $W(x)$, $k \in \mathbb{N} \cup \{0\}$.*

Proof. We use induction on the length $l(w)$. The basis of induction: $l(w) = 1$, i.e. $w \in Z \cap W$. Let now $l(w) = n > 1$. Then $w = [w_1, w_2]$. Since $\bar{w} > x$ we have $\bar{w}_1 > x$. As w_1 is a regular monomial and by the induction hypothesis, w_1 is a monomial in the elements of Z (if $x \in X_-$ then $w_1 = 2^{-k_1} u_1$ where u_1 is a monomial in the elements of W, $k_1 \in \mathbb{N} \cup \{0\}$). If $\bar{w}_2 > x$ then, by analogy with the case of w_1, we have the desired form of presentation for w_2, and in this case our proof is complete. Let now $\bar{w}_2 \leq x$. Since x is the least letter and w_2 is a regular monomial, $w_2 = x$. Furthermore, if $w_1 = [w_{11}, w_{12}]$ then it follows from the regularity of w that $\bar{w}_{12} \leq \bar{w}_2$ and, therefore, $w_{12} = x$. Repeating this procedure we see that $w = y(\text{ad}' x)^m$, $y \in X \setminus x$, $m \in \mathbb{N}$, i.e. $w \in Z$. If $x \in X_-$ and $m = 2t$ then $w = 2^{-t} v$, where $v = y(\text{ad}'[x, x])^t$; and if $m = 2t + 1$, then $w = 2^{-t} v_1$, where $v_1 = [y, x](\text{ad}'[x, x])^t$. This completes the proof of the lemma. □

§ 3. The freeness of subalgebras and its corollaries   55

3.4. Corollary. *Let x be the least element in the well-ordered set X, w an s-regular monomial, $w \neq x$, $l_x(w) = t \neq 0$. Then:*
1) *if either $x \in X_+$ or $w \neq [x, x]$ with $x \in X_-$, then w is a monomial in the elements of the set $Z = Z(x)$, $l_Z(w) = l_x(w) - t < l_x(w)$;*
2) *if $x \in X_-$ then $w = 2^{-k}u$, where $k \in \mathbb{N} \cup \{0\}$ and u is a monomial in the elements of the set $W = W(x)$, $l_W(w) < l_x(w)$.* □

Observe that if in Claim 2 of Corollary 3.4 we have $m(w_1) = m(w_2)$ then it is possible to have $l_W(w_1) \neq l_W(w_2)$. Indeed, let $y_1 > y_2 > x$, $w_1 = [[y_1, x], [y_2, x]]$ and $w_2 = [[[y_1, x], x], y_2]$. Then w_1, w_2 are regular monomials, $m(w_1) = m(w_2)$, but $l_{W(x)}(w_1) = 2 \neq 3 = l_{W(x)}(w_2)$.

3.5. Lemma. *Consider a value of the word f on the elements of the sets Z and W: $u = f(z_i)$, $v = f(w_i)$, $z_i \in Z$, $w_i \in W$. Then the leading terms of u and v (relative to the ordering $\rhd$) have the following form: $u_0 = f(\bar{z}_i)$, $v_0 = 2^l f(\bar{w}_i)$, where $l \in \mathbb{N} \cup \{0\}$.*

Proof. We remark that $[x, x] = 2xx$, $[y, x] = yx - \varepsilon(y, x)xy$, where $yx \rhd xy$ since $y > x$. Substituting z_i and w_i, respectively, into f and recalling Lemma 3.2 and the semigroup property of $\rhd$, we have proved the claim of the lemma. □

3.6. Lemma. *Suppose that the values of some words f and g in some fixed elements $a_1, \ldots, a_n$ of $\bar{Z}$ (or $\bar{W}$) are equal. Then the words f and g coincide.*

Proof. Let
$$f(a_1, \ldots, a_n) = a_{i_1} \ldots a_{i_k} = a_{j_1} \ldots a_{j_l} = g(a_1, \ldots, a_n).$$

We show that $k = l$, $i_q = j_q$ for $1 \leq q \leq k$. The proof will be given by induction on the length $l_x(a_{i_1}, \ldots, a_{i_k}) = t$. It is clear that our assertion is valid for $t = 1$. Let now $t = n > 1$.

The first case: in $a_{i_1} \ldots a_{i_k} = a_{j_1} \ldots a_{j_l}$ we have a subword xy, where $y \in X \setminus x$. Consider the first such subword from the left. By the definition of elements of Z (respectively, of W) it is clear that a_i does not contain xy, $y \in X \setminus x$ for any i. Hence,
$$a_{i_1} \ldots a_{i_p} = a_{j_1} \ldots a_{j_r} \quad \text{and} \quad a_{i_{p+1}} \ldots a_{i_k} = a_{j_{r+1}} \ldots a_{j_l}$$
for some p and r. By the induction hypothesis, we have verified our statement.

The second case: in $a_{i_1} \ldots a_{i_k} = a_{j_1} \ldots a_{j_l}$ we have no subword xy, where $y \in X \setminus x$. If $a_{i_1} \ldots a_{i_k}$ contains $y \in X \setminus x$, then
$$a_{i_1} \ldots a_{i_k} = y_1 \ldots y_r x^m, \quad m \in \mathbb{N} \cup \{0\}.$$

For $\bar{Z}$ we have $y_1 \ldots y_r x^m = y_1 \ldots y_{r-1} \overline{(y_r)} (\mathrm{ad}' x)^m$ and it is clear that this presentation is unique. Therefore $k = l$, $a_{i_s} = a_{j_s}$. For $\bar{W}$ we have

$$y_1 \ldots y_r x^m = \begin{cases} y_1 \ldots y_{r-1} w_1 w^t, & \text{where } 2t = m - 1,\ w_1 = y_r x,\ w = xx; \\ y_1 \ldots y_r w^t, & \text{where } 2t = m,\ w = xx. \end{cases}$$

It is obvious that this presentation is also unique, and therefore $k = l$, $a_{i_s} = a_{j_s}$.

If $a_1, \ldots, a_k$ does not contain any $y \in X \setminus x$, then $a_{i_1} \ldots a_{i_k} = x^m$. In the case of $\bar{Z}$ it is impossible. For $\bar{W}$ it follows that $m = 2t$. Since the presentation $x^m = w^t$, $w = xx$, is unique, we have $k = l = t$, $i_1 = \cdots = i_k = j_1 = \cdots = j_l$. □

3.7. Lemma. *Let $L(X)$ be a free colour Lie superalgebra, $x \in X$, and let $B(x)$ be a subalgebra of $L(X)$, generated by the set $Z(x)$ (for $x \in X_-$ let $C(x)$ be a subalgebra of $L(X)$, generated by the set $W(x)$). Then Z and W are free sets of generators of colour Lie superalgebras $B(x)$ and $C(x)$ respectively.*

Proof. By Lemma 2.4, it is sufficient to show that Z and W are free sets of generators of associative subalgebras $\mathrm{Ass}(Z)$ and $\mathrm{Ass}(W)$ in $A(X)$, respectively generated by Z and W (it is sufficient to show that different associative words in the elements from Z and from W are linearly independent in $\mathrm{Ass}(Z)$ and $\mathrm{Ass}(W)$ respectively).

Assume, on the contrary, that $\sum_{i=1}^n \alpha_i f_i(b_1, \ldots, b_m) = 0$, where $0 \neq \alpha_i \in K$, f_i is a word of $b_i \in Z$ (respectively, $b_i \in W$), $b_i \neq b_j$ for $1 \leq i, j \leq n$, $i \neq j$. By Lemma 3.5

$$\sum_{i=1}^n \alpha_i f_i(b_1, \ldots, b_m) = \sum_{i=1}^n \beta_i f_{i,0} + \sum_{i=1}^n \sum_{j=1}^k \delta_j f_{i,j} = 0,$$

where $0 \neq \beta_i = \alpha_i$ for $b_1, \ldots, b_m \in Z(x)$ and $0 \neq \beta_i = 2^{q_i} \alpha_i$, where $q_i \in \mathbb{N} \cup \{0\}$, for $b_1, \ldots, b_m \in W(x)$; $f_{i,0} = f_i(\bar{b}_1, \ldots, \bar{b}_m)$, $\delta_j \in K$, $f_{i,j} \in S(X)$, $m(f_{i,j}) = m(f_{i,0})$, $f_{i,0} \rhd f_{i,j}$ for all $i, j, j \neq 0$. We extend the partial ordering $\rhd$ to a total ordering on $S(X)$. By Lemma 3.6, all words $f_{i,0}$ are pairwise distinct. Let $f_{i_0,0}$ be the greatest word among them, and therefore, among all $f_{i,j}$. Thus $\beta_{i_0} f_{i_0,0}$ does not cancel in our linear combination. This contradiction completes the proof of the lemma. □

3.8. Definition. We say that a subset M of G-homogeneous elements of $L(X)$ is *independent* if M is a set of free generators of the subalgebra, generated by M in $L(X)$.

As we have shown, each of the sets $Z(x)$ and $W(x)$ is independent.

§3. The freeness of subalgebras and its corollaries

3.9. Definition. A subset $M = \{a_i\}$ of G-homogeneous elements in $L(X)$ is called *reduced* if for any i the leading part a_i^0 of the element a_i does not belong to the subalgebra generated by the set $\{a_j^0, j \neq i\}$.

3.10. Definition. Let $S = \{s_\alpha, \alpha \in I\}$ be a G-homogeneous subset of $L(X)$. We say that a mapping $\omega: S \to L(X)$ is an *elementary transformation* if $\omega(s_\alpha) = s_\alpha$ for all $\alpha \in I \setminus \beta$, $\omega(s_\beta) = \lambda s_\beta + \omega(s_{\alpha_1}, \ldots, s_{\alpha_t})$, where

$$0 \neq \lambda \in K, \alpha_1, \ldots, \alpha_t \neq \beta, d(\omega(s_{\alpha_1}, \ldots, s_{\alpha_t})) = d(s_\beta),$$

$$\omega(y_{\alpha_1}, \ldots, y_{\alpha_t}) \in L(y_1, y_2, \ldots), \qquad d(y_i) = d(s_i).$$

Such an elementary transformation is called *triangular* if $l_X(\omega(s_{\alpha_1}, \ldots, s_{\alpha_t})) \leq l_X(s_\beta)$.

3.11. Lemma. *Elementary transformations of the set X of free generators induce automorphisms of the free colour Lie superalgebra $L(X)$.*

Proof. Let ω be an elementary transformation as above. We consider an elementary transformation ω' such that $\omega'(s_\alpha) = s_\alpha$ for $\alpha \neq \beta$ and $\omega'(s_\beta) = \lambda^{-1} s_\beta - \lambda^{-1} \omega(s_{\alpha_1}, \ldots, s_{\alpha_t})$. It is clear that the composition of ω' and ω gives us the identity transformation of S. In the case $S = X$ let $\bar\omega$ and $\bar\omega'$ be the corresponding endomorphisms of the algebra $L(X)$. Then the endomorphisms $\bar\omega\bar\omega'$ and $\bar\omega'\bar\omega$ are the identity transformations of X. Hence, $\bar\omega$ is an automorphism. □

3.12. Lemma. *Let B be a G-homogeneous subalgebra of $L(X)$. Then there exists a reduced subset M in B such that M generates B.*

Proof. We construct M as a union $\bigcup_{j=0}^\infty M_j$, where M_j consists of G-homogeneous elements of length j, if non-empty. We set $M_0 = \emptyset$. Then, by induction, we suppose that $M_0, \ldots, M_i$ have been defined. Let B_i be the subalgebra of B, generated by $\bigcup_{j=0}^i M_j$ and

$$S_i = \{b \in B_i | l(b) \leq i + 1\}.$$

Consider the G-homogeneous subsets of $\{b \in B | l(b) = i + 1\}$ which are linearly independent modulo the subspace S_i, and let M_{i+1} be a maximal one. It is clear that M generates the subalgebra B.

Let $f \in \bigcup_{j=0}^{i+1} M_j$. Suppose that the leading part f^0 of f belongs to the subalgebra generated by the set $\{h^0, h \in M, h \neq f\}$. It is evident that f^0, in fact, belongs to the subalgebra, generated by the set $\{h^0, h \in \bigcup_{j=0}^{i+1} M_j, h \neq f\}$.

Therefore,
$$f^0 = \alpha_1 f_1^0 + \cdots + \alpha_k f_k^0 + \omega(f_{k+1}^0, \ldots, f_s^0),$$

where $f_1, \ldots, f_k \in M_{i+1}$ and $f_{k+1}, \ldots, f_s \in \bigcup_{j \leq i} M_j$, $l(w(f_{k+1}^0, \ldots, f_s^0)) < i + 1$. Considering relations among elements we have

$$f = \alpha_1 f_1 + \cdots + \alpha_k f_k + \omega(f_{k+1}, \ldots, f_s) + g,$$

where g is an element with $l(g) \leq i$. Hence $c \in B$ and therefore $c \in S_i$. Thus we have a contradiction to the definition of M_{i+1}. So, M is reduced. □

3.13. Lemma. *Let H be a reduced set of l-homogeneous elements,*

$$H = \{h_1, \ldots, h_q\}, \quad 0 \neq f \in L(Y), \quad f(h_1, \ldots, h_q) = 0,$$

and let ω be an elementary transformation of H, $\omega(h_i) = h_i$ for $i \neq 1$,

$$\omega(h_1) = h_1 + s(h_2, \ldots, h_q), \quad s \in L(Y), \quad l(s(h_2, \ldots, h_q)) = l(h_1).$$

Then $\omega(H) = \{\omega(h_1), h_2, \ldots, h_q\}$ is a reduced set and there exists $0 \neq g \in L(Y)$ such that $g(\omega(h_1), h_2, \ldots, h_q) = 0$.

Proof. Since all elements h_i are l-homogeneous, we may assume that all monomials in $h_2, \ldots, h_q$, in the presentation of $s(h_2, \ldots, h_q)$, have length $l_X(h_1)$. As H is reduced, the leading part $\omega(h_1)^0 = \omega(h_1)$ does not belong to the subalgebra generated by elements h_i, $i \neq 1$. Let now, for example, $h_q = \gamma(\omega(h_1), h_2, \ldots, h_{q-1})$, $\gamma \in L(Y)$. One may assume that all monomials in $\omega(h_1)$, $h_2, \ldots, h_{q-1}$ have the length $l_X(h_q)$. Then by l-homogeneity we have

$$h_q = \gamma(h_1 + \lambda h_q + s'(h_2, \ldots, h_{q-1}), h_2, \ldots, h_{q-1}),$$

where $\lambda \in K$ and $l(h_1) = l(h_q)$. Hence

$$h_q = \gamma(h_1 + \lambda h_q, h_2, \ldots, h_{q-1}) + s''(h_2, \ldots, h_{q-1})$$

and again by l-homogeneity we see that

$$h_q = \alpha(h_1 + \lambda h_q) + s'''(h_2, \ldots, h_{q-1}),$$

where $\alpha \in K$, i.e.

$$h_1 = -\alpha^{-1}((\lambda - 1)h_q + s'''(h_2, \ldots, h_{q-1}))$$

for $\alpha \neq 0$, and

$$h_q = s'''(h_2, \ldots, h_{q-1})$$

for $\alpha = 0$. In both cases we arrive at a contradiction to the fact that H is reduced.

Now we are going to prove the last assertion of our lemma. Consider the following elementary transformation ω of the free generating set $Y = \{y_1, \ldots, y_q\}$: $\omega(y_1) = y_1 + s(y_2, \ldots, y_q)$ and $\omega(y_i) = y_i$ if $i \neq 1$ (see the statement of our lemma). By Lemma 3.11

$$\omega(Y) = \{\omega(y_1), y_2, \ldots, y_q\}$$

is a free generating set of the algebra $L(Y)$. Then

$$0 \neq f(y_1, \ldots, y_q) = g(\omega(y_1), y_2, \ldots, y_q) \in L(\omega(Y))$$

and, therefore, $g(\omega(h_1), h_2, \ldots, h_q) = f(h_1, \ldots, h_q) = 0$, completing the proof. $\square$

3.14. Lemma. *Let M be as in Lemma 3.12, $\tilde{M} = \{b_1, \ldots, b_q\} \subset M$, $0 \neq f \in L(Y)$, $f(b_1, \ldots, b_q) = 0$. Then there exists a finite reduced multihomogeneous subset $M_1 = \{c_1, \ldots, c_q\}$ of $L(X)$ such that:*
1) $l_X(c_i) = l_X(b_i)$, $1 \leq i \leq q$;
2) $g(c_1, \ldots, c_q) = 0$ for some $0 \neq g \in L(Y)$.

Proof. We set $\varphi(y_i) = l_X(b_i)$, $1 \leq i \leq q$. Let $\tilde{f}$ be the leading part of f relative to φ. Then

$$0 \neq \tilde{f} \in L(Y), \qquad \tilde{f}(b_1^0, \ldots, b_q^0) = 0,$$

where b_i^0 is the leading part of b_i. Since $\tilde{M}$ is reduced the set $\{b_1^0, \ldots, b_q^0\}$ is also reduced.

Consider the following ordering on the set of monomials: first by length; then by multidegree, i.e. lexicographically for rows $(\alpha_1, \ldots, \alpha_n)$, where $m(u) = \alpha_1 x_1 + \cdots + \alpha_n x_n$ and $\alpha_i = l_{x_i}(u)$.

We shall reduce the set $\{b_1^0, \ldots, b_q^0\}$ with the help of elementary transformations, described in Lemma 3.13, to a reduced set $\{c_1', \ldots, c_q'\}$ such that for any j the leading part $\tilde{c}_j'$ (relative to multidegree) does not belong to the subalgebra generated by $\{\tilde{c}_i', i \neq j\}$. If, for example, $\tilde{b}_1^0 = g(\tilde{b}_2^0, \ldots, \tilde{b}_q^0)$, then we consider the following elementary transformation ω:

$$\omega(b_i^0) = b_i^0 \quad \text{for } i \neq 1 \quad \text{and} \quad \omega(b_1^0) = b_1^0 - g(b_2^0, \ldots, b_q^0).$$

It is clear, since the set $\{b_1^0,\ldots,b_q^0\}$ is reduced, that $\omega(b_1^0) \neq 0$, $l_X(\omega(b_1^0)) = l_X(b_1^0)$. By Lemma 3.13, repeating such transformations, after a finite number of steps we come to the desired set $\{c_1',\ldots,c_q'\}$ such that there exists $0 \neq h = h(y_1,\ldots,y_q) \in L(Y)$ with $h(c_1',\ldots,c_q') = 0$. Now we set

$$\hat{\varphi}(y_i) = (\alpha_1^i,\ldots,\alpha_q^i) = m(\tilde{c}_i').$$

Let $\hat{h}$ be the leading part in an s-regular presentation of h relative to the weight $\hat{\varphi}$. It is clear that $\hat{h}(\tilde{c}_1',\ldots,\tilde{c}_q') = 0$, $0 \neq \hat{h} \in L(Y)$, $\tilde{c}_i'$ is multihomogeneous,

$$l_X(\tilde{c}_i') = l_X(b_i), \qquad i = 1,\ldots,q.$$

Now writing $c_i = \tilde{c}_i'$, $i = 1,\ldots,q$, $g = \hat{h}$ we have proved our assertion. $\square$

3.15. Theorem. *Let K be a field, char $K \neq 2, 3$. Then any reduced subset of a free colour Lie superalgebra $L(X)$ is independent, and, therefore, any G-homogeneous subalgebra of $L(X)$ is free.*

Proof. Let M be a reduced subset in $L(X)$. Assume, on the contrary, that for some finite subset $\tilde{M} = \{b_1,\ldots,b_q\} \subset M$ there exists a nontrivial relation

$$0 \neq f(y_1,\ldots,y_q) \in L(Y), \qquad d(y_i) = d(b_i), \qquad f(b_1,\ldots,b_q) = 0.$$

By Lemma 3.14 we may assume that $\tilde{M}$ is multihomogeneous. If $l_k = l_X(b_k) > 1$ for some $b_k \in \tilde{M}$, then, $\tilde{M}$ being multihomogeneous, there exists $x \in X$ such that $l_x(b_k) \neq 0$, $\beta x \neq b_m$, $1 \leq m \leq q$, $0 \neq \beta \in K$. Changing the ordering of X, if necessary, we may assume that x is the least element of X. Consider now the s-regular presentation of b_i for all i. Observe that an element βx, $0 \neq \beta \in K$, cannot encounter among the summands of the s-regular presentation of b_k because of l-homogeneity of b_k and $l_k > 1$. Since b_m, $b_m \neq b_k$, is multihomogeneous and $\beta x \neq b_m$, $0 \neq \beta \in K$, the elements βx, $0 \neq \beta \in K$, cannot encounter in the s-regular presentation of b_m, $m \neq k$. By Corollary 3.4, if $x \in X_+$ then all elements b_i are polynomials in the elements of $Z = Z(x)$, $l_Z(b_k) = l_X(b_k) - l_x(b_k) < l_k$ and $l_Z(b_m) \leq l_X(b_m) = l_m$ for $b_m \neq b_k$, $1 \leq m \leq q$. By Corollary 3.4, the elements b_i, $1 \leq i \leq q$, are also l-homogeneous relative to Z. Since $f(b_1,\ldots,b_q) = 0$, by Lemma 3.14 one can assume that $\tilde{M}$ is multihomogeneous relative to Z.

If now $x \in X_-$ then, by Corollary 3.4, all elements b_i are polynomials in the set $W = W(x)$, $l_W(b_k) < l_k$ and $l_W(b_m) \leq l_m$ for $b_m \neq b_k$, $1 \leq m \leq q$. Now we apply elementary transformations to the set $\{b_1,\ldots,b_q\}$ and come to a reduced set $C = \{c_1,\ldots,c_q\}$ (relative to the generators in W) with a nontrivial relation. Next we consider triangular elementary transformations (relative to W) of the following form: if, for example, $b_1^0 = -g(b_2^0,\ldots,b_q^0)$, $g \in L(T)$, b_i^0

is the leading part of b_i in the alphabet W, then we apply the triangular elementary transformation ω, where $\omega(b_i) = b_i$ for $i \neq 1$ and $\omega(b_1) = b_1 + g(b_2, \ldots, b_q)$. Observe that this transformation is also a triangular elementary transformation relative to X, i.e. $l_X(\omega(b_1)) = l_X(b_1)$, and therefore by Lemma 3.13 we observe that $\{\omega(b_1), b_2, \ldots, b_q\}$ is X-reduced with a nontrivial relation. After a finite number of such transformations we come to a set C of the desired form which is X-reduced and W-reduced simultaneously. By Lemma 3.14, one can assume that C is W-multihomogeneous.

Thus, after a finite number of such procedures we may assume that $l_i = 1$ for $1 \leq i \leq q$, i.e. $b_i = \alpha_i x_i$, $0 \neq \alpha_i \in K$, $x_i \in X$ for $1 \leq i \leq q$, and the set $\{b_i, 1 \leq i \leq q\}$ is reduced. This is in contradiction with the existence of a nontrivial relation between the elements b_i, $1 \leq i \leq q$, and we have proved that any reduced set is independent.

Finally we consider a G-homogeneous subalgebra B of the free colour Lie superalgebra $L(X)$. By Lemma 3.12 we choose a reduced generating set M for B. As we saw above, M is independent. Therefore $B = L(M)$ is the free colour Lie superalgebra on M. □

3.16. Remark. We give an example of a nonfree subalgebra in $L(X)$ if char $K = 3$. Indeed, let $X = X_- = \{y\}$. Then we consider the subalgebra B of $L(X)$ generated by the monomials $[y, y]$ and $[[y, y], y]$. It is clear that this subalgebra cannot be generated by one element. Otherwise, we have the relation $[[y, y], [[y, y], y]] = 0$, and therefore B is not free. Again, if we add the identity $[[a, a], a] = 0$ with $d(a) \in G_-$ to the defining set of identities in the definition of colour Lie superalgebras then in the case char $K = 3$ the assertion of Theorem 3.15 holds.

In the case when K is a commutative ring, the statement of the theorem does not hold even for Lie algebras. Indeed, let $K = \mathbb{Z}$, $L(x, y)$ be the free Lie ring. Consider the subring H of $L(x, y)$, generated by the elements $2x, y, [x, y]$. Then H is not a free Lie ring (we see that in the abelian group $H/[H, H]$ there is a nonzero element $[x, y]$ of order 2).

Note that the case when K is a ring will be discussed in Section 6.

Now we pass to the corollaries of the theorem about freeness of subalgebras.

3.17. Proposition. *Let M be a reduced subset of $L(X)$. Then the leading part of any element, which is a Lie polynomial in the elements of M, is a Lie polynomial in leading parts of elements in M.*

Proof. The assertion follows from the independence of M. □

3.18. Generators of the derived algebra. Let X be a G-graded set, $|X| \geq 2$. Suppose that X is totally ordered in such a way that from $x \in X_+$, $y \in X_-$ it follows that $y > x$. The derived algebra $L^2(X) = [L(X), L(X)]$ is a free colour Lie superalgebra. As a free set of generators we can choose left-normed monomials $[x_1 x_2 \ldots x_{s-1} x_s] = [[x_1 x_2 \ldots x_{s-1}], x_s]$ where $x_i \in X$, $s \geq 2$, $x_1 > x_2 \leq x_3 \leq \cdots \leq x_s$ and $x_i \neq x_{i+1}$ for $x_i \in X_-$. Also if $x_1, \ldots, x_s \in X_-$ then $x_1 \geq x_2 < x_3 < \cdots < x_s$.

3.19. Theorem (an analogue of Schreier's formula). *Let K be a field, char $K \neq 2, 3$, $L = L(X) = L_+ \oplus L_-$ the free colour Lie superalgebra, rank $L = |X| = N < \infty$. Let M be a G-homogeneous subalgebra, $M = M_+ \oplus M_-$, $M_+ = L_+$, $\dim L_-/M_- = s < \infty$. Then rank $M = 2^s(N - 1) + 1$.*

Proof. We argue by induction on s. The basis of induction: $s = 0$. In this case $M = L$, rank $M = N$. Suppose now that our assertion holds for all $s < n$ and let now $s = n$. We choose a reduced subset H of generators of M. Consider the set of leading parts $H^0 = \{h^0, h \in H\}$. By Theorem 3.15 the set H^0 is independent and $|H^0| = \text{rank } L(H^0) = \text{rank } M$.

By the Lemma in Subsection 1.3 of Chapter 1, we have $\dim L_+/M_+ = \dim L_+/L_+(H^0) = 0$, i.e. $L_+(H^0) = L_+$. Thus it is sufficient to prove the formula for rank $L(H^0) = |H^0|$. We may suppose that $L(H^0) \subset L(W(z))$ for some $z \in X_-$ (changing, if necessary, the bases of subspaces, generated by X_g, $g \in G$). Any such transformation is a composition of elementary transformations and therefore it induces an automorphism of $L(X)$. Since $|W(z)| = 2(N - 1) + 1$, by the induction hypothesis, we have the desired result. $\square$

3.20. Corollary. *Let K be a field, char $K \neq 2, 3$, $L = L_+ \oplus L_-$ the finitely generated colour Lie superalgebra, and let $M = M_+ \oplus M_-$ be a G-homogeneous subalgebra of L such that $M_+ = L_+$ and $\dim L_-/M_- < \infty$. Then M is a finitely generated colour Lie superalgebra.*

Proof. It is sufficient to represent L as a quotient-algebra of some free colour Lie superalgebra $L(X)$ of finite rank and to consider the full inverse image of M in $L(X)$. $\square$

3.21. Theorem. *Let $|X_-| \geq 1$, $L(X) = L_+ \oplus L_-$ be a free colour Lie superalgebra, $M = M_+ \oplus M_-$ a G-homogeneous subalgebra in $L(X)$ such that $M_- = L_-$ and $\dim L_+/M_+ = s < \infty$. Then rank $M = \infty$, $M = L(Y)$ and $|Y_-| = \infty$.*

Proof. We take a reduced subset H of generators of M. Consider the set of leading parts $H^0 = \{h^0, h \in H\}$. By Theorem 3.15 H^0 is independent and rank $M = |H^0|$. By the Lemma in Subsection 1.3 of Chapter 1 and Proposition 3.17 we have $\dim L_+/M_+ = \dim L_+/L_+(H^0)$ and $\dim L_-/M_- = \dim L_-/L_-(H^0) = 0$, i.e. $L_-(H^0) = L_-$.

Thus it is sufficient to prove our assertion for $L(H^0)$. The proof will be carried out by induction on s. The basis of induction: $s = 1$. As in the proof of Theorem 3.19, we may assume that $L(H^0) \subset L(Z_X(z))$ for some $z \in X$. Since $L(H^0)$ is an l-homogeneous subalgebra of codimension one in $L(X)$, we have $L(H^0) = L(Z_X(z))$, $|Z_X(z)| = \infty$ and $|(Z_X(z))_-| = \infty$. Suppose now that $\dim L_+/L_+(H^0) = s > 1$. As above, we may assume that $L(H^0) \subset L(Z_X(z))$ for some $z \in X_+$ and $L_-(H^0) = L_-(Z_X(z)) = L_-$, $|(Z_X(z))_-| = \infty > 1$, $\dim L_+(Z_X(z))/L_+(H^0) = s - 1$. By the induction hypothesis our statement follows. □

3.22. Lemma. *Let A, B be G-homogeneous subalgebras in $L(X)$. Suppose that B is a free factor of some G-homogeneous subalgebra $\tilde{B}$ of $L(X)$. Then the subalgebra $A \cap B$ is a free factor of the subalgebra $A \cap \tilde{B}$.*

Proof. Let M, $M \cup N$, R be reduced independent subsets, $B = L(M)$, $\tilde{B} = L(M \cup N)$ and $A \cap B = L(R)$. Then $A \cap \tilde{B} = L(R) \oplus H$ where H is a linear subspace of elements $h \in A \cap \tilde{B}$ such that in its s-regular presentation in the generators from $M \cup N$ we have each s-regular monomial containing some element in N. It is obvious that H is a subalgebra. Let $H = L(S)$ where S is an independent subset. By triangular elementary transformations we transform $R \cup S$ to a reduced subset T. By our definition of the subsets R and S and by Theorem 3.15 we have T reduced, $T = R \cup S'$ and $A \cap \tilde{B} = L(R \cup S')$. This completes the proof. □

3.23. Lemma. *Let H be a G-homogeneous subalgebra of finite rank in $L(X)$. Then H is a free factor of some G-homogeneous subalgebra of finite codimension in $L(X)$.*

Proof. We choose a reduced set $M = \{m_1, \ldots, m_l\}$ of free generators for H. By the Lemma in Subsection 1.3 of Chapter 1, Proposition 3.17 and Theorem 3.15, it is sufficient to find a G-homogeneous set R of l-homogeneous elements such that $M^0 \cup R$ is a reduced set and $\dim L(X)/L(M^0 \cup R) < \infty$ where $M^0 = \{m_1^0, \ldots, m_l^0\}$ is the set of leading parts. Let $k = \max_{1 \leq i \leq l} l_X(m_i)$ and let B be the subalgebra of $L(X)$ which is the linear span of all s-regular monomials w with $l(w) > k$, R_1 a reduced l-homogeneous set of generators for B. We consider the l-homogeneous set $M^0 \cup R_1$ and transform it by triangular elementary transformations to the desired form. Since $l(y) > k$ for any $y \in B$, our transformations have the following form: $m_i^0 \to m_i^0$, $1 \leq i \leq l$, $r_j \to r_j'$, $l(r_j') = l(r_j)$, $d(r_j') = d(r_j)$ for $r_j' \neq 0$ and either $r_j'^0 = r_j'$ or $r_j' = 0$. We transform analogously the sets $R_1' = \{r_j', r_j \in R_1\}$, R_1'' and so on. Using such transformations, we obtain a reduced l-homogeneous set $M^0 \cup R$ generating the same subalgebra as $M^0 \cup R_1$. Since this subalgebra contains the subalgebra B of finite codimension in $L(X)$, it is also of finite codimension in $L(X)$. □

3.24. Lemma. *Let $|X_-| \geq 1$ and let $M = M_+ \oplus M_-$ be a G-homogeneous subalgebra in $L = L(X) = L_+ \oplus L_-$ such that $M_+ \neq L_+$ and $M_- = L_-$. Then rank $M = \infty$.*

Proof. Assume to the contrary that rank $M < \infty$. Then by Lemma 3.23 we may include M as a free factor in a subalgebra $\hat{M}$ of finite codimension in $L(X)$. It is clear that $(\hat{M})_- = L_-$. By Theorem 3.21, $\hat{M} = L(Y)$ and $|Y_-| = \infty$. But $|Y_-| = \dim \hat{M}_-/[\hat{M}, \hat{M}]_- = \dim L_-/[\hat{M}, \hat{M}]_- = \infty$. Since $M \subset \hat{M}$, we have $[M, M] \subset [\hat{M}, \hat{M}]$. As $M_- = \hat{M}_- = L_-$, we have $\dim M_-/[M,M]_- = \dim L_-/[M,M]_- \geq \dim L_-/[\hat{M}, \hat{M}]_- = \infty$. Thus rank $M = \infty$ and therefore we have a contradiction. □

3.25. Theorem (an analogue of P.M. Cohn's theorem on generators of the automorphism group of a free Lie algebra of finite rank). *The set of elementary automorphisms generates the automorphism group of a free colour Lie superalgebra of finite rank.*

Proof. Let $|X| < \infty$ and $L(X) = L(Y)$. By the invariance of the rank we have $|X| = |Y|$ (see 2.10). We transform $Y = \{y_1, \ldots, y_n\}$ by elementary transformations to some X-reduced set D in the following way: if, for example, $y_n^0 = g(y_1^0, \ldots, y_{n-1}^0), 0 \neq g \in L(T)$ then $\omega(y_i) = y_i$ for $i \neq n$ and $\omega(y_n) = y_n - g(y_1, \ldots, y_{n-1}) \neq 0$ (by the independence of Y). Thus, by finitely many steps, we come to a set $D = \{d_1, \ldots, d_n\}$. We have

$$x_i = g_i(d_1, \ldots, d_n), \qquad i = 1, \ldots, n, g_i \in L(T). \tag{1}$$

Since D is reduced, by Corollary 3.17 the leading part of $g_i(d_1, \ldots, d_n)$ is a Lie polynomial in leading parts d_i^0, $1 \leq i \leq n$. Therefore, it follows from (1) that d_i is a linear combination of generators $x_i \in X$. By a non-degenerate linear transformation we can transform D into X. It is clear that this transformation is a composition of several elementary automorphisms. □

3.26. Proposition. *Any free Lie superalgebra with a countable set of free generators can be embedded in a 2-generated Lie superalgebra.*

Proof. Consider $X_+ = \emptyset$, $X_- = \{y_1, y_2\}$ and $L(X)$. Observe that $W_X(y_1) = \{y_2, [y_2, y_1], [y_1, y_1]\}$. By Lemma 3.7, $W_X(y_1)$ is an independent set. Let $T = Z_{W_X(y_1)}([y_1, y_1]) = \{x, x(\text{ad}'[y_1, y_1])^n, x \in W_X(y_1) \setminus [y_1, y_1], n \in \mathbb{N}\}$. By Lemma 3.7, T is an independent set. That

$$T_+ = \{[y_2, y_1], [y_2, y_1](\text{ad}'[y_1, y_1])^n, n \in \mathbb{N}\}, \qquad |T_+| = \infty,$$

$$T_- = \{y_2, y_2(\text{ad}'[y_1, y_1])^m, m \in \mathbb{N}\}, \qquad |T_-| = \infty.$$

This completes the proof. □

Observe that, in the proof, it was also possible to consider $X_+ = \{x\}$, $X_- = \{y\}$, $X = X_+ \cup X_-$, $T = Z_X(y) = \{x, x(\text{ad}' y)^n, n \in \mathbb{N}\}$ with $T_+ = \{x, x(\text{ad}' y)^{2m}, m \in \mathbb{N}\}$, $|T_+| = \infty$, $T_- = \{x(\text{ad}' y)^{2m-1}, m \in \mathbb{N}\}$, $|T_-| = \infty$.

Remark. By the composition technique developed in Section 3 of Chapter 3 and some of A.I. Shirshov's results [Shirshov, 1956, 1958] it is possible to prove that any countably generated Lie superalgebra can be embedded in a suitable 2-generated Lie superalgebra over the same field. For colour Lie superalgebras this fact holds if the grading group G is 2-generated. For the details see [Mikhalev, 1990].

3.27. Proposition. *Any nonzero ideal $H = H_+ \oplus H_-$ of the free colour Lie superalgebra $L(X) = L_+ \oplus L_-$ such that $H \neq 0$, $|X| > 1$ and $H_+ \neq L_+$, is a free colour Lie superalgebra of infinite rank.*

Proof. Since $H_+ \neq L_+$, there exists a G-homogeneous element $a \in L_+ \setminus H_+$. As H is an ideal of $L(X)$ and $a \in L_+$, $E = H \oplus Ka$ is a subalgebra in $L(X)$ and $[E, E] \subset H$. Therefore we may choose a free generating set Y for the subalgebra E such that $a \in Y$ and $Y \setminus a \subset H$. But the set $Z_Y(a)$ is a free generating set for H and $|Z_Y(a)| = \infty$. By the invariance of the rank for free colour Lie superalgebras and by Theorem 3.15 we observe that rank $H = \infty$. □

Remark that the condition $H_+ \neq L_+$ in Proposition 3.27 is essential. Indeed, let $X_+ = \{x\}$, $X_- = \{y\}$, $L(X) = L_+ \oplus L_-$, $H = L(W_X(y)) = L(x, [x, y], [y, y]) = H_+ \oplus H_-$. Then H is a proper ideal of $L(X)$ with rank $H = 3 < \infty$ but $H_+ = L_+$.

3.28. Theorem. *Let K be a field, char $K \neq 2, 3$, $|X| > 1$, $|X_-| \geq 1$, and let $L(X) = L_+ \oplus L_-$ be a free colour Lie superalgebra. Then L_+ is a free colour Lie algebra of infinite rank (we give the explicit form for a set of free generators for L_+); in particular, if $L(X)$ is a free Lie superalgebra, then L_+ is a free Lie algebra of infinite rank.*

Proof. It is sufficient to consider only the case $|X| < \infty$. Let $X_- = \{y_1, \ldots, y_s\}$ with $s \geq 1$ for $|X_+| \geq 1$ and $s \geq 2$ for $X_+ = \emptyset$, $y_1 < \cdots < y_s$ and $y_s < x$ for all $x \in X_+$. We extend this ordering to a total ordering on X. We compare words from $S(X)$ at first by their lengths and then lexicographically. We introduce a partial order on $\Gamma(X)$ by setting $u < v$ if $\bar{u} < \bar{v}$ in $S(X)$. Let $Y_1 = W_X(y_1)$. By Lemma 3.7 Y_1 is an independent set. It is obvious that $C_1 = L(Y_1)$ is an l-homogeneous subalgebra of $L(X)$. Let z be the least element among the $u \in Y_1$ such that

$$l_X(u) = \min_{\substack{\omega \in Y_1 \\ d(\omega) \in G_-}} (l_X(\omega)).$$

Now we consider the l-homogeneous subalgebra $C_2 = L(Y_2)$, where $Y_2 = W_{Y_1}(z)$. Note that our partial ordering on $\Gamma(X)$ induced the total ordering on Y_1. Repeating this procedure, we construct Y_n and choose the least generator v among all odd elements of Y_n (at first by length, and then lexicographically). Furthermore, if we consider the set $Y_{n+1} = W_{Y_n}(v)$ then, by Lemma 3.7, Y_{n+1} is an independent subset. Let $C_{n+1} = L(Y_{n+1})$. By induction on n we show that it is possible to choose such an element v and at the same time to have $|Y_n| > 1$ and $|(Y_n)_-| \geq 1$. The basis of induction: $n = 1$, i.e. the subalgebra C_1. In this case our assertion follows from the conditions $|X| > 1$, $|X_-| \geq 1$ and the explicit form for the elements of $W_X(y_1)$. By the induction hypothesis for $n - 1$ we have $|Y_{n-1}| > 1$, $|(Y_{n-1})_-| \geq 1$, and therefore there exists the desired element $v \in Y_{n-1}$. Now applying the above procedure we construct the set $Y_{n-1}(v)$. It is clear that we have $|Y_n| > 1$ and $|(Y_n)_-| \geq 1$. Observe that from $|Y_{n-1}| < \infty$ it follows that $|Y_n| < \infty$. Since $|X| < \infty$, we have $|Y_n| < \infty$ for all $n \in \mathbb{N}$. Set $l_n(Y_n) = \min_{u \in (Y_n)_-} l_X(u)$. Since $|Y_n| < \infty$ for all $k \in \mathbb{N}$ there exists $N \in \mathbb{N}$ such that $l_N(Y_N) > k$. It is obvious that $L(X) \supset C_1 \supset C_2 \supset \cdots \supset C_n \supset \cdots$, $C_i \supset L_+$ for all i. We show that $L_+ = (C_i)_+$ for all $i \in \mathbb{N}$. Let now x be the least element of X_- and w an s-regular monomial, $w \neq x$. Then $w = 2^{-k}u$ with $k \in \mathbb{N} \cup \{0\}$ and u a monomial in the elements of $W_X(x)$ (see Corollary 3.4). Since y_1 is the least element of X, $y_1 \in X_-$, we have $w \neq y_1$ for any even s-regular monomial $w \in L_+$, and therefore $(C_1)_+ = L_+$. Similarly, we have $(C_t)_+ = (C_{t+1})_+$ for all $t \in \mathbb{N}$. Observe that $X_+ \subset (Y_1)_+ \subset \cdots \subset (Y_n)_+ \subset \cdots$. Let $M = \bigcup_{i=1}^{\infty}(Y_i)_+$. Considering the construction of Y_n and the conditions $|Y_{n-1}| > 1$, $|(Y_{n-1})_-| \geq 1$, we find that $|Y_n| - |Y_{n-1}| \geq 1$, and therefore $|M| = \infty$. Since any finite subset of M belongs to some independent set Y_N, M is independent. Now we need the following:

3.29. Lemma. *Let* $|X| > 1$, $|X_-| \geq 1$, $|X_-| = \{y_1, \ldots, y_s\}$, $y_1 < \cdots < y_s$. *Then for any* $u \in W_X(y_i)$, *where* $i > 1$, $u \neq y_j$ *for all* $j = 1, \ldots, i - 1$, *there exists* $u' \in W_{Y_{i-1}}(y_i)$ *such that* $u' = \alpha u$ *in* $L(X)$ *with* $0 \neq \alpha \in K$.

Proof. We consider the following possibilities:
1) $u \in X$. Since $u \neq y_j$ for $j = 1, \ldots, i - 1$, $u' = u \in W_{Y_{i-1}}(y_i)$.
2) $u = [x, y_i]$, $x \in X_+$. Then $u' = u \in W_{Y_{i-1}}(y_i)$.
3) $u = [y_j, y_i]$, $j \geq i$. Then $u' = u \in W_{Y_{i-1}}(y_i)$.
4) $u = [y_j, y_i]$, $j < i$. Then $u' = [y_j, y_i] = \alpha[y_i, y_j]$, $0 \neq \alpha \in K$, $u = [y_i, y_j] \in W_{Y_{i-1}}(y_i)$.
This completes the proof. $\square$

Now we continue the proof of our theorem. By induction on the length we show that for any s-regular monomial $w \in L_+$ we have $w \in L(M)$. The basis of induction: $w \in X_+$ (in this case $w \in (Y_1)_+$). Let now $l_X(w) = N$. Suppose that $l_{y_1}(w) \neq 0$. By Corollary 3.4, $w = 2^{-k}u$, where u is a monomial in the elements

§3. The freeness of subalgebras and its corollaries   67

from $W_X(y_1) = Y_1$ and $l_X(w) < l_{Y_1}(w)$. The monomial u is a linear combination of s-regular monomials with the length $l_{Y_1}(u)$. By the induction hypothesis we see that $w \in L(\tilde{M})$, where $\tilde{M}$ is the set constructed in the same way as M, but starting with $W_X(y_1)$. Therefore $w \in L(M)$. If now $l_{y_1}(w) = 0$, then we consider $l_{y_2}(w)$, and so on. In the case $l_{y_i}(w) = 0$ for all i, we have $w \in L(X_+) \subset L((Y_1)_+)$. Let i_0 be the least index i such that $l_{y_i}(w) \neq 0$. Then we take $W_{Y_{i_0-1}}(y_{i_0})$. As above $w = 2^{-t}u$ where $t \in \mathbb{N}$, u is a monomial in $W_X(y_{i_0})$ and $l_{W_X(y_{i_0})}(u) < l_X(w)$. By the induction hypothesis and Lemma 3.29, we have $w \in L(\tilde{M})$, where $\tilde{M}$ is constructed as M, but starting with $W_{Y_{i_0-1}}(y_{i_0})$. Therefore $w \in L(M)$.

Thus, we have constructed the independent set M of multihomogeneous elements (in particular, the set M is reduced) such that $|M| = \infty$ and $L_+ = L(M)$. It is clear that rank $L_+ = \infty$. This completes the proof of the theorem. □

3.30. Theorem. *Let K be a field, char $K \neq 2, 3$, $|X_-| \geq 1$, and let A, B be G-homogeneous subalgebras of finite rank in the free colour Lie superalgebra $L(X)$, $A = A_+ \oplus A_-$, $B = B_+ \oplus B_-$. If $A_- = B_- \neq \{0\}$, then $A = B$.*

Proof. We assume to the contrary that $A_+ \neq B_+$. Consider the subalgebra $M = M_+ \oplus M_-$ of $L(X)$, generated by $A \cup B$. It is clear that rank $M < \infty$. We show that $M_- = A_-$. Let $A = L(H)$ and $B = L(P)$, $|H| < \infty$, $|P| < \infty$. Taking into account our assumptions, we see that $H_- \neq \emptyset$, $P_- \neq \emptyset$. It is clear that the subalgebra M is generated by $H \cup P$ and any element of M_- is a linear combination of regular monomials w_i in the elements of $H \cup P$ (possibly, there are many such presentations). We show by induction on the length $l = l_{H \cup P}(w)$ that all $w_i \in A_-$. The basis of induction: $l = 1$ (and then $w \in H_- \cup P_- \subset A_-$). Let now $l = n > 1$. Then $w = [w_{11}, w_{12}]$, where $d(w) \in G_-$. We may assume that $d(w_{11}) \in G_-$, $d(w_{12}) \in G_+$. By the induction hypothesis we see that $w_{11} \in A_-$. Under the induction hypothesis we show again by induction on $l_{H \cup P}(w_{12})$ for even w_{12} that $[w_{11}, w_{12}] \in A_-$. The basis for this induction: $l_{H \cup P}(w_{12}) = 1$ (then it is evident that $[w_{11}, w_{12}] \in A_-$). Suppose that $[w_{11}, w'_{12}] \in A_-$ for all even monomials w'_{12} such that $l(w'_{12}) < l(w_{12})$. Let now $w_{12} = [w^1_{12}, w^2_{12}]$. Consider two possible cases.

1) The monomials w^1_{12}, w^2_{12} are odd. Then by the assumption of the first induction we have $w \in A_-$ since $w_{11}, w^1_{12}, w^2_{12} \in A_-$.

2) The monomials w^1_{12}, w^2_{12} are even. Then

$$[w_{11}, [w^1_{12}, w^2_{12}]] = \alpha_1[[w_{11}, w^1_{12}], w^2_{12}] + \alpha_2[[w_{11}, w^2_{12}], w^1_{12}],$$

where $0 \neq \alpha_1, \alpha_2 \in K$, and applying twice the second induction hypothesis we have $w = [w_{11}, w_{12}] \in A_-$.

Thus, $M_- = A_-$ and we have, in our situation, that $|X| < \infty$, $|X_-| \geq 1$, and $L(X) = L_+ \oplus L_-$ is a free colour Lie superalgebra, $M = M_+ \oplus M_-$ is a G-homogeneous subalgebra in L such that $M_- = L_-$, $M_+ \neq L_+$ and rank $M < \infty$. By Lemma 3.24 we know that rank $M = \infty$ and therefore we have come to a contradiction. □

3.31. Remarks.
1) In Theorem 3.30 the assumption rank $A < \infty$ and rank $B < \infty$ is essential. Indeed, let $X_+ = \{x\}$, $X_- = \{y\}$, $A = L(X)$, $B = L(Z_X(x))$. Since $x \in A_+$ and $x \notin B_+$, we see that $A_+ \neq B_+$; but at the same time $A_- = B_-$.
2) The assumption $A_- \neq 0$ is also essential. Indeed, we may take any two distinct subalgebras of finite rank in L_+.
3) In general, from $A_+ = B_+$ it does not follow that $A = B$. Let $X_+ = \{x\}$, $X_- = \{y\}$, $A = L(X)$, $B = L(W_X(y))$, where $W_X(y) = \{x, [x, y], [y, y]\}$. Then rank $A < \infty$, rank $B < \infty$ and $A_+ = B_+$. Since $y \in A_-$ and $y \notin B_-$, we have at the same time $A_- \neq B_-$.

3.32. Remark.
If a and b are G-homogeneous linearly independent elements in the free colour Lie superalgebra $L(X)$, $a \neq [b,b]$ and $b \neq [a,a]$, then the colour Lie superalgebra generated by a and b is a free colour Lie superalgebra with free generators $\{a,b\}$.

3.33. Theorem.
Let $L(X)$ be a free colour Lie superalgebra over a field K, $a, w_1, \ldots, w_k$ G-homogeneous elements in $L(X)$, B the subalgebra in $L(X)$ generated by $W = \{w_1, \ldots, w_n\}$. Then there exists an algorithm (which amounts to solving systems of linear equations over K) to solve the problem if a belongs to B.

Proof. It is sufficient to consider the case when $|X| < \infty$. First using triangular elementary transformations we pass from W to a reduced set W'. If, for example, the leading term (by the length) w_1^0 of w_1 belongs to the subalgebra generated by $\{w_2^0, \ldots, w_n^0\}$ then $w_1^0 = \sum a_i g_i(w_2^0, \ldots, w_n^0)$ where $a_i \in K$, g_i is a monomial, $l_X(g_i(w_2^0, \ldots, w_n^0)) = l_X(w_1^0)$ for all i (the number of such monomials g_i is finite). Rewriting left and right sides of this equality in s-regular monomials relatively X, we have a system of linear equations over K in variables $\{a_i\}$. Choosing such $\{a_i\}$ we set $w_1' = w_1 - \sum a_i g_i(w_2, \ldots, w_n)$. Continuing such process, after a finite number of steps we come to a reduced set W' generating B.

Thus we may assume that W is a reduced set. If $a \in B$ then by Proposition 3.17 we have $a^0 = \sum a_i g_i(w_1^0, \ldots, w_n^0)$ where $a_i \in K$, $l_X(g_i(w_1^0, \ldots, w_n^0)) = l_X(a^0)$. As above, the problem of existence of such presentation is reduced to a system of linear equations. Again, if it is possible to find such $\{a_i\}$ then we consider the element $a' = a - \sum a_i g_i(w_1, \ldots, w_n)$, $l_X(a') < l_X(a)$. Therefore after a finite number of steps we find the solution of the problem if $a \in B$, thus finishing the proof. □

3.34. Remark. In some cases it is important to consider arbitrary subalgebras of Lie superalgebras. I.B. Volichenko proved in 1987 the following theorem. A is a (non-homogeneous) subalgebra of a Lie superalgebra if, and only if, the following conditions hold:
1) A is Lie-admissible, i.e. A is a Lie algebra with respect to the product defined by the bracket (not superbracket) $[a,b] = ab - ba$;
2) the subalgebra A^{JJ} generated by all Jordan and Jacobi elements (i.e. the elements of the from $a \circ b = ab + ba$ and $j(a,b,c) = a(bc) + c(ab) + b(ca)$, respectively) belong to the anticenter of A, in other words $ax + xa = 0$ for any $a \in A^{JJ}$, $x \in A$;
3) $a(xy) = (ax)y + x(ay)$ for any $a \in A^{JJ}$, $x, y \in A$.

§4. Bases and subalgebras of free colour Lie p-superalgebras

In this section for a free colour Lie p-superalgebra $L^p(X)$ we construct a linear basis (Lemma 4.3), prove a theorem about the freeness of subalgebras (Theorem 4.15) and as a corollary we obtain a theorem on a generating set of the automorphism group of $L^p(X)$, where $|X| < \infty$ (Theorem 4.17) and the rank formula for the subalgebras of finite codimension in $L^p(X)$ (Theorem 4.20 is an analogue of Schreier's Formula).

In all that follows in this section we assume that K is a field with char $K = p > 3$.

4.1. Definition. Let $X = \bigcup_{g \in G} X_g$ be a G-graded set and $A(X)$ the free G-graded associative algebra. We say that an element $w \in A(X)$ is an *as-regular* monomial relative to the alphabet X if $w = \pi(\tilde{u})$ where u is an s-regular monomial relative to X, and, respectively, a *ps-regular* monomial, if either w is an as-regular monomial or $w = (\ldots((\pi(\tilde{v}))^p)^p \ldots)^p = (\pi(\tilde{v}))^{p^k}$, where $k \in \mathbb{N}$,
$$\underbrace{\phantom{(\ldots((\pi(\tilde{v}))^p)^p \ldots)^p}}_{k \text{ times}}$$
and v is an s-regular monomial with $d(v) \in G_+$ (see Definition 1.2 and Lemma 2.4).

Observe that, by definition, ps-regular monomials are just the associative presentations of s-regular monomials and their p-th powers for even s-regular monomials.

4.2. Definition. On the set $PS[X]$ of ps-regular monomials we consider the following mapping $\Psi: PS[X] \to S(X)$: for $\pi(\tilde{u})$ we set $\Psi(\pi(\tilde{u})) = \bar{u}$; for $(\pi(\tilde{u}))^{p^s}$ we set $\Psi((\pi(\tilde{u}))^{p^s}) = (\bar{u})^{p^s}$.

Observe that if we consider the operation $a \to a^p$ for even homogeneous elements $a \in [A(X)]$ then we have a colour Lie p-superalgebra which will be denoted by $[A(X)]^p$. Let $L^p(X)$ be the subalgebra of $[A(X)]^p$ generated by X (see Subsection 1.9 from Chapter 1).

4.3. Lemma. *The set of ps-regular monomials forms a linear basis in the colour Lie p-superalgebra $L^p(X)$.*

Proof. a) By Theorem 2.6, Lemma 2.4 and the definition of the operation $a \to a^p$, we see that any element of $L^p(X)$ is a linear combination of ps-regular monomials in the elements of X.
b) By Lemma 2.4, $A(X)$ is the universal enveloping algebra for the free colour Lie superalgebra $L(X)$ (see Definition 1.9 from Chapter 1 and Subsection 2.1 from Chapter 3). By Theorem 2.6, Lemma 2.4 and the Poincaré–Birkhoff–Witt Theorem (Theorem 2.2 of Chapter 3) the set of ps-regular monomials is linearly independent since it is part of a basis in $A(X)$. $\square$

4.4. Proposition. *The colour Lie p-superalgebra $L^p(X)$ constructed above is a free colour Lie p-superalgebra with free generating set $X = \bigcup_{g \in G} X_g$ (i.e. X generates the algebra $L^p(X)$ and any mapping $\varphi: X \to R$, where R is any colour Lie p-superalgebra with the same G and ε, can be extended to a homomorphism $\overline{\varphi}: L^p(X) \to R$ of colour Lie p-superalgebras).*

Proof. Any element a in $L^p(X)$ can be written as $a = \sum_i \lambda_i (\pi(\tilde{\omega}_i))^{p^{t_i}} + \sum_j \beta_j \pi(\tilde{v}_j)$, where $\lambda_i, \beta_j \in K$, ω_i, v_j are s-regular monomials, $d(\omega_i) \in G_+$, $t_i \in \mathbb{N}$ for all i, j. It is clear that the mapping φ can be extended to a homomorphism $\varphi_1: [X] \to R$ of colour Lie superalgebras. Now we set $\overline{\varphi}(a) = \sum \lambda_i (\varphi_1(\pi(\tilde{\omega}_i)))^{[p]^{t_i}} + \sum \beta_j \varphi_1(\pi(\tilde{v}_j))$. $\square$

4.5. Definition. The number rank $L^p(X) = |X|$ is called the *rank* of the free colour Lie p-superalgebra. It is clear that the rank of $L^p(X)$ does not depend on the choice of a set of free generators.

The maximal length of ps-regular monomials in the presentation of $a \in L^p(X)$ is called the length $l_X(a)$. If $a = \sum_{i \in I} \alpha_i \omega_i$, where $0 \neq \alpha_i \in K$, ω_i are ps-regular monomials, then $a^0 = \sum_{i \in \Omega} \alpha_i \omega_i$ with $\Omega = \{i \in I, l(\omega_i) = l(a)\}$ is called the leading part of $a \in L^p(X)$. We say that an element $a \in L^p(X)$ is *l-homogeneous* if $a = a^0$. Analogously we define the length $l_x(a)$ for $x \in X$ and say, that an element $a \in L^p(X)$ is *multihomogeneous* if a is homogeneous relative to all generators $x \in X$.

4.6. Definition. We say that a subset $M = \{a_i\}$ of G-homogeneous elements of $L^p(X)$ is *reduced* if for all i the leading part a_i^0 does not belong to the subalgebra generated by the set $\{a_j^0, j \neq i\}$ and, respectively, *independent* if M is a set of free generators of the subalgebra which it generates.

As in 3.1, we introduce the semigroup partial ordering $\rhd$ on $S(X)$ by the number of inversions.

4.7. Definition. Let $x \in X_+$ and $z \in X_-$. We use the notation $PZ(x) = \{y, x^p, [y, x], \ldots, y(\mathrm{ad}'\, x)^{p-1} \mid y \in X \setminus x\} \subset L^p(X)$, $W(z) = \{y, [y, z], [z, z] \mid y \in X \setminus z\} \subset L^p(X)$.

For $u \in \Gamma(X)$ we again denote by $\bar{u} \in S(X)$ the word which is obtained by forgetting the brackets. Let $\overline{x^p} = x^p$ and $\overline{PZ}(x) = \{\bar{w}, w \in PZ(x)\}$, $\overline{W}(z) = \{\bar{w}, w \in W(z)\}$.

4.8. Lemma. *The following assertions hold*:
1) *Let $x \in X_+$ be the least element in the totally ordered set $X = \bigcup_{g \in G} X_g$ and $u \in PZ(x)$. Then $u = \bar{u} + \sum \alpha_i u_i$ where u_i are associative words, $m(u_i) = m(u)$, $0 \neq \alpha_j \in K$, $\bar{u} \rhd u_i$ for all i.*
2) *Let $z \in X_-$ be the least element of the totally ordered set $X = \bigcup_{g \in G} X_g$ and $u \in W(z)$. Then $u = \alpha \bar{u} + \sum \alpha_i u_i$, where u_i are associative words, $m(u_i) = m(u)$, $0 \neq \alpha_i \in K$, $\alpha = 1$ or $\alpha = 2$ and $\bar{u} \rhd u_i$ for all i.*

Proof. Our assertions follow from Lemma 3.2 and the explicit form of elements of the set $W(z)$. □

4.9. Lemma. *Consider the values of an associative word f in the elements of the sets $PZ(x)$ and $W(z)$. Let $u = f(z_i)$ and $v = f(w_i)$ where $z_i \in PZ(x)$ and $w_i \in W(z)$, respectively. Then, relative to the ordering $\rhd$, the leading parts u_0 and v_0 of the elements u and v, respectively, have the form $u_0 = f(\bar{z}_i)$ and $v_0 = 2^l f(\bar{w}_i)$, where $l \in \mathbb{N} \cup \{0\}$.* □

(The proof follows immediately by Lemma 4.8 and the semigroup property of the ordering $\rhd$.)

4.10. Lemma. *Suppose that the values of the words f and g on some fixed subset of elements of $\overline{PZ}(x)$ (or $\overline{W}(z)$) are the same. Then the words f and g coincide.* □

(The proof can be given along the lines of the proof of Lemma 3.7.)

4.11. Lemma. *The subsets $PZ(x)$ and $W(z)$ are free generating sets of the G-graded associative subalgebras which they generate in $A(X)$.*

Proof. The existence of a non-trivial associative relation among the elements of $PZ(x)$ (or $W(z)$) gives us a contradiction to the assertions of Lemmas 4.9 and 4.10. □

4.12. Corollary. *The subsets $PZ(x)$ and $W(z)$ are independent.* □

4.13. Lemma. *Let x be the least element of the totally ordered set X, w a ps-regular monomial on X such that $w \neq x$ and $l_x(w) \neq 0$. Then:*
1) *if $x \in X_+$ then $w = u$ in $L^p(X)$, where u is a monomial in elements of the set $PZ(x)$ and $l_{PZ(x)}(w) < l_X(w)$;*
2) *if $x \in X_-$, then $w = 2^{-k} u$ in $L^p(X)$, with $k \in \mathbb{N} \cup \{0\}$, u a monomial in elements of the set $W(x)$ and $l_{W(x)}(w) < l_X(w)$.*

Proof. Observe that $l_X(x^{p^s}) = p^s$ and $l_{PZ(x)}(x^{p^s}) = s$. Let $l = pt + s, t \in \mathbb{N}, 1 \leq s < p$. Then $y(\text{ad}'\,x)^l = [y(\text{ad}'\,x)^s, x^{p^t}]$ in $L^p(X)$. Therefore, the elements of $Z(x)$ can be written as monomials in the elements of the set $PZ(x)$ without increasing the length (relative to X). By Corollary 3.4, we have our assertions 1) and 2). □

We remark that if, in Lemma 4.13, claim 1), we assume $m(w_1) = m(w_2)$, then it is possible that $l_{PZ(x)}(w_1) \neq l_{PZ(x)}(w_2)$. Indeed, let $y_1 > y_2 > x$, $w_1 = [y_1(\text{ad}'\,x)^{p-1}, [y_2, x]]$ and $w_2 = [y_1(\text{ad}'\,x)^p, y_2]$. Then w_1, w_2 are regular monomials and $m(w_1) = m(w_2)$, but $l_{PZ(x)}(w_1) = 2 \neq 3 = l_{PZ(x)}(w_2)$. The corresponding remark for the Claim 2) has been given after Corollary 3.4.

4.14. Definition. Let $S = \{s_\alpha, \alpha \in I\}$ be a G-homogeneous subset of the free colour Lie p-superalgebra $L^p(X)$. The mapping $\omega: S \to L^p(X)$ such that $\omega(s_\alpha) = s_\alpha$ for all $\alpha \in I \setminus \beta$ and $\omega(s_\beta) = \lambda s_\beta + \omega(s_{\alpha_1}, \ldots, s_{\alpha_t})$, where $0 \neq \lambda \in K$, $\alpha_1, \ldots, \alpha_t \neq \beta$, $\omega(y_{\alpha_1}, \ldots, y_{\alpha_t}) \in L^p(Y)$, $d(y_i) = d(s_i)$, $d(\omega(s_{\alpha_1}, \ldots, s_{\alpha_t})) = d(s_\beta)$, is called an *elementary transformation* of S. An elementary transformation ω is called *triangular* if $l_X(\omega(s_{\alpha_1}, \ldots, s_{\alpha_t})) \leq l_X(s_\beta)$.

4.15. Theorem. *Let K be a field, $p = \text{char}\,K > 3$, $L^p(X)$ the free colour Lie p-superalgebra. Then any reduced subset of $L^p(X)$ is independent, and therefore, any G-homogeneous subalgebra H of $L^p(X)$ is a free colour Lie p-superalgebra.*

Proof. As in the proof of Theorem 3.15, let $\tilde{M}$ be a reduced subset of $L^p(X)$ and $\{a_1, \ldots, a_k\} \subset \tilde{M}$. Suppose that we have a nonzero element $g \in L^p(Y)$, where $Y = \{y_1, \ldots, y_k\}$, $d(y_j) = d(a_j)$ for all $j = 1, \ldots, k$, and $g(a_1, \ldots, a_k) = 0$. Now we need the following result.

4.16. Lemma. *For a free colour Lie p-superalgebra $L^p(X)$ the following holds:*
1) *Let $M' = \{e_1, \ldots, e_q\}$ be a reduced subset of l-homogeneous elements of $L^p(X)$, $0 \neq f \in L(Y)$, $f(e_1, \ldots, e_q) = 0$, and let ω be an elementary transformation of M', $\omega(e_i) = e_i$ for $i \neq 1$, $\omega(e_1) = e_1 + s(e_2, \ldots, e_q)$, $s \in L(Y)$, $l(s(e_2, \ldots, e_q)) = l(e_1)$. Then $\omega(M') = \{\omega(e_1), e_2, \ldots, e_q\}$ is a reduced subset and there exists $0 \neq g \in L(Y)$ such that $g(\omega(e_1), e_2, \ldots, e_q) = 0$.*

2) Let $M = \{a_1, \ldots, a_k\}$ be a reduced subset of $L^p(X)$ with a non-trivial relation. Then there exists a reduced subset $M_2 = \{b_1, \ldots, b_k\}$ of multihomogeneous elements with a non-trivial relation and $l(a_i) = l(b_i), 1 \leq i \leq k$. □

(The proof can be performed by analogy with the proof of Lemmas 3.13 and 3.14.)

Now we continue the proof of Theorem 4.15. By Lemma 4.16 we may assume that there is a non-trivial relation among multihomogeneous elements of a reduced subset $M_2 = \{b_1, \ldots, b_k\}$ and, moreover, $l_X(b_j) = l_X(a_j)$ for $1 \leq j \leq k$. Since M_2 is reduced and multihomogeneous we observe that there exists an element $x \in X$, such that $b_i \neq x$ for all $i = 1, \ldots, k$ and $l_x(b_{i_0}) \neq 0$ for some i_0, $1 \leq i_0 \leq k$. Reordering X, if it is necessary, we have that x is the least element in X. By Corollary 4.12 and Lemma 4.13, we derive that all b_i can be written as Lie polynomials in the free generators of $PZ(x)$ in the case $x \in X_+$, and of $W(x)$ in the case $x \in X_-$. Observe that $l_{PZ(x)(W(x))}(b_i) \leq l_X(b_i)$ for all i and $l_{PZ(x)(W(x))}(b_{i_0}) < l_X(b_{i_0})$. By analogy with the proof of Theorem 3.15, using Lemma 4.16 we show that the set $M_2 = \{b_1, \ldots, b_k\}$ can be reduced by triangular elementary transformations to a reduced set $C = \{b'_1, \ldots, b'_k\}$ in the alphabet $PZ(x)$ (respectively, $W(x)$) with a non-trivial relation and $l_X(b'_i) = l_X(b_i)$, $l_{PZ(x)(W(x))}(b'_i) \leq l_{PZ(x)(W(x))}(b_i)$. By Lemma 4.16 we may assume that C is a multihomogeneous set. Repeating this procedure we see that there exists a non-trivial relation among the elements of length one from a reduced multihomogeneous subset of a free colour Lie p-superalgebra. This contradiction shows that any reduced subset of $L^p(X)$ is independent.

Now, as in the proof of Lemma 3.12, we choose a reduced generating set M of H. As above, M is independent. This completes the proof. □

4.17. Theorem. *The elementary transformations generate the automorphism group of the free colour Lie p-superalgebra of finite rank.* □

(The proof is given by analogy with the proof of Theorem 3.25 with Theorem 4.15 as a basis.)

4.18. Corollary. *Let M be a reduced subset of $L^p(X)$ and a an element of the subalgebra, generated by M. Then the leading part a^0 of a belongs to the subalgebra generated by the set M^0 of leading parts of elements from M.* □

4.19. Remark. If a and b are G-homogeneous elements of $L^p(X)$, $a \neq [b,b]$, $b \neq [a,a]$, $a_0 \neq b_0^{p^k}$, $b_0 \neq a_0^{p^l}$, where a_0 and b_0 are the leading ps-regular monomials in the ps-regular presentations of a and b respectively, then the set $\{a,b\}$ is a reduced subset of $L^p(X)$ (if $x \in X_+$, $a = x$, $b = x^p + x$, then $[a,b] = 0$ and therefore the set $\{a,b\}$ is not independent).

4.20. Theorem. *Let K be a field, $p = \operatorname{char} K > 3$, $|X| = N < \infty$. Suppose that $H = H_+ \oplus H_-$ is a G-homogeneous subalgebra of the free colour Lie p-superalgebra $L^p(X)$, $\dim L_+^p/H_+ = t < \infty$ and $\dim L_-^p/H_- = s < \infty$. Then $\operatorname{rank} H = 2^s p^t(N-1) + 1$.*

Proof. We proceed by induction on $t + s$. The basis of induction: $t = 0, s = 0$ (and then $H = L^p(X)$, $\operatorname{rank} H = N$). Suppose now that our assertion is true for $t + s < n$ and let $t + s = n$. We choose in H a reduced independent generating subset M. Let now $M^0 = \{m^0, m \in M\}$ be the set of leading parts for M. By Theorem 4.15, M^0 generates the free colour Lie p-superalgebra $L^p(M^0)$ and $\operatorname{rank} L^p(M^0) = \operatorname{rank} H = |M|$. By the Lemma in Subsection 1.3 of Chapter 1, we have that $\dim L_+^p/H_+ = \dim L_+^p/L_+^p(M^0)$ and $\dim L_-^p/H_- = \dim L_-^p/L_-^p(M^0)$. Therefore, it is sufficient to prove our formula for $\operatorname{rank} L^p(M^0)$. Changing, if necessary, bases of the linear spans of X_g, $g \in G$, we may assume that either $L^p(M^0) \subset L^p(PZ(x))$ or $L^p(M^0) \subset L^p(W(z))$ for some $x \in X_+$, $z \in X_-$. Since $|PZ(x)| = p(N-1) + 1$ and $|W(z)| = 2(N-1) + 1$, by induction hypothesis we have our assertion. $\square$

4.21. Corollary. *Let K be a field, $p = \operatorname{char} K > 3$, and let $L = L_+ \oplus L_-$ be a finitely generated colour Lie superalgebra, M its G-homogeneous subalgebra, $M = M_+ \oplus M_-$, $\dim L_+/M_+ < \infty$, $\dim L_-/M_- < \infty$. Then M is also a finitely generated colour Lie p-superalgebra.*

Proof. It is sufficient to consider L as a quotient algebra of some free colour Lie p-superalgebra $L^p(X)$ of finite rank and to take the inverse image of M in $L^p(X)$. This completes the proof. $\square$

4.22. Remark. The statement of Theorem 3.30 does not hold for $L^p(X)$. Indeed, let $X_+ = \{x\}$, $X_- = \{y\}$, $A = L^p(X)$ and $B = L^p(PZ(x))$. Then $\operatorname{rank} A < \infty$, $\operatorname{rank} B < \infty$, $A_- = B_- \neq 0$, but $A_+ \neq B_+$ because of $x \in A_+$ and $x \notin B_+$.

4.23. Theorem. *If $a, w_1, \ldots, w_k$ are G-homogeneous elements of the free colour Lie p-superalgebra $L^p(X)$ over a field K then there exists an algorithm (which reduces to systems of linear equations over K) to solve the problem if a belongs to the subalgebra in $L^p(X)$ generated by elements $w_1, \ldots, w_k$.*

Proof. We may apply the same arguments as in the proof of Theorem 3.33 using Corollary 4.18. $\square$

§5. The lattice of finitely generated subalgebras

Our next aim is to prove that the intersection of two subalgebras of finite rank in the free colour Lie superalgebra $L(X)$ (and also in the free colour Lie p-superalgebra $L^p(X)$) is of finite rank, provided that K is a field and $p = \operatorname{char} K > 3$ (Theorems 5.6 and 5.4).

In all that follows in this section, K is a field and $p = \operatorname{char} K > 3$.

5.1. Lemma. *Let A and B be G-homogeneous subalgebras of $L^p(X)$. If B is a free factor of some G-homogeneous subalgebra $\tilde{B}$ in $L^p(X)$, then the subalgebra $A \cap B$ is a free factor of $A \cap \tilde{B}$.* □

(The proof can be given by analogy with the proof of Lemma 3.22, using the results of the preceding section.)

5.2. Corollary. *Let A and B be G-homogeneous subalgebras of $L^p(X)$. If A and B are free factors of some G-homogeneous subalgebras $\tilde{A}$ and $\tilde{B}$ of $L^p(X)$ respectively, then the subalgebra $A \cap B$ is a free factor of the subalgebra $\tilde{A} \cap \tilde{B}$.* □

5.3. Lemma. *Let $|X| < \infty$ and let H be a G-homogeneous subalgebra of finite rank in $L^p(X)$. Then H is embeddable as a free factor in some G-homogeneous subalgebra $\tilde{H}$ of finite codimension in $L^p(X)$.* □

(The proof is analogous to the proof of Lemma 3.23, using the results of the preceding section.)

5.4. Theorem. *Let K be a field, $p = \operatorname{char} K > 3$, $L^p(X)$ the free colour Lie p-superalgebra, and let A and B be some G-homogeneous finitely generated subalgebras of $L^p(X)$. Then the subalgebra $A \cap B$ is also finitely generated.*

Proof. We may assume that $|X| < \infty$. By Lemma 5.3 we may embed A and B as free factors in some subalgebras $\tilde{A}$ and $\tilde{B}$ of finite codimension in $L^p(X)$.

By Corollary 5.2, $A \cap B$ is a free factor of $\tilde{A} \cap \tilde{B}$ which is a subalgebra of finite codimension in $L^p(X)$. By Theorem 4.20 we have $\operatorname{rank}(\tilde{A} \cap \tilde{B}) < \infty$ and therefore $\operatorname{rank}(A \cap B) < \infty$. □

5.5. Definition. Let $PS[X]$ be ordered in such a way that $u_1 > u_2$ if, and only if, $\Psi(u_1) > \Psi(u_2)$ (see Subsection 4.2). The leading ps-regular monomial $\hat{a}$ in the presentation of $a \in L^p(X)$ is called the *leading term* of a.

5.6. Theorem. *Let K be a field, $p = \mathrm{char}\, K > 3$, $L(X)$ a free colour Lie superalgebra, and let A and B be some G-homogeneous finitely generated subalgebras of $L(X)$. Then their intersection $A \cap B$ is also finitely generated.*

Proof. We may assume that $|X| < \infty$. We first consider the case where K is a perfect field. We consider the free colour Lie superalgebra $L(X)$ embedded in the free colour Lie p-superalgebra $L^p(X)$. By Theorem 3.15 there exist G-homogeneous generating sets M and N of subalgebras A and B respectively with $|M| < \infty$, $|N| < \infty$. By Theorem 5.4, $L^p(M) \cap L^p(N) = L^p(H)$ and $|H| < \infty$. Any element $h \in H$ can be written as

$$h = \sum \alpha_{i0} s_{i0} + \sum \alpha_{i1} s_{i1}^p + \cdots + \sum \alpha_{im} s_{im}^{p^m}$$
$$= \sum \beta_{i0} \sigma_{i0} + \sum \beta_{i1} \sigma_{i1}^p + \cdots + \sum \beta_{im} \sigma_{im}^{p^m}, \tag{1}$$

where s_{ij} are as-regular monomials in the alphabet M, $s_{ij}^{p^j}$ are ps-regular monomials in the alphabet M; σ_{ij} are as-regular monomials in the alphabet N, $\sigma_{ij}^{p^j}$ are ps-regular monomials in the alphabet N; $\alpha_{ij}, \beta_{ij} \in K$. Then

$$\sum_{j\geq 1} \sum_i \alpha_{ij} s_{ij}^{p^j} - \sum_{j\geq 1} \sum_i \beta_{ij} \sigma_{ij}^{p^j} \in L(X).$$

Since K is perfect, for the element

$$f = \sum_{j\geq 1} \sum_i \sqrt[p]{\alpha_{ij}}\, s_{ij}^{p^{j-1}} - \sum_{j\geq 1} \sum_i \sqrt[p]{\beta_{ij}}\, \sigma_{ij}^{p^{j-1}}$$

we have

$$f^p - \left(\sum_j \sum_i \alpha_{ij} s_{ij}^{p^j} - \sum_j \sum_i \beta_{ij} \sigma_{ij}^{p^j} \right) \in L(X),$$

and therefore $f^p \in L(X)$. But $\hat{f}^p = (\hat{f})^p$ (see Subsection 5.5). By Lemma 2.4, Theorem 2.6 and Lemma 4.3 we have $f = 0$. Thus

$$\sum_{j\geq 1} \sum_i \sqrt[p]{\alpha_{ij}}\, s_{ij}^{p^{j-1}} = \sum_{j\geq 1} \sum_i \sqrt[p]{\beta_{ij}}\, \sigma_{ij}^{p^{j-1}}.$$

Continuing the procedure we come to $\sum_i \sqrt[p^m]{\alpha_{im}}\, s_{im} = \sum_i \sqrt[p^m]{\beta_{im}}\, \sigma_{im}$ and thus $h_1 = \sum_i \sqrt[p^m]{\alpha_{im}}\, s_{im} \in A \cap B \subset L^p(M) \cap L^p(N)$. Now we add h_1 to H and change h to $h_2 = h - (h_1)^{p^m}$. Then, in the presentation (1) for the element h_2, we pass from m to $m - 1$. After a finite number of steps we come to $m = 0$ for h and then also for all other elements of H. Thus we have a G-homogeneous set H_1 such that $|H_1| < \infty$, H_1 generates the subalgebra $L^p(M) \cap L^p(N)$ and $H_1 \subset$

§ 5. The lattice of finitely generated subalgebras   77

$A \cap B \subset L(X)$. For all $f \in A \cap B$ we have $f = \sum_{j=1}^{k} \sum_i \gamma_{ij} s_{ij}^{p^j} + \sum_i \gamma_{i0} s_{i0}$, where s_{ij} are as-regular monomials in the elements of H_1, $s_{ij}^{p^j}$ are ps-regular monomials in the elements from H_1, $\gamma_{ij} \in K$. Then for the element $g = \sum_{j=1}^{k} \sum_i \sqrt[p]{\gamma_{ij}} s_{ij}^{p^{j-1}}$ we have $g^p - f \in L(X)$, and therefore $g^p \in L(X)$ and $g = 0$. But the element $g^p - f$ belongs to the linear span of the set $\{s_{ij}\}$. Consequently, f belongs to the colour Lie superalgebra which is generated by the finite G-homogeneous set H_1. Thus, in this case, the proof is complete.

Now we consider the general case. Let $\bar{K}$ be the algebraic closure of the field K. We embed $L(X)$ as a K-subspace in the algebra $L_{\bar{K}} = \bar{K} \otimes_K L(X)$ over the perfect field $\bar{K}$. By Theorem 3.15 we find free G-homogeneous generating sets M and N for the algebras A and B: $A = L(M)$; $B = L(N)$; $A_{\bar{K}} = L_{\bar{K}}(M)$; $B_{\bar{K}} = L_{\bar{K}}(N)$. By the above, $A_{\bar{K}} \cap B_{\bar{K}} = L(H)$ and $H = \{h_j, 1 \leq j \leq r\}$. Now we present h_j as a linear combination of s-regular monomials in some elements from M and N respectively: $h_j = \sum_i \varphi_{ij} s_i = \sum_i \psi_{ij} \sigma_i$, where $\varphi_{ij}, \psi_{ij} \in \bar{K}$. In the K-subspace, generated by φ_{ij}, ψ_{ij}, we choose a basis $\{q_l, 1 \leq l \leq t\}$. Let $\varphi_{ij} = \sum_l \alpha_{ijl} q_l$ and $\psi_{ij} = \sum_l \beta_{ijl} q_l$, where $\alpha_{ijl}, \beta_{ijl} \in K$. Then

$$h_j = \sum_{i,l} \alpha_{ijl} q_l s_i = \sum_{i,l} \beta_{ijl} q_l \sigma_i.$$

Now we write s_i and σ_i as linear combinations of the elements of the basis $\{\omega_k\}$ consisting of s-regular monomials in the elements of X: $s_i = \sum_k \gamma_{ik} \omega_k$ and $\sigma_i = \sum_k \delta_{ik} \omega_k$, where $\gamma_{ik}, \delta_{ik} \in K$, $1 \leq k \leq q$. Then $\sum_{i,l,k} \alpha_{ijl} q_l \gamma_{ik} \omega_k = \sum_{i,l,k} \beta_{ijl} q_l \delta_{ik} \omega_k$. Since $\{\omega_k\}$ is a basis both in $L(X)$ and in $L_{\bar{K}}$, we have $\sum_{i,l} \alpha_{ijl} q_l \gamma_{ik} = \sum_{i,l} \beta_{ijl} q_l \delta_{ik}$ for $1 \leq k \leq q$. Since $\{q_l\}$ is linearly independent over K, we have $\sum_i \alpha_{ijl} \gamma_{ik} = \sum_i \beta_{ijl} \delta_{ik}$ for $1 \leq k \leq q$ and $1 \leq l \leq t$. Therefore

$$\sum_i \alpha_{ijl} s_i = \sum_{i,k} \alpha_{ijl} \gamma_{ik} \omega_k = \sum_{i,k} \beta_{ijl} \delta_{ik} \omega_k = \sum_i \beta_{ijl} \sigma_i$$

for $1 \leq l \leq t$. Replacing every element h_j, $1 \leq j \leq r$ by a finite set of G-homogeneous elements $h_{jl} = \sum_i \alpha_{ijl} s_i$, where $1 \leq l \leq t$, we come to the set $H = \{h_{jl}, 1 \leq j \leq r, 1 \leq l \leq t\} \subset A \cap B$ which generates the subalgebra $A_{\bar{K}} \cap B_{\bar{K}}$ containing $A \cap B$. Thus H is a finite G-homogeneous generating set of the subalgebra $A \cap B$. This completes the proof. □

5.7. Remark. Let K be a field, $\operatorname{char} K = 0$, $X = \bigcup_{g \in G} X_g$, $X = \{x_1, \ldots, x_{n+1}\}$, $d(x_{n+1}) \in G_+$, $L(X)$ the free colour Lie superalgebra and $x_{rk} = x_r(\operatorname{ad}' x_{n+1})^k$, $1 \leq r \leq n$, $k = 0, 1, 2, \ldots$; $\{x_{rk}\} = \{Z(x_{n+1})\}$ (see Definition 3.1 and Lemma 3.7). Consider the subalgebra H_t, generated by the set $\{x_{rk}, 1 \leq r \leq n, 0 \leq k \leq t\}$. Let B be a G-homogeneous subalgebra of $L(X)$ which does not contain x_{n+1} such that $\operatorname{rank} B < \infty$. Then $B \cap H_t$ is a subalgebra of finite rank.

§6. Free colour Lie super-rings

In this section we consider subrings of the free colour Lie (p-)super-rings. Theorem 6.4 is an analogue of Witt's theorem on subrings of free Lie rings. We give a description of identical relations between left (right) normed monomials (Proposition 6.5 is an analogue of the corresponding theorem by P.M. Cohn).

In all that follows in this section K is a commutative associative ring with 1, K^* is the group of its invertible elements, char $K \neq 2$.

6.1. Definition. A colour Lie K-superalgebra L is called a *colour Lie K-super-ring*, if $[x,x] = 0$ and $[y,[y,y]] = 0$ for all homogeneous elements $x \in L_+$ and $y \in L_-$ (in the case $K = \mathbb{Z}$ we have a colour Lie super-ring).

Observe that we do not assume that $2, 3 \in K^*$.

If K is a ring, char $K = p > 2$, then in a similar manner we define the notion of a colour Lie p-super-ring over K (see Definition 1.10 from Chapter 1).

The G-homogeneous subring B of a colour Lie (p-)super-ring over K is called isolated if $B = \{a \in L | \exists 0 \neq \alpha \in K, \alpha a \in B\}$.

6.2. Free colour Lie super-rings. Let L be a colour Lie superalgebra over K, I the ideal of L generated by all elements of the form $[x,x]$ and $[y,[y,y]]$, where x, y are homogeneous elements from L, $x \in L_+$, $y \in L_-$. Then the algebra L/I is a colour Lie K-super-ring.

Let $X = \bigcup_{g \in G} X_g$ be a G-graded set and $A(X)$ the free G-graded associative K-algebra, $[A(X)] \supset [X]$ the free colour Lie superalgebra (see Lemma 2.4). Then the algebra $[X]/I$ is the *free colour Lie K-super-ring* (here I is defined as above).

Let now $L^p(X)$ be the free colour Lie p-superalgebra over K and I as above the ideal of $L^p(X)$ generated by all elements of the form $[a,a]$ and $[b,[b,b]]$ where a and b are homogeneous elements from $L^p(X)$, $d(a) \in G_+$ and $d(b) \in G_-$. Then the algebra $L^p(X)/I$ is the *free colour Lie p-super-ring* over K.

The universal properties of free super-rings $[X]/I$ and $L^p(X)/I$ are analogous to the corresponding universal properties of the free superalgebras $[X]$ and $L^p(X)$.

6.3. Proposition. *The following assertions hold:*
1) *The set of s-regular monomials supplies us with a basis of $[X]/I$ (as a free K-module).*
2) *If char $K = p$, then the set of ps-regular monomials is a basis of $L^p(X)/I$ (as a free K-module).* □

The proof can be given by analogy with the proof of Theorem 2.6 and Lemma 4.3 (in place of the condition 2, 3 ∈ K^* we use the conditions $[a, a] = 0$ and $[b, [b, b]] = 0$ for $d(a) \in G_+$ and $d(b) \in G_-$).

6.4. Theorem. *Let K be a commutative domain over which all projective modules are free, L the free colour Lie (p-)super-ring over K, B an isolated subring of L which is a direct summand of the K-module L. Then the ring B is a free colour Lie (p-)super-ring over K.*

Proof. We say that a G-homogeneous subset M of nonzero elements from L is reduced if for each $a \in M$ no $\alpha a'$ belongs to the subalgebra generated by $\{b' | b \in M, b \neq a\}$ where a' and b' are the leading parts of a and b and $0 \neq \alpha \in K$. Using the fact that B is an isolated subring of L and applying A.G. Kurosh's method (see Lemma 3.12 and [Kurosh, 1947]) we arrive at a reduced G-homogeneous generating set M for the subring B. Now we embed K in its field of quotients $Q(K)$. Let $L(M)$ be a colour Lie (p-)superalgebra generated by M in the free colour Lie (p-)superalgebra $L_Q = L \otimes Q(K)$ over $Q(K)$. Since M is reduced in L, we see that M is reduced in L_Q. Applying Theorems 3.15 and 4.15 we have the assertion of our theorem. □

6.5. Remark. Under the hypothesis of Theorem 6.4, except that B need not be isolated, we assume that B is φ-homogeneous relative to some weight $\varphi: X \to \mathbb{N}$. Then B is a free colour Lie (p-)super-ring over K. The proof can be given by analogy with the proof of Theorem 2.4.3 from [Bahturin, 1987a] with the use of the results of this chapter.

At last we give a description of the identical relations between left (right) normed monomials in free colour Lie super-rings.

Let δ, $\delta': A(X) \to [X]/I$ be linear mappings such that $\delta(x_1, \ldots, x_n) = [x_1, \ldots, x_n]$ is the right normed monomial and $\delta'(x_1, \ldots, x_n) = [x_1, \ldots, x_n]'$ is the left normed monomial for all $x_1, \ldots, x_n \in X$.

As we have remarked, any element from $[X]/I$ is a linear combination of left (right) normed monomial.

6.6. Proposition (see [Mikhalev, 1991b]). *For the mappings δ, δ' we have*
1) $\delta(a) = 0$ *if, and only if,* $a = \sum_i w_i(\delta(a_i)b_i + \varepsilon(a_i, b_i)\delta(b_i)a_i)$;
2) $\delta'(a) = 0$ *if, and only if,* $a = \sum_i (c_i \delta'(d_i) + \varepsilon(c_i, d_i)d_i \delta'(c_i))\omega_i$,
where a_i, b_i, c_i, d_i are homogeneous elements from $A(X)$, w_i, $\omega_i \in A(X)$. □

This proposition gives us an analogue of the following theorem due to P.M. Cohn (Theorem 10.1 from [Cohn, 1951]): let $L(X)$ be a free Lie ring, then $\delta'(a) = 0$ iff $a = \sum_i u_i \delta'(u_i) + \sum_j v_j \delta'(v_j)w_j$, where u_i, v_j, $w_j \in A(X)$.

Comments to Chapter 2

The main source of the theory of free Lie algebras lies in the papers of P. Hall [Hall P., 1933], W. Magnus [Magnus, 1935, 1937], and E. Witt [Witt, 1937] devoted to the study of free groups. The construction of a linear basis for free Lie algebras has been suggested in the article [Hall M., 1950] of M. Hall (so-called "Hall basis"). The theorem about freeness of subalgebras in free Lie algebras has been proved by A.I. Shirshov [Shirshov, 1953] and in free Lie rings and p-algebras by E. Witt [Witt, 1956]. Later A.I. Shirshov has given some other constructions of free linear bases [Shirshov, 1958, 1962c]. There are also other variations of the bases (see for example, [Michel, 1975], [Schützenberger, 1971], [Ufnarovsky, 1990], [Viennot, 1978]). The theorem on the finiteness of the rank for the intersection of two subalgebras of finite rank is due to G.P. Kukin [Kukin, 1972b]. The main results of Chapter 2 are contained in the papers of A.A. Mikhalev [Mikhalev, 1985, 1986, 1988, 1990, 1991b, 1992a, 1992b], giving an extensive development of A.I. Shirshov's techniques for free colour Lie p-superalgebras.

An independent approach to the theorems on bases and subalgebras for free Lie superalgebras over a field is due to A.S. Shtern [Shtern, 1986]. Even earlier the theorem on subalgebras of free $\mathbb{Z}$-graded Lie superalgebras had been mentioned by J. Lemaire [Lemaire, 1974]. Various forms of the dimension formula for the homogeneous components of free Lie superalgebras have been suggested by I.L. Kantor [Kantor, 1984], A.A. Mikhalev [Mikhalev, 1986], A.I. Molev and L.M. Tsalenko [Molev, Tsalenko, 1986].

Chapter 3

Composition techniques in the theory of Lie superalgebras

In this chapter we present some techniques concentrating around the so-called Diamond Lemma (Theorem 1.2) and with the help of this lemma we prove the PBW-theorems (Poincaré-Birkhoff-Witt) for bases of (restricted) universal enveloping algebras (Theorems 2.2, 2.5). We also consider filtrations on universal enveloping algebras and their associated graded algebras (Proposition 2.8). We describe the structure of the colour Hopf algebra on the universal enveloping algebra (see point 2.9) and its primitive elements (Theorems 2.10, 2.11 are in fact "colour" analogues of Friedrichs' Criterion). We show that free colour Lie (p-)superalgebras are finitely separable (Theorem 2.12).

In §3 we prove the Diamond Lemma for colour Lie superalgebras and for p-superalgebras (Theorem 3.10).

In §4 we use the Diamond Lemma to construct bases of the free product with amalgamated subalgebra for colour Lie superalgebras and p-superalgebras (Theorem 4.4, 4.8).

§1. The Diamond Lemma for associative rings

1.1. Let K be a commutative and associative ring with 1, X a set, $S(X)$ the free semigroup with 1 on X, $A(X)$ the free associative K-algebra on X.

We consider a set S of pairs of the form $\sigma = (W_\sigma, f_\sigma)$, where $W_\sigma \in S(X)$ and $f_\sigma \in A(X)$. For any $\sigma \in S$ and $A, B \in S(X)$ let $R_{A\sigma B}$ denote the endomorphism of the K-module $A(X)$ that acts trivially on all elements of $S(X)$ other than $AW_\sigma B$ and at the same time $R_{A\sigma B}(AW_\sigma B) = Af_\sigma B$.

We call this set S a *reduction system* and the K-endomorphisms $R_{A\sigma B}$: $A(X) \to A(X)$ *reductions*. We say that an element $a \in A(X)$ is *irreducible* if a does not contain monomials of the form $AW_\sigma B$, $\sigma \in S$. The K-submodule of all irreducible elements of $A(X)$ will be denoted by $A(X)_{irr}$.

Suppose that for $a \in A(X)$ we have a finite sequence $R_1, \ldots, R_n$ of reductions such that $R_n \ldots R_1(a) \in A(X)_{irr}$. The element $R_n \ldots R_1(a)$ will be called a *normal form* of a. The set of elements $a \in A(X)$ with unique normal form $R_S(a)$ will be denoted $A(X)_{red}$.

Consider an overlap ambiguity (σ, τ, A, B, C) with $\sigma, \tau \in S$ and $A, B, C \in S(X)\setminus 1$ such that $W_\sigma = AB$ and $W_\tau = BC$. We shall say the overlap ambiguity (σ, τ, A, B, C) is resolvable if there exist compositions of reductions R and R', such that $R(f_\sigma C) = R'(Af_\tau)$.

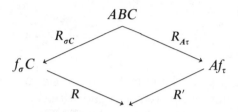

Now consider an inclusion ambiguity (σ, τ, A, B, C) with $\sigma, \tau \in S$ and $A, B, C \in S(X)$ such that $W_\sigma = B$ and $W_\tau = ABC$. We shall say the inclusion ambiguity is resovable if $Af_\sigma C$ and f_τ can be reduced to a common element.

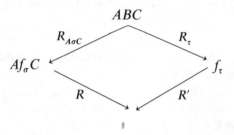

We say that a partial ordering $\leq$ on $S(X)$ is a semigroup partial ordering if $B < B'$ with $B, B' \in S(X)$ implies $ABC < AB'C$ for all $A, C \in S(X)$ and that it is compatible with S if for all $\sigma \in S$ the element f_σ is a linear combination of words V such that $V < W_\sigma$.

Let I be the two-sided ideal of $A(X)$ generated by the elements $W_\sigma - f_\sigma$, $\sigma \in S$. It is clear that as a K-module I is spanned by the elements $A(W_\sigma - f_\sigma)B$ with $A, B \in S(X)$.

1.2. Theorem (see [Bergman, 1978, Theorem 1.2] and [Bokut', 1976]). *Let S be a reduction system for the free associative algebra $A(X)$ and $\leq$ a semigroup partial ordering on $S(X)$ compatible with S and satisfying the descending chain condition (DCC). Then the following conditions are equivalent:*
1) *All ambiguities of S are resolvable.*
2) $A(X) = A(X)_{red}$.
3) *A set of representatives in $A(X)$ for the elements of the algebra $R = A(X)/I$ defined in terms of generators X and relations $W_\sigma = f_\sigma (\sigma \in S)$ is given by the K-submodule $A(X)_{irr}$ (i.e. $A(X) = A(X)_{irr} \oplus I$).*
When these conditions hold, R may be identified with the K-module $A(X)_{irr}$, made into a K-algebra by the multiplication $a \times b = R_S(ab)$.

Proof. Since we consider the S-compatible partial ordering with DCC every element of $A(X)$ can be reduced to a normal form under a finite sequence of reductions.

Assuming 2) the mapping R_S will be a projection of $A(X)$ onto $A(X)_{irr}$, its kernel being contained in I because every reduction replaces an element a by an element $a + b$ with $b \in I$; its kernel is contained in I since $R_S(A(W_\sigma - f_\sigma)B) = R_S(AW_\sigma B) - R_S(Af_\sigma B) = 0$ for all A, B, σ (all elements of $A(X)$ have unique normal form), just proving 3).

Now assume 3) holds and suppose that $a \in A(X)$ can be reduced to distinct $b, b' \in A(X)_{irr}$. Then $b - b' \in A(X)_{irr} \cap I = 0$. Thus 2) and 3) are equivalent.

It is clear that 2) $\Rightarrow$ 1). Now we want to show that 1) $\Rightarrow$ 2). At the beginning observe that $A(X)_{red}$ is a K-submodule of $A(X)$. Let $f, g \in A(X)_{red}$, $\alpha, \beta \in K$ and R be any finite composition of reductions. Then we can find finite compositions of reductions R' and R'' such that $(R'R)(f) = R_S(f)$ and $(R''R'R)(g) = R_S(g)$.

Therefore

$$(R''R'R)(\alpha f + \beta g) = \alpha R''R_S(f) + \beta R_S(g) = \alpha R_S(f) + \beta R_S(g) \in A(X)_{irr}.$$

Thus the element $\alpha f + \beta g$ has unique normal form. Moreover we have proved that the mapping $R_S: A(X)_{red} \to A(X)_{irr}$ is K-linear.

It remains to prove $S(X) \subset A(X)_{red}$. If w is a minimal element in $S(X)$ then it is obvious that $w \in A(X)_{red}$. Suppose now that w is not a minimal element in $S(X)$ and that for any $v \in S(X)$ with $v < w$ we have $v \in A(X)_{red}$. We prove that $w \in A(X)_{red}$. For, given any two reductions $R_{U\sigma V'}$ and $R_{U'\tau V}$ each acting nontrivially on w, from the preceding inductive hypothesis we derive that

$$R_{U\sigma V'}(w), \quad R_{U'\tau V}(w) \in A(X)_{red}.$$

We would like to show that

$$R_S(R_{U\sigma V'}(w)) = R_S(R_{U'\tau V}(w)).$$

Without loss of generality we may assume that $l(U) < l(U')$.
Consider three possible cases.

Case 1. We have an overlap ambiguity $w = UABCV$, $W_\sigma = AB$, $W_\tau = BC$. By 1) this ambiguity is resolvable, i.e. the elements $f_\sigma C$ and Af_τ can be reduced each to one and the same element f. Therefore the elements $R_{U\sigma V'}(w) = Uf_\sigma CV$ and $R_{U'\sigma V}(w) = UAf_\tau V$ can be reduced to one element $g = UfV$. Thus

$$R_S(Uf_\sigma CV) = R_S(g) = R_S(UAf_\tau V).$$

Case 2. We have an inclusion ambiguity. By analogy with the first case, the present case can be handled with the solvability of inclusion ambiguities.

Case 3. The subwords W_σ and W_τ are disjoint in w (i.e. $w = AW_\sigma BW_\tau C$). Then the elements $R_{U_\sigma V'}(w) = Af_\sigma BW_\tau C$ and $R_{U'_\tau V}(w) = AW_\sigma Bf_\tau C$ can be reduced to one element $f = Af_\sigma Bf_\tau C$, and therefore the normal form of each of these elements is equal to $R_S(f)$. This completes the proof of the theorem. □

1.3. Remark. For any ideal I of the algebra $A(X)$ there exists a system of reductions S satisfying the conditions of Theorem 1.2 (this ideal I is generated by elements $W_\sigma - f_\sigma, \sigma \in S$). Such a system of generators of I will be called a complete system of relations. Observe that a finitely generated ideal may have an infinite complete system of relations. For example if $X = \{x, y\}$, $x > y$ and if I is generated by $x^2 - yx$, then a complete system of relations for I is $\{W_\sigma - f_\sigma | \sigma \in S\}$, where $S = \{(xy^{n+1}x, y^{n+2}x) | n = -1, 0, 1, 2, \ldots\}$.

Further we remark that for any ideal I there exists a complete system of relations S such that all elements $W_\sigma - f_\sigma$ are not reducible with the help of reductions $W_\tau \to f_\tau$, where $\tau \in S$, $\tau \neq \sigma$ (such a complete system of relations is unique and sometimes it is called the *Gröbner Basis*, see for example [Computer Algebra, 1983], [Latyshev, 1988], [Ufnarovsky, 1990]).

Let K be a field, char $K \neq 2, 3$, Y a set and let L be a colour Lie superalgebra with G-homogeneous basis $E = \{e_i, i \in Y\}$, $\lambda_i: G \to K^*$ a homomorphisms for $i = 1, \ldots, n$, $T = \{t_i, i = 1, \ldots, n, d(t_i) = 0\}$, $X = \{x_i, i \in Y, d(x_i) = d(e_i)\}$. Let $A(T \cup X)$ be the free G-graded associative algebra and I be the two-sided ideal in $A(T \cup X)$, generated by the following sets of elements:

$\{t_i t_j - t_j t_i, 1 \leq i, j \leq n\}$; $\{t_i x - \lambda_i(g) x t_i - (\lambda_i(g) - 1)x, \ 1 \leq i \leq n, \ x \in L_g, g \in G\}$; $\{ab - \varepsilon(a, b)ba - [a, b]$ for G-homogeneous elements $a, b \in A(X)\}$.

By Theorems 1.2 and 2.2 the cosets with representatives of the form $t_1^{\lambda_1} \ldots t_n^{\lambda_n} x_1 \ldots x_k$, $\lambda_i \geq 0$, $k \in \mathbb{N} \cup 0$, $x_i \leq x_{i+1}$, $x_i \neq x_{i+1}$ with $x_i \in X$, form a linear basis of the quotient algebra $A(T \cup X)/I$. (This algebra is a semidirect product of the polynomial ring $K[t_1, \ldots, t_n]$ and the universal enveloping algebra $U(L)$.)

§ 2. Universal enveloping algebras

In this section we use the Diamond Lemma (Theorem 1.2) to prove PBW-theorems for (restricted) universal enveloping algebras (Theorems 2.2, 2.5). We consider filtrations on universal enveloping algebras and their graded algebras (Proposition 2.8). We also describe the colour Hopf algebra structure on universal enveloping algebras (see Subsection 2.9) and their primitive elements (Theorems 2.10 and 2.11 give us analogues of Friedrichs' Criterion). We show that free colour Lie (p-)superalgebras are finitely separable (Theorem 2.12).

2.1. Recall some definitions from Chapter 1. Let K be a commutative associative ring with 1, L a colour Lie superalgebra, U a G-graded associative algebra with 1, $\delta: L \to [U]$ a homomorphism of colour Lie superalgebras. We say that the algebra U is the *universal enveloping algebra* of L (we denote it by $U(L)$) if for any homomorphism σ of the colour Lie superalgebra L into a colour Lie superalgebra $[R]$ (with the same ε and G) for some G-graded associative algebra R with 1 there exists a unique homomorphism θ of G-graded associative algebras with 1 of U into R such that the following diagram is commutative.

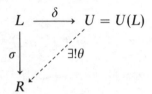

It is clear that the universal enveloping algebras are determined uniquely (up to isomorphisms of G-graded associative algebras with 1).

In Lemma 2.4 of Chapter 2 it was proved that $\pi: L(X) \to A(X)$ is an embedding and $\pi(L(X)) = [X]$, $A(X) = U(L(X))$. Suppose now that $2, 3 \in K^*$ and that L is a colour Lie superalgebra which is a free K-module with a G-homogeneous basis $X = X_+ \cup X_-$. Let $A(X)$ be the free G-graded associative algebra, I be the two-sided ideal in $A(X)$ generated by elements of the form $ab - \varepsilon(a,b)ba - [a,b]$ for all homogeneous $a, b \in L$. Consider the canonical mapping $\delta: L \to A(X)/I = U(L)$. It is clear that $U(L)$ is the universal enveloping algebra of L.

Let $\leq$ be a total ordering of X. Since the multiplication $[,]$ is K-linear, the elements $yx - \varepsilon(y,x)xy - [y,x]$ with $x < y \in X$ and $yy - \frac{1}{2}[y,y]$ with $y \in X_-$ generate the above ideal I.

2.2. Theorem (Poincaré-Birkhoff-Witt). *Let $2, 3 \in K^*$. The universal enveloping algebra $U(L)$ constructed above is a free K-module with a G-homogeneous basis consisting of 1 and all monomials of the form $\delta(x_1)\ldots\delta(x_n)$ where $n \in \mathbb{N}$, $x_i \in X$, $x_i \leq x_{i+1}$ and $x_i \neq x_{i+1}$ if $x_i \in X_-$ for all $i = 1, \ldots, n-1$ (in particular, δ is an embedding).*

Proof. Let S be the reduction system in $A(X)$ consisting of the pairs $\sigma_{yx} = (yx, \varepsilon(y,x)xy + [y,x])$ for all $y > x$ in X and $\sigma_{yy} = (yy, \frac{1}{2}[y,y])$ for all $y \in X_-$. It is clear that the ideal I is generated by all elements $W_\sigma - f_\sigma$, $\sigma \in S$. Define the disordering index of an element $x_1 \ldots x_n \in S(X)$ as the number of pairs (i, j) such that $i < j$ but $x_i > x_j$. Partially order $S(X)$ by setting $A < B$ if A is of smaller length than B, or if $m(A) = m(B)$ (i.e. A and B have the same multidegrees), but A has smaller disordering index than B. It is clear that this ordering is a semigroup ordering, it is S-compatible and satisfies DCC.

Consider possible ambiguities.

Case 1. $(\sigma_{zy}, \sigma_{yx}, z, y, x)$ with $z > y > x \in X$. To resolve such an ambiguity relative to $\leq$ we must study the element

$$R_{\sigma_{zy}x}(zyx) - R_{z\sigma_{yx}}(zyx) = (\varepsilon(z, y)yzx + [z, y]x) - (\varepsilon(y, x)zxy + z[y, x]).$$

To reduce this element further we apply reductions $R_{y\sigma_{zx}}$, $R_{\sigma_{zx}y}$, $R_{\sigma_{yx}z}$, $R_{x\sigma_{zy}}$.

Thus, we have

$$\varepsilon(z, y)(\varepsilon(z, x)(\varepsilon(y, x)zyx + [y, x]z) + y[z, x]) +$$

$$[z, y]x\varepsilon(y, x)(\varepsilon(z, x)(\varepsilon(z, y)xyz + x[z, y]) + [z, x]y) - z[y, x] =$$

$$[[z, y], x] - [z, [y, x]] + \varepsilon(z, y)[y, [z, x]] \equiv 0 (\text{mod } I).$$

Case 2. $(\sigma_{zy}, \sigma_{yy}, z, y, y)$ with $z \in X$, $y \in X_-$, $z > y$.
Consider the element

$$R_{\sigma_{zy}y}(zyy) - R_{z\sigma_{yy}}(zyy) = \varepsilon(z, y)yzy + [z, y]y - \tfrac{1}{2}z[y, y].$$

We apply now the reductions $R_{y\sigma_{zy}}$ and $R_{\sigma_{yy}z}$ to this element. Then we have $[[z, y], y] - \tfrac{1}{2}[z, [y, y]] \equiv 0 (\text{mod } I)$.

Case 3. $(\sigma_{yy}, \sigma_{yz}, y, y, z)$ with $z \in X$, $y \in X_-$, $y > z$.

We consider this situation by analogy with the second case. It is obvious that the ambiguity $(\sigma_{yy}, \sigma_{yy}, y, y, y)$ is resolvable. Remark that in our system of reductions S there are no inclusion ambiguities.

The claim of our theorem now follows from Theorem 1.2 and the fact that the irreducible words have the form described in the statement of our theorem. □

2.3. Remark. If K is a field and char $K = 3$, the algebra $U(L)$ constructed above is the universal enveloping algebra, but it is possible that the mapping δ is not an embedding. For example, if $X = X_- = \{y\}$, $L(X)$ is the free Lie K-superalgebra, $A(X)$ is the free associative algebra, then $A(X) = U(L(X))$, but $[y, [y, y]] \neq 0$ and $\delta([y, [y, y]]) = 0$; nonzero elements from Theorem 2.2 form a basis. The assertion of Theorem 2.2 remains true if in addition we assume that $[y, [y, y]] = 0$ for all homogeneous elements $y \in L_-$.

§2. Universal enveloping algebras

2.4. Definition. Let K be a field, char $K = p > 3$, L a colour Lie p-superalgebra, U a G-graded associative K-algebra with 1, $\delta: L \to [U]^p$ a homomorphism of colour Lie p-superalgebras. We say that U is the *restricted universal enveloping algebra* for the colour Lie p-superalgebra L (notation $u(L)$) if for any homomorphism $\sigma: L \to [R]^p$ of colour Lie p-superalgebras, where R is a G-graded associative K-algebra with 1 (with the same ε and G), there exists a unique homomorphism $\theta: U \to R$ of G-graded associative K-algebras with 1 such that the following diagram is commutative.

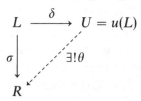

It is clear that the restricted universal enveloping algebra is defined uniquely (up to an isomorphism of G-graded associative K-algebras with 1).

Let K be a field, char $K = p > 3$, L a colour Lie p-superalgebra with G-homogeneous linear basis X. Let $A(X)$ be a free G-graded associative K-algebra and let I be the two sided ideal of $A(X)$ generated by the elements $ab - \varepsilon(a,b)ba - [a,b]$ (where $a, b \in L$ are homogeneous) and $a^p - a^{[p]}$ (where $a \in L_g, g \in G_+$).

Let $\delta: L \to A(X)/I$ be the canonical mapping. It is clear that $A(X)/I = u(L)$ is the restricted universal enveloping algebra for L, $u(L) \cong U(L)/J$ where $U(L)$ is the universal enveloping algebra for L constructed above and J is the two-sided ideal of $U(L)$ generated by elements $a^p - a^{[p]}$ ($a \in L_g, g \in G_+$).

Let $\leq$ be a total order on X. Since $[\ ,\ ]$ is a K-linear multiplication we see that the elements $yx - \varepsilon(y,x)xy - [y,x]$, where $y > x \in X$, and $x^p - x^{[p]}$, where $x \in X_g, g \in G_+$, generate the ideal I.

2.5. Theorem (Poincaré-Birkhoff-Witt). *Let K be a field, char $K > 3$. The monomials of the form $\delta(x_1)^{\lambda_1} \ldots \delta(x_n)^{\lambda_n}$, where $x_i \in X$, $x_1 < \cdots < x_n$, $0 \leq \lambda_i \leq p - 1$ for $x_i \in X_+$ and $\lambda_i = 0, 1$ for $x_i \in X_-$, form a linear basis of $u(L)$.*

Proof. Let S be the reduction system in $A(X)$ consisting of the pairs $\sigma_{yx} = (yx, \varepsilon(y,x)xy + [y,x])$ for all $y > x$, $y, x \in X$, $\sigma_{yy} = (yy, \frac{1}{2}[y,y])$ for all $y \in X_-$, and $\sigma_{x^p} = (x^p, x^{[p]})$ for all $x \in X_+$. As in the proof of Theorem 2.2 we consider the semigroup partial ordering first by length and then by the disordering index.

As we have remarked this semigroup ordering is S-compatible and has DCC.

Consider now all possible ambiguities. We do not have any inclusion ambiguities because the pairs σ_{yy} are defined for $y \in X_-$ and the pairs σ_{x^p} are defined for $x \in X_+$. Looking at the overlap ambiguities we have the following possible cases.

Case 1. Overlap ambiguities between σ_{xy}, σ_{yy} and σ_{yz}. These ambiguities are resolvable as was shown considering the Cases 1–3 in the proof of Theorem 2.2.

Case 2. $(\sigma_{yx}, \sigma_{x^p}, y, x, x^{p-1})$ with $y > x$, $y \in X$, $x \in X_+$. Consider the element $R_{y\sigma_{x^p}}(yx^p) - R_{\sigma_{yx}x^{p-1}}(yx^p) = yx^{[p]} - (xy - [y,x])x^{p-1}$.

Applying to this element the reductions $R_{x\sigma_{yx}x^{p-2}}, R_{x^2\sigma_{yx}x^{p-3}}, \ldots, R_{x^{p-1}\sigma_{yx}}$ and then $R_{\sigma_{x^p y}}$, taking into account char $K = p$ and the properties of the operation $[p]$, we have $[y, x^{[p]}] - \underbrace{[[\ldots[[y,x],x],\ldots],x]}_{p \text{ times}} \equiv 0 (\mathrm{mod}\, I)$.

Case 3. $(\sigma_{x^p}, \sigma_{xy}, x^{p-1}, x, y)$ with $x > y$, $y \in X$, $x \in X_+$. By analogy with the Case 2 similar arguments work in this new situation.

Observe that we do not have any overlap ambiguities between σ_{yy} and σ_{x^p} because $y \in X_-$ and $x \in X_+$. It is obvious that all ambiguities between x^p and x^p are resolvable.

Thus, using now Theorem 1.2 and the fact that irreducible words have the form described in the statement of our theorem, we complete the proof. $\square$

2.6. Remarks. 1. Let L be a colour Lie superalgebra over a field K, char $K = p > 3$, E a G-graded basis of L_+ and assume that for any $e_i \in E$ the derivation $(\mathrm{ad}\, e_i)^p$ is inner (i.e. there exists an element $e_i^{[p]} \in L_{pg}$, where $g = d(e_i) \in G$, such that $(\mathrm{ad}\, e_i)^p = \mathrm{ad}\, e_i^{[p]}$). Then there is a unique mapping $a \to a^{[p]}$, defined on the G-homogeneous elements of L_+ and coinciding on basic elements with the above mentioned mapping which gives us the structure of colour Lie p-superalgebra on L.

2. There is a colour Lie superalgebra which does not have colour Lie p-superalgebra structures. For example, if $K = F_{p^n}$ is the finite field of p^n elements, $n > 1$, L a linear space over K with a basis $\{v_g, g \in K\}$, then we may consider the multiplication with $[v_g, v_h] = (g - h)v_{g+h}$ on basic elements. Thus L is a Lie algebra which is not compatible with any further structure of a Lie p-algebra.

2.7. Colour tensor product. Let Q and R be G-graded associative algebras over a commutative ring K with 1, G an abelian group, $\varepsilon \colon G \times G \to K^*$ an alternating bilinear form. Consider the tensor product $Q \otimes_K R$ as a K-module and define multiplication on it by setting $(a \otimes b)(c \otimes d) = \varepsilon(b,c) ac \otimes bd$ for homogeneous elements $a, c \in Q$, $b, d \in R$ (for other elements use linearity). This algebra $Q \hat{\otimes} R$ is G-graded with $d(a \otimes b) = d(a) + d(b)$ for homogeneous elements $a \in Q$, $b \in R$ and it is associative since

$$(a \otimes b)((c \otimes g)(e \otimes f)) = \varepsilon(b, ce)\varepsilon(g, e) ace \otimes bgf = \varepsilon(b,c)\varepsilon(b,e)\varepsilon(g,e) ace \otimes bgf$$

$$= \varepsilon(b,c)\varepsilon(bg, e) ace \otimes bgf = ((a \otimes b)(c \otimes g))(e \otimes f)$$

for homogeneous elements $a, c, e \in Q$, $b, g, f \in R$. (We call $Q \hat{\otimes} R$ the colour tensor product of G-graded associative algebras Q and R.)

Consider now the filtration (see Subsection 1.3 from Chapter 1) on the universal enveloping algebra $U(L)$ of the colour Lie superalgebra L defined by $U^{-1} = \{0\}$, $U^0 = K \cdot 1$ and for $n > 0$. Let U^n denote the K-submodule generated by all monomials of length $\leq n$ (see Theorem 2.2). As usual (see Subsection 1.3 in Chapter 1) the associated graded module $\operatorname{gr} U(L) = \bigoplus_{n=-1}^{\infty} U^{n+1}/U^n$ with multiplication $(a + U^{n-1})(b + U^{m-1}) = ab + U^{n+m-1}$, where $a \in U^n$, $b \in U^m$, is a G-graded associative algebra. It is clear that $\operatorname{gr} U(L)$ is generated by all elements $a + U^0$, $a \in L$. For G-homogeneous elements a and b we have $(a + U^0)(b + U^0) = ab + U^1 = \varepsilon(a,b)ba + [a,b] + U^1 = \varepsilon(a,b)ba + U^1$ because $[a,b] \in L \subset U^1$.

Now let X, Y be G-graded sets, $X = X_+$, $Y = Y_-$. Recall, from Chapter 1, Subsection 1.9 that $K^\varepsilon[X]$ is our notation for the algebra of colour polynomials in variables from X and that $\Lambda^\varepsilon(Y)$ is the same for the colour Grassmann algebra.

2.8. Proposition. *Let K be a field, $\operatorname{char} K \neq 2, 3$.*
1) *Suppose that L is a colour Lie superalgebra with a G-homogeneous basis $E = E_+ \cup E_-$ and $U(L)$ is its universal enveloping algebra. Then*

$$\operatorname{gr} U(L) \cong K^\varepsilon[E_+] \hat{\otimes} \Lambda^\varepsilon(E_-).$$

2) *Let $\operatorname{char} K = p > 3$. If L is a colour Lie p-superalgebra and $u(L)$ is its restricted enveloping algebra (observe that the filtration on $u(L)$ and associative graded algebra $\operatorname{gr} u(L)$ are constructed by analogy with the use of Theorem 2.5) then, denoting a G-homogeneous linear basis in L by $E = E_+ \cup E_-$, we have*

$$\operatorname{gr} u(L) \cong (K^\varepsilon[E_+]/I) \hat{\otimes} \Lambda^\varepsilon(E_-).$$

Here $K^\varepsilon[E_+]/I$ is the algebra of reduced colour polynomials and the ideal I is generated by elements e_i^p, $e_i \in E_+$.
3) *The universal enveloping algebra $U(L)$ of a colour Lie algebra L (i.e. $L = L_+$) has no zero divisors.*
4) *Let $\dim L < \infty$. Then the algebra $U(L)$ is left and right Noetherian.*
5) *If $L = L_+$, then the Jacobson radical $J(U(L))$ is equal to zero.*

Proof. The assertions 1) and 2) follow immediately from Theorem 2.2 and 2.5 respectively.

3) Let now $a \in U = U(L)$, $a \in U^n$, $a \notin U^{n-1}$. The element $\bar{a} = a + U^{n-1}$ is homogeneous of degree n in $\operatorname{gr} U$ (we call it the leading term of a, setting $\bar{0} = 0$). If $a, b \in U$ then either $\overline{ab} = 0$ or $\overline{ab} = \bar{a}\bar{b}$. We show that if $\operatorname{gr} U$ does

not have zero division then U also does not have zero divisors. Let $b, c \in U$ and $b \neq 0, c \neq 0$. Then $\bar{b} \neq 0, \bar{c} \neq 0$ and, therefore, $\bar{b}\bar{c} \neq 0$. But $\bar{b}\bar{c} = \overline{bc}$ so $bc \neq 0$.

By assertion 1) of our proposition we have $\operatorname{gr} U \cong K^\varepsilon[X]$, where X is a G-graded basis of L, $X = X_+$ and $K^\varepsilon[X]$ is the ring of colour polynomials. The ring $K^\varepsilon[X]$ does not have zero divisors because the leading term of the product of two elements is equal to the product of their leading terms (up to a K-multiplier). Thus from the above remark we have that for $L = L_+$ the algebra $U(L)$ does not have zero divisors.

4). We show that if $\operatorname{gr} U$ is left (right) Noetherian then the ring U is also left (right) Noetherian. Let V be a left ideal of U. Then the set of finite sums of leading terms of elements from V is a left homogeneous ideal $\bar{V}$ of $\operatorname{gr} U$. Being left Noetherian is equivalent to having a finite system of generators for any left ideal. Suppose that, by our assumption, $\bar{a}_1, \ldots, \bar{a}_t$ generate $\bar{V}$. Let $\bar{a}_i$ be the leading term of $a_i \in V$, $1 \le i \le t$. We show that a_i, $1 \le i \le t$, generate V, i.e. for any $a \in V$ we have $a = \sum b_i a_i$ for some $b_i \in V$. Using induction on the degree we may assume that our statement is proved for each element a with the degree of its leading term $\hat{a}$ smaller than n. Assume that the degree of $\bar{a}$ is equal to n. It is clear that $\bar{a} = \sum \bar{b}_j \bar{a}_j$, where $\bar{b}_j \in \operatorname{gr} U$ is homogeneous and $\bar{b}_j \bar{a}_j$ is homogeneous of degree n, $\bar{b}_j$ is the leading term of b_j, the leading term of $\sum b_j a_j$ is equal to $\bar{a}$. Then $c = a - \sum b_j a_j \in U^{(n-1)}$ and therefore c is in the left ideal generated by a_j. By the inductive assumption we prove our assertion.

At the beginning, we consider the case $L = L_+$. As we have remarked, in our case $\operatorname{gr} U \cong K^\varepsilon[X]$ is the ring of colour polynomials, where $X = X_+$. We now use the Hilbert Basis Theorem in the following form.

Theorem ([McConnell, Robson, 1987, Chapter 1, Theorem 2.10]). *Let R be a left Noetherian ring and S be an over-ring generated by R and an element x such that $xR + R = Rx + R$. Then S is left Noetherian.* □

Let now $X = X_+ = \{x_1, \ldots, x_n\}$, $S = K^\varepsilon[X]$, $x = x_n$ and $R = K^\varepsilon[X \setminus x_n]$. By the inductive assumption the ring R is left Noetherian. If $r = \sum_{g \in G} \alpha_g r_g$, $\alpha_g \in K$, $r_g \in R_g$ then $x \cdot r = \sum \alpha_g x_n \cdot r_g = \sum \alpha_g \varepsilon(d(x_n), g) r_g \cdot x_n = (\sum \alpha_g \varepsilon(d(x_n), g) r_g) x$ and $r \cdot x = \sum \alpha_g r_g \cdot x_n = \sum \alpha_g \varepsilon(g, d(x_n)) x_n \cdot r_g = x(\sum \alpha_g \varepsilon(g, d(x_n)) r_g)$.

By the Hilbert Basis Theorem we see that $K^\varepsilon[X]$ is left Noetherian. As we mentioned above, $U(L)$ is also left Noetherian.

Consider now the general case: $L = L_+ \oplus L_-$. As K is a field and $U(L_+)$ is a left Noetherian subalgebra of finite codimension in $U(L)$ (see the **PBW Theorem 2.2**), we see that $U(L)$ is itself left Noetherian.

5). We may define the Jacobson radical $J = J(R)$ of an associative ring R with 1 as the largest left ideal consisting of quasi-regular elements r (i.e. $1 + r$

is invertible in R). It is clear that in our case the set of all invertible elements of $U(L)$ together with the zero element is equal to the field K. Therefore either $J \cap K = 0$ or $J \cap K = K$. Since $1 \notin J$, $J \cap K = \{0\}$. Therefore, if $z \in J$ then $1 - z \in K$ and $z \in K$. Thus $z \in J \cap K = 0$, hence $J = 0$. This completes the proof of our proposition. □

Remark. Note that some statements of Proposition 2.8 are valid in case char $K = 3$. For example, if L is a finite-dimensional colour Lie superalgebra over a field of characteristic 3 then $U(L)$ is right and left Noetherian.

2.9. A Hopf algebra structure on the universal enveloping algebra. Let K be a field, char $K \neq 2, 3$, L a colour Lie superalgebra, $U(L)$ its universal enveloping algebra, $U(L) \hat{\otimes} U(L)$ the colour tensor product (see Subsection 2.7). We define the *diagonal mapping* $\delta \colon L \to U(L) \hat{\otimes} U(L)$ by setting $\delta(a) = a \otimes 1 + 1 \otimes a$ for $a \in L$.

Assuming $\delta(1) = 1 \otimes 1$, by Theorem 2.2 this mapping has a unique extension to a homomorphism $\delta \colon U(L) \to U(L) \hat{\otimes} U(L)$ of G-graded associative algebras, called the *co-product* on the universal enveloping algebra $U(L)$.

It is clear that the co-product is co-associative, i.e. the following diagram is commutative:

$$\begin{array}{ccc} U(L) & \xrightarrow{\delta} & U(L) \hat{\otimes} U(L) \\ \delta \downarrow & & \downarrow \delta \otimes id \\ U(L) \hat{\otimes} U(L) & \xrightarrow{id \otimes \delta} & U(L) \hat{\otimes} U(L) \hat{\otimes} U(L) \end{array}$$

where $id = id_{U(L)}$.

We consider now the unique linear homomorphism $e \colon U(L) \to K$ such that $e(L) = 0$ and $e(1) = 1$. Then e is the co-unit of $U(L)$, i.e. $(e \otimes id) \circ \delta = id = (id \otimes e) \circ \delta$ (identifying $K \hat{\otimes} U(L)$ and $U(L) \hat{\otimes} K$ canonically with $U(L)$). Thus, $U(L)$ is a colour co-algebra with co-unit. Since δ and e are homomorphisms and $\delta(1) = 1 \otimes 1$, $e(1) = 1$, we see that $U(L)$ with multiplication and co-multiplication is a colour bialgebra. Observe that $U(L)$ is co-commutative, i.e. for the unique automorphism τ of the G-graded associative algebra $U(L) \hat{\otimes} U(L)$ with $\tau(x \otimes y) = \varepsilon(x, y)y \otimes x$, $x, y \in U(L)$ homogeneous, we have $\tau \circ \delta = \delta$.

Consider now the linear mapping $\mu \colon U(L) \hat{\otimes} U(L) \to U(L)$ induced by the multiplication in $U(L)$. We define the mapping $\bar{e} \colon U(L) \to U(L)$ by setting $\bar{e}(x) = e(x) \cdot 1_U$ for all $x \in U(L)$. Let $\eta \colon U(L) \to U(L)$ be the linear mapping such that $\eta(1) = 1$, $\eta(x) = -x$ for all $x \in L$ and $\eta(xy) = \varepsilon(x, y)\eta(y)\eta(x)$ for all homogeneous $x, y \in U(L)$. It is clear that $\eta^2 = id_U$ and $\mu \circ (\eta \otimes id_U) \circ \delta = \mu \circ$

$(id_U \otimes \eta) \circ \delta = \bar{e}$. Thus the constructed linear mapping η is an antipode, and our colour bialgebra $U(L)$ with the antipode η is a *co-commutative colour Hopf algebra*.

Analogously, we may check that in the case of a field K with char $K = p > 3$ the restricted universal enveloping algebra of a colour Lie p-superalgebra L also has the structure of a co-commutative colour Hopf algebra.

For a colour Hopf algebra we define the set of all its *primitive elements* by $\mathscr{P}(H) = \{a \in H | \delta(a) = a \otimes 1 + 1 \otimes a\}$.

2.10. Theorem. *Let K be a field, char $K = 0$. Then the set of primitive elements $\mathscr{P}(U(L))$ in the universal enveloping algebra $U(L)$ of the colour Lie superalgebra L coincides with L. In particular, $\mathscr{P}(A(X)) = [X]$ where $A(X) = U(L(X))$ is the free G-graded associative algebra with 1 and $[X] = L(X)$ is the free colour Lie superalgebra.*

Proof. Observe that

$[a \otimes 1 + 1 \otimes a, b \otimes 1 + 1 \otimes b]$

$= a \otimes 1 \cdot b \otimes 1 + 1 \otimes a \cdot b \otimes 1 + a \otimes 1 \cdot 1 \otimes b + 1 \otimes a \cdot 1 \otimes b$

$\quad - \varepsilon(a,b)(b \otimes 1 \cdot a \otimes 1 + 1 \otimes b \cdot a \otimes 1 + b \otimes 1 \cdot 1 \otimes a + 1 \otimes b \cdot 1 \otimes a)$

$= ab \otimes 1 + \varepsilon(a,b)b \otimes a + a \otimes b + 1 \otimes ab$

$\quad - \varepsilon(a,b)(ba \otimes 1 + \varepsilon(b,a)a \otimes b + b \otimes a + 1 \otimes ba)$

$= ab \otimes 1 - \varepsilon(a,b)1 \otimes ba + 1 \otimes ab - \varepsilon(a,b)ba \otimes 1$

$= (ab - \varepsilon(a,b)ba) \otimes 1 + 1 \otimes (ab - \varepsilon(a,b)ba)$

$= [a,b] \otimes 1 + 1 \otimes [a,b]$.

Therefore the set of primitive elements $\mathscr{P}(U(L))$ is a colour Lie superalgebra, $L \subset \mathscr{P}(U(L))$. Let $e_1, e_2, \ldots$ be a homogeneous basis of L_+ and $f_1, f_2, \ldots$ a homogeneous basis of L_-. By Theorem 2.2 the elements of the form

$$e_1^{k_1} e_2^{k_2} \ldots e_m^{k_m} f_1^{l_1} \ldots f_n^{l_n}, \text{ where } m, n \in \mathbb{N}, k_i \geq 0, l_i = 0, 1, e_i^0 = 1, f_i^0 = 1, \quad (1)$$

are a basis of $U(L)$). Therefore, the elements of the form

$$e_1^{k_1} e_2^{k_2} \ldots e_m^{k_m} f_1^{l_1} \ldots f_n^{l_n} \otimes e_1^{s_1} e_2^{s_2} \ldots e_p^{s_p} f_1^{t_1} \ldots f_q^{t_q} \quad (2)$$

form a basis of $U(L) \hat{\otimes} U(L)$.

Applying the diagonal mapping δ to elements of the form (1), we see that

$$\delta(e_1^{k_1}e_2^{k_2}\ldots e_m^{k_m}f_1^{l_1}\ldots f_n^{l_n}) = (e_1 \otimes 1 + 1 \otimes e_1)^{k_1}\ldots(e_m \otimes 1 + 1 \otimes e_m)^{k_m}$$

$$\times (f_1 \otimes 1 + 1 \otimes f_1)^{l_1}\ldots(f_n \otimes 1 + 1 \otimes f_n)^{l_n}$$

$$= e_1^{k_1}\ldots e_m^{k_m}f_1^{l_1}\ldots f_n^{l_n} \otimes 1$$

$$+ k_1\alpha_1 e_1^{k_1-1}e_2^{k_2}\ldots e_m^{k_m}f_1^{l_1}\ldots f_n^{l_n} \otimes e_1$$

$$+ k_2\alpha_2 e_1^{k_1}e_2^{k_2-1}\ldots e_m^{k_m}f_1^{l_1}\ldots f_n^{l_n} \otimes e_2 \quad (3)$$

$$+ \cdots + k_m \cdot \alpha_m e_1^{k_1}\ldots e_m^{k_m-1}f_1^{l_1}\ldots f_n^{l_n} \otimes e_m$$

$$+ \beta_1 l_1 e_1^{k_1}\ldots e_m^{k_m}f_2^{l_2}\ldots f_n^{l_n} \otimes f_1 + \cdots$$

$$+ \beta_n l_n e_1^{k_1}\ldots e_m^{k_m}f_1^{l_1}\ldots f_{n-1}^{l_{n-1}} \otimes f_n + A$$

where $k_1,\ldots,k_m \in \mathbb{N}, \alpha_1,\ldots,\alpha_m, \beta_1,\ldots,\beta_n \in K^*$ and A is a linear combination of elements of the form (2) with $\sum_{i=1}^{p} s_i + \sum_{j=1}^{q} t_j \geq 2$.

All summands in (3), except for the first one and A, are not involved in similar presentations for any other element of the form (1). Therefore, an element $\delta(a) \in U(L) \hat{\otimes} U(L)$ is a linear combination of elements of the form $e_1^{k_1}\ldots e_m^{k_m}f_1^{l_1}\ldots f_n^{l_n} \otimes 1$ and $1 \otimes e_1^{s_1}\ldots e_p^{s_p}f_1^{t_1}\ldots f_q^{t_q}$ if, and only if, in any presentation of a in the basis (1) we have only elements with $\sum_{i=1}^{m} k_i + \sum_{j=1}^{n} l_j = 1$, i.e. $a \in L$. This completes the proof of the theorem. □

Remark that Theorem 2.10 also holds in the situation when K is a commutative ring which is torsion-free as a $\mathbb{Z}$-module and L is a free K-module.

2.11. Theorem. *Let K be a field,* char $K = p > 3$, L *a colour Lie superalgebra, M a colour Lie p-superalgebra, $U(L)$ the universal enveloping algebra, $u(M)$ the restricted universal enveloping algebra. Then*
1) $\mathscr{P}(u(M)) = M$;
2) $\mathscr{P}(U(L))$ *coincides with the colour Lie p-superalgebra generated by L in $U(L)$.*

Proof. 1) For any $a \in \mathscr{P}(U(M))$ we have $(a \otimes 1 + 1 \otimes a)^p = a^p \otimes 1 + 1 \otimes a^p$ because $\varepsilon(d(a), d(a)) = 1$ and char $K = p$. Therefore $M \subset \mathscr{P}(u(M))$. For the opposite inclusion we consider (3) under the assumptions of our theorem (with $0 \leq k_i < p$). Since $a \in \mathscr{P}(u(M))$, $\sum_{i=1}^{m} k_i + \sum_{j=1}^{n} l_j = 1$ for all elements of the presentation for $a \in u(M)$, i.e. $a \in M$. This completes the proof in Case 1.

2) Again consider (3) in our situation.

If $a \in \mathscr{P}(U(L))$ then each k_i either divides p or is equal to 1. Furthermore, let $k_{i_1} = ps_1, \ldots, k_{i_l} = ps_l$ ($i_1 < \cdots < i_l$). Then $\delta(e_{i_1}^{ps_1} \ldots e_{i_l}^{ps_l}) = (e_{i_1}^p \otimes 1 + 1 \otimes e_{i_1}^p)^{s_1} \ldots (e_{i_l}^p \otimes 1 + 1 \otimes e_{i_l}^p)^{s_l}$. A similar argument (instead of e_i we consider e_i^p) works and we have $p|s_i$ hence $k_i = p^{t_i}$, $t_i \in \mathbb{N} \cup 0$. Thus we see that an element $a \in U(L)$ is primitive (i.e. $a \in \mathscr{P}(U(L))$) if, and only if, in its presentation in the basis of the form (1) we have that only one of the coefficients k_i, l_i is nonzero and if $k_j \neq 0$ then $k_j = p^{t_j}$, $t_j \in \mathbb{N} \cup 0$ (i.e. a belongs to the colour Lie p-superalgebra generated by L in $U(L)$). This completes the proof of the theorem. □

2.12. Finite separability. A G-graded algebra L is called *finitely separable* if for any G-homogeneous $a \in L$ and a finitely generated G-homogeneous subalgebra B, $a \notin B$, there exists a finite-dimensional G-graded H and a homomorphism $\varphi: L \to H$ such that $\varphi(a) \notin \varphi(B)$.

Theorem. *Let $L(X)$ be a free colour Lie superalgebra over a field K (in the case char $K = p > 3$ we consider also a free colour Lie p-superalgebra $L^p(X)$). Then the algebra $L = L(X)$ (respectively $L = L^p(X)$) is finitely separable.*

Proof. It is sufficient to prove the statement in the case where $|X| < \infty$. At the beginning we consider the free colour Lie superalgebra $L = L(X)$. The free G-graded associative algebra $A(X)$ is the universal enveloping algebra for L. Consider a well ordering of X and its extension to the free monoid $S(X)$ (first by length and then by lexicographical order). For $b \in A(X)$ we denote by $\hat{b}$ its leading term and by $l(b)$ its length. We say that a subset $W = \{w_1, w_2, \ldots, w_k\} \subset A(X)$ is r-reduced if the leading words $\hat{w}_i$ have the identity as their coefficients in w_i and $\hat{w}_i \neq \hat{w}_j t$ for all $t \in S(X)$, $1 \leq i, j \leq k$.

First, we show that each G-homogeneous finitely generated right ideal J in $A(X)$ has an r-reduced generating set. Indeed, let the right ideal J be generated by $W = \{w_1, w_2, \ldots, w_k\}$ and, for example, $\hat{w}_1 = \hat{w}_2 t$, $t \in S(X)$. Then we consider the element $w_1' = w_1 - aw_2 t$ where $a \in K$ is such that $\hat{w}_1' < \hat{w}_1$. After a finite number of such transformations, multiplying, if necessary, by elements from K, we come to an r-reduced set W' generating J. It is clear that if $b \in J$ then $\hat{b} = \hat{w}_i t$ for some $w_i \in W'$ and $t \in S(X)$ and that the set of words $u \in S(X)$ such that $\hat{u} \neq \hat{w}_i t$, $w_i \in W'$, $t \in S(X)$, gives us a linear basis of the K-space $A(X)/J$.

Second, we remark that if J is a finitely generated G-homogeneous right ideal in $A(X)$, $a \notin J$, then there exists a G-homogeneous finitely generated right ideal J' in $A(X)$ such that $J \subset J'$, $a \notin J'$ and $\dim_K A(X)/J' < \infty$. Indeed, let $W = \{w_1, w_2, \ldots, w_k\}$ be an r-reduced generating set of the right ideal J. We may assume that $\hat{a} \neq \hat{w}_i t$, $w_i \in W$, $t \in S(X)$ (otherwise we consider the element $a' = a - aw_i t$ where $a \in K$ is such that $\hat{a}' < \hat{a}$ and $a' \notin J$). Let

$N = \max(l(a), l(w_1), \ldots, l(w_k))$. Consider the set $\{w_{k+1}, \ldots, w_l\}$ of all words in $S(X)$ such that $l(w_s) = N + 1$ for all $k + 1 \le s \le l$ and $\hat{w}_{k+i} \ne \hat{w}_j t$ for all $t \in S(X), 1 \le i \le l - k, 1 \le j \le k$. The set $\{w_1, \ldots, w_l\}$ is r-reduced and for the right ideal J' of $A(X)$ generated by this set we have $J \subset J'$ and $a \notin J'$.

Let now J be the right ideal of $A(X)$ generated by B. We prove $L \cap J = B$. Let $E = \{e_1, e_2, \ldots\}$ be a G-homogeneous ordered K-linear basis of B. We extended E to a G-homogeneous ordered K-linear basis $E \cup C$ of L such that $C = \{c_1, c_2, \ldots\}$ and $e_i < c_j$ for all i, j. By Theorem 2.2 the set of all monomials $e_{i_1} \ldots e_{i_m} c_{j_1} \ldots c_{j_n}$ where $e_{i_1} \le \cdots \le e_{i_m}$, $c_{j_1} \le c_{j_n}$, $i_s \ne i_{s+1}$ with $d(e_{i_s}) \in G_-$, $j_s \ne j_{s+1}$ with $d(c_{j_s}) \in G_-$, is a K-linear basis of $A(X)$. Since B is a subalgebra of $L(X)$ these monomials with $m \ge 1$ give us a K-linear basis of J. Therefore $L \cap J = B$. As $a \in L$ and $a \notin B$, we have $a \notin J$. Any G-homogeneous generating set $\{u_1, \ldots, u_k\}$ of B generates J as a right ideal. Thus there exists a G-homogeneous right ideal J' of $A(X)$ such that $\dim_K A(X)/J' < \infty$, $J \subset J'$ and $a \notin J'$.

Consider the finite-dimensional right $U(L)$-module $M = U(L)/J'$ ($U(L) = A(X)$). Let $\mathrm{Ann}(M)$ be the annihilator of M and $R = U(L)/\mathrm{Ann}(M)$. Then R is a finite-dimensional G-graded associative algebra, M is a faithful R-module and $\mathrm{Ann}(M) \subset J'$ since $1 \in U(L)$. Let $\varphi: U(L) \to R$ be the canonical epimorphism. It is clear that $H = [\varphi(L)]$ is a finite dimensional colour Lie superalgebra. Suppose now that $\varphi(a) \in \varphi(B)$. Then $\varphi(a)$ belongs to the right ideal in R generated by $\{\varphi(u_1), \ldots, \varphi(u_k)\}$. Therefore there exist elements $f_1, \ldots, f_k \in U(L)$ such that $a = \sum_{i=1}^k \varphi(u_i)\varphi(f_i)$, i.e. $a - \sum_{i=1}^k u_i f_i \in \mathrm{Ann}(M) \subset J'$ and thus $a \in J'$. This contradiction completes the proof of our theorem in the case $L = L(X)$.

In the case $\mathrm{char}\, K = p$ and $L = L^p(X)$ the proof can be given in the same way, using Theorem 2.5 and the fact that the free G-graded associative algebra $A(X)$ is the restricted universal enveloping algebra for L. This completes the proof of the theorem. □

§ 3. The Composition Lemma

In this section we introduce the notion of composition for colour Lie superalgebras (see Definition 3.7) and for colour Lie p-superalgebras (see Definition 3.8). The main results are some versions of the Composition Lemma in all these cases (Theorem 3.10). As a corollary we obtain the algorithmical solution of the equality problem for colour Lie superalgebras and for colour Lie p-superalgebras with a homogeneous (relative to the weight of generators) set of defining relations (Theorem 3.11) and various forms of the Freiheitssatz (Proposition 3.12, Remark 3.13).

Recall the definitions of regular words and monomials given in Definition 1.2 of Chapter 2.

3.1. Definition. Consider the weight function $\varphi\colon X \to \mathbb{N}$, let $\varphi(x_1\ldots x_n) = \sum_{i=1}^n \varphi(x_i)$ for $x_1, \ldots, x_n \in X$ and let $\hat{a}$ be the leading term in $a \in A(X)$ first relative to the weight φ and then lexicographically.

3.2. Lemma. *Let u, v be regular words.*
1) *If $u = avb$ ($a, b \in S(X)$) then on u there is an arrangement of brackets $[u] = [a[v]b]$ such that $[v]$ is a regular monomial, $\widehat{[u]} = u$ in $L(X)$ (in $L^p(X)$).*
2) *Let $u = avvb$, $d(v) \in G_-$, $a, b \in S(X)$. Then on u there is an arrangement of brackets $[u]$ such that $[u] = [a[v,v]b]$ where $[v,v]$ is an s-regular monomial, $\widehat{[u]} = 2u$ in $L(X)$ (in $L^p(X)$).*
3) *Let $u = av^{p^t}b$, $d(v) \in G_+$. Then on u there is an arrangement of brackets $[u]$ such that $[u] = [a[v]^{p^t}b]$ where $[v]$ is a regular monomial, $\widehat{[u]} = u$ in $L^p[X]$.*
4) *Let $u = av^{2p^t}b$, $d(v) \in G_-$. Then there is an arrangement $[u]$ of brackets on u such that $[u] = [a[v,v]^{p^t}b]$ where $[v,v]$ is an s-regular monomial, $[\hat{u}] = 2^{p^t}u$ in $L^p[X]$.*

Proof. We show the first statement by induction on the length $l(u)$ of u. The starting point of the induction when $u = v$ is evident. If $l(v) = 1$ then $[u]$ is the arrangement of brackets on u such that $[u]$ is a regular monomial and by Lemma 2.4 of Chapter 2 our statement 1) is true. Suppose that we have proved statement 1) for all u with $l(u) < n$. Let $l(u) = n > 1$ and let x be the least possible letter from X which occurs in u. In the proof of Lemma 1.8 in Chapter 2 we have remarked that the new letters $x_i x^k$ had the same lexicographical order as that of the letters $x_i \in X$ (in Lemma 3.6 of Chapter 2 we proved that the $x_i x^k$ were free generators of the associative subalgebra of $A(X)$ which they generate). Since u, v are regular words and $l(u) > 1$, $l(v) > 1$, the initial letters of the words u and v are different from x. Writing down u in the new letters $x_i x^k$ we have $l_{\{x_i x^k\}}(u) < l_X(u)$. Let v' be a subword in the new letters in u of the form $v' = vx^l$ where $l \geq 0$ is such that either to the right of v' in u we have a letter which is different from x, or v' is the end of u. Since v is a regular word and $v > x$, by Lemma 1.7 of Chapter 2 we have that v' is a regular word. By the inductive hypothesis we have the desired statement for the subword v' of u in the new letters $x_i x^k$. We consider an arrangement of brackets on $x_i x^k$ of the form $x_i (\operatorname{ad}' x)^k$. As $x_i (\operatorname{ad}' x)^k$ is a regular monomial we have $\widehat{x_i(\operatorname{ad}' x)^k} = x_i x^k$. Let $[v']$ be the arrangement of brackets on v' such that $[v']$ is a regular monomial in letters $x_i x^k$ and on $x_i x^k$ we have the following arrangement of brackets: $x_i(\operatorname{ad}' x)^k$. Now we consider another arrangement of brackets on v': $[v']' = [v](\operatorname{ad}' x)^l$ where $[v]$ is a regular monomial. Since v is a regular word and $v > x$, $\widehat{[v']'} = v' = \widehat{[v']}$. Now we impose brackets on u in the following way: on v' as $[v']'$; all other brackets are in accordance with the inductive hypothesis; on x_i^k as $x_i(\operatorname{ad}' x)^k$. Thus we have proved the first statement of our lemma.

§3. The Composition Lemma

Statement 2) of our lemma also will be proved by induction on $l(u)$. The basis of induction: a) $u = vvx$, where $x \in X$; b) $u = yvv$, where $y \in X$. In the case a) taking into account that u is a regular word we have $vvx > x$ and therefore $v > x$. Thus for the arrangement of brackets $[u] = [[v,v], x]$, where $[v,v]$ is an s-regular monomial, we find that $\widehat{[u]} = 2u$. In the case b), by analogy with a), for $u = yvv$, $y \in X$, we have $[u] = [y, [v,v]]$.

Let $v \in X$. If $v = x$ is the least letter in u, then $[u]$ is a regular monomial and there is a letter $x_i x^k$ in u with $k \geq 2$. Remark that $[x_i x^k] = x_i (\text{ad}' x)^k = \frac{1}{2}[x_i(\text{ad}' x)^{k-2}, [x, x]]$ and therefore we have our statement. If $v \neq x$ then we express v in the letters $x_i x^k$. As a subword v' of u in new letters $x_i x^k$ we again consider $v' = vvx^l$, where $l \geq 0$ is such that either to the right of v' in u we have a letter which is different from x, or v' is the end of u. In the case $l = 0$ we may use the inductive hypothesis. If $l > 0$ then the word v' is regular. Let $[v'] = [v,v](\text{ad}' x)^l$. By 1) we may consider the induced arrangement on u, and therefore we have proved our statement. Suppose that 2) holds for all u with $l(u) < n$. Let $l(u) = n > 1$, $l(v) > 1$ and let x be the least letter from X in u. Since u and v are regular words, their initial letters are different from x. Therefore we may rewrite u and v in new letters $x_i x^k$. Consider now the subword $v' = vvx^l$ of u such that either to the right of v' in u we have a letter which is different from x, or v' is the end of u. If $l = 0$ then by the inductive hypothesis with $[x_i x^k] = x_i(\text{ad}' x)^k$ we have our claim.

If $l \neq 0$ then v' is a regular word. Let $[v'] = [v,v](\text{ad}' x)^l$ where $[v,v]$ is an s-regular monomial. Then $\widehat{[v']} = 2v'$. If $u = av'b'$ then we put $[u] = [a[v']b']$, where we consider the induced arrangement of brackets on u with the above mentioned arrangement of brackets on v' (see Statement 1)). Thus we have proved the second statement.

For the third statement we argue again by induction on $l(u)$. The basis of induction is as follows: a) $u = v^{p^t} x$, $x \in X$ (u is a regular word, therefore $v^{p^t} > x$ and $v > x$ and for $[u] = [[v]^{p^t}, x] = (\text{ad}[v])^{p^t}(x)$ we have $\widehat{[u]} = u$); b) $u = yv^{p^t}$, $y \in X$ (then $yv^{p^t} > v$, therefore $y > v$ and for $[u] = [y, [v]^{p^t}] = y(\text{ad}'[v])^{p^t}$ we have $\widehat{[u]} = u$).

Let $v \in X$. If $v = x$ is the least letter in u, then arranging brackets on u in such a way that $[u]$ is a regular monomial we have that there is a letter $x_i x^k$ in u with $k \geq p^t$. Remark that $[x_i x^k] = x_i(\text{ad}' x)^k = [x_i(\text{ad}' x)^{k-p^t}, x^{p^t}]$. This proves our statement. If $v \neq x$ then we rewrite the word u using letters $x_i x^k$. As a subword v' of u in letters $x_i x^k$ we again consider $v' = v^{p^t} x^l$, $l \geq 0$, such that either to the right of v' in u we have a letter which is different from x, or v' is the end of u. If $l = 0$ then we use the inductive hypothesis on $l(u)$. If $l > 0$ then v' is a regular word, let $u = av'b'$, and we consider the following arrangement of brackets on v': $[v'] = ([v]^{p^t})(\text{ad}' x)^l$. By Statement 1) of our lemma we have the induced arrangement of brackets on u, thus we derive our statement.

Suppose now that we have proved the third statement for all u with $l(u) < n$. Let $l(u) = n > 1$, $l(v) > 1$ and let x be the least letter from X in u. As above we write the words u and v in letters $x_i x^k$, $v' = v^{p^t} x^l$. If $l = 0$ we use the inductive hypothesis and obtain our statement. If $l > 0$ we impose brackets on v' in the following way: $[v'] = ([v]^{p^t})(\mathrm{ad}'\, x)^l$. By Statement 1) of our lemma we consider the induced arrangement of brackets on u, and therefore we have our third statement. The proof of the fourth statement is similar to that of the third one. This completes the proof of the lemma. $\square$

3.3. Lemma. *Let e_1, e_2, $e_3 \in S(X)\backslash 1$ and $e_1 e_2$, $e_2 e_3$ be regular words. Then the word $e_1 e_2 e_3$ is regular.*

Proof. Let $e_1 e_2 e_3 = w_1 w_2$. a) If w_1 is a subword of $e_1 e_2$ then $e_1 e_2 = w_1 v$, $v \in S(X)\backslash 1$. Since $e_1 e_2$ is a regular word, by Lemma 1.4 of Chapter 2 we have that $e_1 e_2 > v$. As $l(v) < l(e_1 e_2)$, we have that $w_1 w_2 = e_1 e_2 e_3 > v e_3 w_1 = w_2 w_1$. b) If w_2 is a subword of e_3 (possibly, $w_2 = e_3$) then $e_3 = u w_2$, $u \in S(X)$. Since $e_1 e_2$ is a regular word, $e_1 e_2 > e_2$. It is clear that $l(e_1 e_2) > l(e_2)$. As $e_2 e_3$ is a regular word we have $e_1 e_2 e_3 > e_2 e_3 = e_2 u w_2 > w_2$. On the other hand, $w_2 > w_2 w_1$. Thus we obtain $w_1 w_2 > w_2 w_1$. This completes the proof. $\square$

3.4. Lemma. *Let w be a regular word. Then there are no e_1, e_2, $e_3 \in S(X)\backslash 1$ such that $w = e_1 e_2 = e_2 e_3$.*

Proof. Since w is a regular word, $e_2 e_3 = e_1 e_2 > e_2 e_1$ and therefore $e_3 > e_1$. On the other hand, $e_1 e_2 = e_2 e_3 > e_3 e_2$ and therefore $e_1 > e_3$. This contradiction proves our lemma. $\square$

3.5. Lemma. *Let u, v be regular words, $u \neq v$, $k, l \in \mathbb{N}$, $u^k = e_1 e_2$, $v^l = e_2 e_3$, $e_1, e_2, e_3 \in S(X)\backslash 1$. Then either $e_2 = u$ and $v = ue$ or $e_2 = v$ and $u = fv$ (e, $f \in S(X)\backslash 1$) or $u = u' e_2$, $v = e_2 v'$, where u', $v' \in S(X)\backslash 1$ (in all cases $e_1 e_2 e_3$ is a regular word). If $u = v$, then $e_2 = u^m$, $m \in \mathbb{N}$.*

Proof. Let $u \neq v$, $u = u_1 u_2$, $v = v_1 v_2$, u_2, $v_1 \in S(X)$, $e_2 = u_2 u^{l'} = v^{k'} v_1$, l', $k' \in \mathbb{N}$. Then we have to consider two cases.

1. $v = u_2 u^{l_1} a$ where $l_1 \geq 0$, $u = ab$. If $u_2 \neq 1$ and $a \neq 1$ then, taking into account that v is a regular word, we have $v > a$ and therefore $u_2 > a$. Since u is a regular word, $u > u_2$. It follows from $u = ab$ that $a > u$; thus we have a contradiction. If $a = 1$ and $u_2 \neq 1$ then $v = u_2 u^{l_1}$. In the case $l_1 > 0$, since v is a regular word, we have $v = u_2 u^{l_1} > u$. But u is also a regular word, therefore $u > u_2$, so that $u > u_2 > u_2 u^{l_1} > v$ which gives a contradiction. If $a \neq 1$, $u_2 = 1$ and $l_1 > 0$ then $v = u^{l_1} a$, $v > a$ (v is a regular word), $a > u$ and $u > u^{l_1} a = v$. Again we have a contradiction. If $u_2 = 1 = a$ then $v = u^{l_1}$ is not a regular word.

§3. The Composition Lemma

2. Similarly, $u = dv^{k_1}v_1$ where $k_1 \geq 0$ and $v = cd$. If $d \neq 1$ and $v_1 \neq 1$ then $u > v_1, v > d > dv^{k_1}v_1 = v$ and thus we have a contradiction. If $d \neq 1, v_1 = 1$ and $k_1 > 0$ then $u = dv^{k_1} > v > d > dv^{k_1}$, again a contradiction. If $d = 1$, $v_1 \neq 1$ and $k_1 > 0$ then $u = v^{k_1}v_1 > v_1 > v > v^{k_1}v_1$. If $d = 1$ and $v = 1$ then the word $u = v^{k_1}$ is not a regular word.

Thus for $e_2 = u$ we have $v = ue$, $e \in S(X) \backslash 1$, $u > ue$ and therefore the word uue is a regular word since the words u and v are regular (see Lemma 1.7 of Chapter 2). It follows that $uue > ue$, therefore the word $u^{k-1}v^l = e_1 e_2 e_3$ is regular. If $e_2 = v$ then $u = fv$, $f \in S(X) \backslash 1$, $fv > v$ and therefore fvv is a regular word. Since $u = fv > fvv$, the word $u^k v^{l-1}$ is regular. If $u = u'e_2$, $v = e_2 v'$, u', $v' \in S(X) \backslash 1$, then it follows from Lemma 3.3 that the word $u'e_2 v'$ is regular and $u = u'e_2 > u'e_2 v' > e_2 v'$, and therefore the word $e_1 e_2 e_3$ is regular. In the case $u = v$ contrary to the assumption it follows that $u = u_2 u_1 = u_1 u_2$ where $u_1, u_2 \in S(X) \backslash 1$, a contradiction (recall that the word u is regular). This completes the proof of the lemma. □

3.6. Lemma. *Let u and v be regular words, $u \neq v$, $k, l \in \mathbb{N}$, v^k is a subword of u^l. Then v^k is a subword of u.*

Proof. Assume to the contrary that v^k is not a subword of u. Then by our hypothesis we have that either $v = u_2 u^{l_1} a$, $l_1 \geq 0$, $u_1 u_2 = u = ab$, $u_2, a \in S(X)$, or $u = dv^{k_1} v_1$, $k_1 \geq 0$, $v_1 v_2 = v = cd$, $v_1, d \in S(X)$.

In the proof of Lemma 3.5 we considered these cases and showed that $l_1 = 0 = k_1$ and either $v = u_2$ or $v = a$, respectively, either $u = d$ or $u = v_1$. By our assumption, v^k is not a subword of u, and therefore in the case $v = u_2$ we have $u = ve$, hence $v > u$. Since u is a regular word we obtain $u > u_2 = v$ which is a contradiction. If $v = a$ (and therefore $v > u$), then we have $u = fv$. Since u is a regular word, we get $u > v$, which is again a contradiction. If either $u = d$ or $u = v_1$, then it follows from the assumption of the lemma that either $v = gu^s$, $s > 0$ and $hg = u$ or $v = u^t w$, $t > 0$ and $wq = u$.

As it has been shown in the proof of Lemma 3.5 these possibilities do not arise. This completes the proof of the lemma. □

3.7. Definition. Let a, b be elements of $L(X)$. Assume that their leading terms a_0 and b_0 in the s-regular presentation (first relative to the weight φ and then lexicographically) have coefficients which are equal to 1 and $\bar{a}_0 = e_1 e_2$, $\bar{b}_0 = e_2 e_3$ where $e_1, e_2, e_3 \in S(X)$. By Lemma 3.5 the word $e_1 e_2 e_3$ is regular. Now we define the element $(a, b) \in L(X)$ which is called the *composition* of the elements $a, b \in L(X)$ (relative to a subword e_2).

1) In the case where $\bar{a}_0, \bar{b}_0$ are regular words, $\bar{a}_0 \neq \bar{b}_0$, we set $(a, b) = [(a)e_3] - [e_1(b)]$. Here $[(a)e_3]$ is obtained from the monomial $[[a_0]e_3]$ with the arrangement of brackets described in Part 1 of Lemma 3.2 by replacing a_0 by a and analogously $[e_1(b)]$ is obtained from $[e_1[b_0]]$ by replacing b_0 by b.

100   3. Composition techniques in the theory of Lie superalgebras

2) In the case where u, $\bar{b}_0$ are regular words, $u \neq \bar{b}_0$, $d(u) \in G_-$, $\bar{a}_0 = uu$, $e_1 = ue_1'$, $u = e_1'e_2$ and $e_1' \in S(X)$, we set $(a,b) = [(a)e_3] - 2[e_1(b)]$. Here $[(a)e_3]$ is the element obtained from the monomial $[[u,u]e_3]$ with the arrangement of brackets described in Part 2 of Lemma 3.2 by replacing $[u,u]$ by a and $[e_1(b)]$ is the element obtained from the monomial $[e_1[b_0]]$ with the arrangement of brackets described in Part 1 of Lemma 3.2 by replacing b_0 by b.

3) In the case where $\bar{a}_0$, u are regular words, $d(u) \in G_-$, $\bar{a}_0 \neq u$, $\bar{b}_0 = uu$, $e_3 = e_3'u$, $u = e_2e_3'$, $e_3' \in S(X)$, we set $(a,b) = 2[(a)e_3] - [e_1(b)]$. The element $[(a)e_3]$ is obtained from the monomial $[[a_0]e_3]$ with the arrangement of brackets described in Part 1 of Lemma 3.2 by replacing a_0 with a and $[e_1(b)]$ is the element obtained from the monomial $[e_1[u,u]]$ with the arrangement of brackets described in Part 2 of Lemma 3.2 by replacing $[u,u]$ with b.

4) In the case where u, v are regular words with $u \neq v$, $d(u), d(v) \in G_-$, $\bar{a}_0 = uu$, $\bar{b}_0 = vv$, $e_1 = ue_1'$, $e_3 = e_3'v$, $u = e_1'e_2$, $v = e_2e_3'$ and $e_1', e_3' \in S(X)$, we set $(a,b) = [e_1(b)] - [(a)e_3]$. The element $[e_1(b)]$ is obtained from the monomial $[e_1[v,v]]$ by replacing $[v,v]$ with b and the element $[(a)e_3]$ is obtained from the monomial $[[u,u]e_3]$ by replacing $[u,u]$ with a (the arrangements of brackets are described in Part 2 of Lemma 3.2).

5) For $a = b$, $\bar{a}_0 = uu$, where u is a regular word and $d(u) \in G_-$ we set $(a,a) = [a,[u]] = [a - [u,u], [u]]$, where $[u]$ is a regular monomial.

3.8. Definition. Let a, b be elements in $L^p(X)$. Suppose that their leading terms a_0 and b_0 in the ps-regular presentation (first relative to the weight φ and then lexicographically) have coefficients which are equal to 1 and $\bar{a}_0 = e_1e_2$, $\bar{b}_0 = e_2e_3$ where $e_1, e_2, e_3 \in S(X)$. By Lemma 3.5 the word $e_1e_2e_3$ is regular. Now we define the element $(a,b) \in L^p(X)$ which is called the *composition* of the elements $a, b \in L^p(X)$ (relative to a subword e_2).

1) In the case where $\bar{a}_0$ and $\bar{b}_0$ are either regular words or squares uu of regular words u with $d(u) \in G_-$ the composition $(a,b) \in L^p(X)$ is defined as in Parts 1)–5) of Definition 3.7.

2) In the case where $\bar{a}_0$, v are regular words, $\bar{a}_0 \neq v$, $\bar{b}_0 = v^{p^t}$, $e_3 = e_3'v^{p^t-1}$, $t \in \mathbb{N}$, $v = e_2e_3'$, $d(v) \in G_+$, we set $(a,b) = [(a)e_3] - [e_1(b)]$. Here $[(a)e_3]$ is obtained from the monomial $[[a_0]e_3]$ with the arrangement of brackets described in Part 1 of Lemma 3.2 by replacing a_0 with a and $[e_1(b)]$ is obtained from the monomial $[e_1[v]^{p^t}]$ with the arrangement of brackets from Part 3 of Lemma 3.2 by replacing $[v]^{p^t}$ with b.

3) In the case where u, $\bar{b}_0$ are regular words, $u \neq \bar{b}_0$, $\bar{a}_0 = u^{p^t}$, $e_1 = u^{p^t-1}e_1'$, $t \in \mathbb{N}$, $u = e_1'e_2$, $d(u) \in G_+$, we set $(a,b) = [(a)e_3] - [e_1(b)]$. We consider the arrangement of brackets on $[[a_0]e_3]$ in accordance with Part 3 and on $[e_1[b_0]]$ in accordance with Part 1 of Lemma 3.2.

4) In the case when u, v are regular words, $u \neq v$, $d(u) \in G_+$, $d(v) \in G_-$, $\bar{a}_0 = u^{p^t}$, $\bar{b}_0 = vv$, $e_1 = u^{p^t-1}e_1'$, $e_3 = e_3'v$, $u = e_1'e_2$, $v = e_2e_3'$, we set $(a,b) =$

§3. The Composition Lemma

$2[(a)e_3] - [e_1(b)]$. We consider the arrangement of brackets on $[[a_0]e_3]$ in accordance with Part 3 and on $[e_1[b_0]]$ in accordance with Part 1 of Lemma 3.2.

5) In the case when u, v are regular words, $u \neq v$, $d(u), d(v) \in G_+$, $\bar{a}_0 = u^{p^s}$, $\bar{b}_0 = v^{p^t}$, $s, t \in \mathbb{N}$, $e_1 = u^{p^s-1}e_1'$, $e_3 = e_3'v^{p^t-1}$, $u = e_1'e_2$ and $v = e_2e_3'$, we set $(a, b) = [(a)e_3] - [e_1(b)]$. Both arrangements of brackets on $[[a_0]e_3]$ and on $[e_1[b_0]]$ are as in Part 3 of Lemma 3.2.

6) In the case when u, v are regular words, $u \neq v$, $d(u) \in G_-$, $d(v) \in G_+$, $\bar{a}_0 = uu$, $\bar{b}_0 = v^{p^t}$, $t \in \mathbb{N}$, $e_1 = ue_1'$, $e_3 = e_3'v^{p^t-1}$, $u = e_1'e_2$, $v = e_2e_3'$, we set $(a, b) = [(a)e_3] - 2[e_1(b)]$. We consider the arrangement of brackets on $[[a_0]e_3]$ in accordance with Part 2 and on $[e_1[b_0]]$ in accordance with Part 3 of Lemma 3.2.

7) For $a = b$ where u is a regular word, $d(u) \in G_+$, $\bar{a}_0 = u^{p^t}$, $t \in \mathbb{N}$, we set $(a, a) = [a, [u]] = [a - [u]^{p^t}, [u]]$, where $[u]$ is a regular monomial.

8) In the case when u, v are regular words, $u \neq v$, $d(u) \in G_-$, $\bar{a}_0 = u^{2p^t}$, $\bar{b}_0 = v^l$, $t, l \in \mathbb{N}$, we set $(a, b) = [(a)e_3] - a[e_1(b)]$ where the arrangements of brackets are as in Lemma 3.2, $0 \neq a \in K$ is such that $\widehat{[(a)e_3]} = a\widehat{[e_1(b)]}$; analogously we define (a, b) for $\bar{b}_0 = u^{2p^t}$, $\bar{a}_0 = v^p$; if $u = v$, $a = b$, then we set $(a, a) = [a, [u]] = [a - [u, u]^{p^t}, [u]]$ where $[u]$ is a regular monomial.

3.9. Definition. A subset $S \in L(X)(L^p(X))$ is called *stable* if the following conditions are satisfied.

a) The coefficients of the leading monomials in s-regular (ps-regular) presentations are equal to 1 for all elements from S.

b) All words $\hat{s}$, where $s \in S$, are distinct and do not contain the other ones as subwords.

c) If $a, b \in S$ and the composition (a, b) does exist (see Definitions 3.7 and 3.8) then either $(a, b) \in S$ (up to a coefficient from K) or, after expressing the elements associatively, we have $(a, b) = \sum \alpha_i A_i s_i B_i$, where $\alpha_i \in K$, $s_i \in S$, A_i, $B_i \in S(X)$, $\alpha_1 A_1 \hat{s}_1 B_1 = \widehat{(a, b)}$, $A_i \hat{s}_i B_i < A_1 \hat{s}_1 B_1 (i \neq 1)$ for all i.

3.10. Theorem (The Composition Lemma). *Let K be a field, char $K \neq 2, 3$, $L(X)$ a free colour Lie superalgebra, $L^p(X)$ a free colour Lie p-superalgebra, S a stable set (see Definition 3.9), id(S) the ideal in $L(X)$ (in $L^p(X)$) generated by S. Then for any element $a \in \text{id}(S)$ the word $\hat{a}$ contains a subword $\hat{b}$, $b \in S$. Furthermore, (p)s-regular monomials u such that $\hat{u}$ does not contain subwords $\hat{b}$, $b \in S$, give us a linear basis of $L(X)/\text{id}(S)$ (respectively, of $L^p(X)/\text{id}(S)$).*

Proof. We prove our statement for elements of the associative ideal ID(S) generated by S in the free associative algebra $A(X)$. Let $a \in \text{ID}(S)$. Then $a = \sum \alpha_i A_i s_i B_i$ where A_i, $B_i \in S(X)$, $s_i \in S$ and $\alpha_i \in K$. Let $A_1 \hat{s}_1 B_1$ be the highest order term among all $A_i \hat{s}_i B_i$. If $A_1 \hat{s}_1 B_1 \neq A_i \hat{s}_i B_i$ for all $i \neq 1$ then our theorem is proved. If for example $\alpha_1' A_1 \hat{s}_1 B_1 = -\alpha_2' A_2 \hat{s}_2 B_2$, $\alpha_1', \alpha_2' \in K$, then we have the following cases.

Case 1. The subwords $\hat{s}_1$ and $\hat{s}_2$ have empty intersection in $A_1\hat{s}_1 B_1$. Then we may assume that $A_1 s_1 B_1 = A s_1 B \hat{s}_2 C$, $A_2 s_2 B_2 = A \hat{s}_1 B s_2 C$ where $A, B, C \in S(X)$. Therefore $\alpha'_1 A_1 s_1 B_1 + \alpha'_2 A_2 s_2 B_2 = \alpha'_1 A s_1 B \hat{s}_2 C + \alpha'_2 A \hat{s}_1 B s_2 C = -\alpha'_1 A s_1 B(\hat{s}_2 - \hat{s}_2)C - \alpha'_2 A(\hat{s}_1 - \hat{s}_1) B s_2 C$, $A\hat{s}_1 B(\widehat{\hat{s}_2 - \hat{s}_2})C < A\hat{s}_1 B\hat{s}_2 > A(\widehat{\hat{s}_1 - \hat{s}_1}) B \hat{s}_2 C$.

Case 2. Since S is stable the words $\hat{s}_1$ and $\hat{s}_2$ are different and they are not subwords of each other. Thus we have to consider only the following cases:
A) $s_1 \neq s_2$, $\hat{s}_1 = e_1 e_2$, $\hat{s}_2 = e_2 e_3$; B) $s_1 = s_2$.

Remark that in $L(X)$ the words $\hat{s}_1, \hat{s}_2$ are regular or may be written in the form uu where u is a regular word, $d(u) \in G_-$. In $L^p(X)$ in addition to the possibilities mentioned above we have u^{p^t}, $t \in \mathbb{N}$, where u is a regular word, $d(u) \in G_+$; v^{2p^t}, $t \in \mathbb{N}$, where v is a regular word, $d(v) \in G_-$.

In case A by Lemma 3.5 all possible overlaps for $\hat{s}_1$ and $\hat{s}_2$ look as in 1)–4) of Definition 3.7 for $L(X)$ and in 1)–6), 8) of Definition 3.8 for $L^p(X)$. Remark that $As_1 e_3 B - A[(s_1)e_3]B = \sum \gamma_i E_i s_1 F_i$, $\gamma_i \in K$, E_i, $F_i \in S(X)$, $E_i \hat{s}_1 F_i < A\hat{s}_1 e_3 B$ for all i; $Ae_1 s_2 B - A[e_1(s_2)]B = \sum \delta_i T_i s_2 H_i$, $\delta_i \in K$, T_i, $H_i \in S(X)$, $T_i \hat{s}_2 H_i < Ae_1 \hat{s}_2 B$ for all i, where the arrangements of brackets are described in Lemma 3.2. Thus we have $\alpha'_1 As_1 e_3 B + \alpha'_2 Ae_1 s_2 B = \alpha'_1 A[(s_1)e_3]B + \alpha'_2 A[e_1(s_2)]B + \sum \alpha'_1 \gamma_i E_i s_1 F_i + \sum \alpha'_2 \delta_i T_i s_2 H_i = \alpha A(s_1, s_2) B + \sum \alpha'_1 \gamma_i E_i s_1 F_i + \sum \alpha'_2 \delta_i T_i s_2 H_i = \sum \beta_i C_i g_i D_i$, α, $\beta_i \in K$, $g_i \in S$, C_i, $D_i \in S(X)$, $C_i \hat{g}_i D_i < A_1 \hat{s}_1 B_1$ for all i (considering the stability of S and $(\widehat{s_1, s_2}) < e_1 e_2 e_3$).

In case B, if $\hat{s}_1 = 2uu$, u is a regular word, $d(u) \in G_-$, then by Lemma 3.5 the overlaps have the form $\alpha'_1(A s_1 uB - A u s_1 B)$ (it is clear that $\alpha'_1 = -\alpha'_2$). Note that if $[u]$ is a regular monomial then by Lemma 2.4 in Chapter 2 $As_1 uB - As_1[u]B = \sum \gamma_i E_i s_1 F_i$, $\gamma_i \in K$, E_i, $F_i \in S(X)$, $E_i \hat{s}_1 F_i < AuuuB$ for all i; $Au s_1 B - A[u] s_1 B = \sum \delta_i T_i s_1 H_i$, $\delta_i \in K$, T_i, $H_i \in S(X)$, $T_i \hat{s}_1 H_i < AuuuB$ for all i. Therefore $As_1 uB - Au s_1 B = As_1[u]B - A[u]s_1 B + \sum \gamma_i E_i s_1 F_i - \sum \delta_i T_i s_1 H_i = A[s_1, [u]]B + \sum \gamma_i E_i s_1 F_i - \sum \delta_i T_i s_1 H_i$ (since $[s_1, [u]] = s_1[u] - \varepsilon(d(s_1), d(u))[u]s_1 = s_1[u] - \varepsilon(2d(u), d(u))[u]s_1 = s_1[u] - (\varepsilon(d(u), d(u)))^2 [u]s_1 = s_1[u] - [u]s_1$). Taking into account the stability of S as above we have $\alpha'_1(A s_1 uB - A u s_1 B) = \sum \beta_i C_i g_i D_i$, $\beta_i \in K$, C_i, $D_i \in S(X)$, $C_i \hat{g}_i D_i < A_1 \hat{s}_1 B_1$ for all i.

Now if $\hat{s}_1 = u^{p^t}$, u is a regular word, $d(u) \in G_+$, then by Lemma 3.5 the overlaps have the form $e_2 = u^l$, $l \in \mathbb{N}$. At first we consider the case $e_2 = u$. Then the overlap has the form $As_1 uB - Au s_1 B$. As above we have $As_1 uB - As_1[u]B = \sum \gamma_i E_i s_1 F_i$, $\gamma_i \in K$, E_i, $F_i \in S(X)$, $E_i \hat{s}_1 F_i < Au^{p^t+1} B$ for all i; $Au s_1 B - A[u]s_1 B = \sum \delta_i T_i s_1 H_i$, $\delta_i \in K$, T_i, $H_i \in S(X)$, $T_i \hat{s}_1 H_i < Au^{p^t+1} B$ for all i, with $[u]$ being a regular monomial. Since $\varepsilon(d(u), d(u)) = 1$, $[s_1, [u]] = s_1[u] - \varepsilon(d(s_1), d(u))[u]s_1 = s_1[u] - \varepsilon(p^t d(u), d(u))[u]s_1 = s_1[u] - (\varepsilon(u, u))^{p^t}[u]s_1 = s_1[u] - [u]s_1$, $As_1 uB - Au s_1 B = A[s_1, [u]]B + \sum \gamma_i E_i s_1 F_i - \sum \delta_i T_i s_1 H_i = \sum \beta_i C_i g_i D_i$, $\beta_i \in K$, C_i, $D_i \in S(X)$, $C_i \hat{g}_i D_i < A_1 \hat{s}_1 B_1 = Au^{p^t+1} B$ for all i (S is stable!).

Let now $e_2 = u^l$, $l > 1$. We have just shown that $s_1 u = u s_1 + \sum \beta_i' C_i' g_i' D_i'$, $\beta_i' \in K$, C_i', $D_i' \in S(X)$, $\widehat{C_i' \hat{g}_i' D_i'} < \widehat{u s_1} = \widehat{s_1 u}$. Therefore $A s_1 u^l B - A u^l s_1 B = \sum \beta_i C_i g_i D_i$, $\beta_i \in K$, C_i, $D_i \in S(X)$, $\widehat{C_i \hat{g}_i D_i} < A \hat{s}_1 u^l B = A u^l \hat{s}_1 B = A u^{p^t + l} B$.

The case $\hat{s}_1 = u^{2p^t}$ with u regular, $d(u) \in G_-$, is considered as above taking into account $[s_1, [u]] = s_1[u] - \varepsilon(2p^t d(u), d(u))[u] s_1 = s_1[u] - (-1)^{2p^t}[u] s_1 = s_1[u] - [u] s_1$ where $[u]$ is a regular monomial.

Thus we have proved that in the case of cancellation of $\alpha_1 A_1 \hat{s}_1 B_1$ in the sum $a = \sum \alpha_i A_i s_i B_i$ it was possible to write $a = \sum \beta_i C_i g_i D_i$, $\beta_i \in K$, C_i, $D_i \in S(X)$, $\widehat{C_i \hat{g}_i D_i} < A_1 \hat{s}_1 B_1$ for all i. It is clear that our statement follows from this fact for each $a \in \mathrm{ID}(S)$ and therefore for each $a \in \mathrm{id}(S) \subset \mathrm{ID}(S)$. This completes the proof of the theorem. □

3.11. Theorem. *There exists an algorithm for solving the equality problem for colour Lie superalgebras and for colour Lie p-superalgebras with a stable or a homogeneous (relative to weight φ) finite set of defining relations.*

Proof. We show the homogeneous case (the stable case is, in fact, a specific case of this situation).

Let $|X| < \infty$, $L(X)$ the free colour Lie superalgebra ($L^p(X)$ the free colour Lie p-superalgebra), and let R be a subset consisting of φ-homogeneous elements in $L(X)$ (in $L^p(X)$). To solve the equality problem means precisely to find an algorithm which gives the possibility to decide whether $a \in L(X)$ ($a \in L^p(X)$) belongs to the ideal $\mathrm{id}(R)$. We will change R in a step by step way with the aim to arrive at a stable set S such that $\mathrm{id}(S) = \mathrm{id}(R)$, using the following transformations.

1) Eliminations of subwords.

If, for example, $r_1, r_2 \in R$ and for $\hat{r}_1, \hat{r}_2$ their corresponding words are equal, then we set $r_1' = r_1 - \alpha r_2$, where $\alpha \in K$ is such that $\hat{r}_1 = \alpha \hat{r}_2$ (and therefore $\hat{r}_1' < \hat{r}_1$). If $\hat{r}_1 = \alpha A \hat{r}_2 B$, $\alpha \in K$, $A, B \in S(X)$, then by Lemmas 3.6 and 3.2 there exists an element $[A(r_2)B]$ (in the case of the algebra $L^p(X)$ it is possible that $[A(r_2)B] = [C(r_2)D]^{p^k}$, where $C, D \in S(X)$), such that $\widehat{[A(r_2)B]} = \beta A \hat{r}_2 B$, $\beta \in K$, and therefore, for $r_1' = r_1 - (\alpha/\beta)[A(r_2)B]$, we have $\hat{r}_1' < \hat{r}_1$. Observe that r_1 and r_2 are φ-homogeneous and thus r_1' is also φ-homogeneous.

2) Joining compositions.

If $r_1, r_2 \in R$ and there exists the composition (r_1, r_2), then we adjoin it to the set R. Remark, that r_1 and r_2 are φ-homogeneous and therefore their composition (r_1, r_2) is also φ-homogeneous. It is clear that $\varphi((r_1, r_2)) > \varphi(r_1)$ and $\varphi((r_1, r_2)) > \varphi(r_2)$.

Consider the following algorithm. At the beginning we take the subset $\{r \in R | \varphi(r) = \min_{a \in R} \varphi(a)\}$. Then we apply all possible transformations of the first type to this subset (observe that applying a transformation of the first type we have either zero or an element with the same value of φ). Then we apply all possible transformations of the second type (all of the corresponding compositions belong to higher order φ-subsets). By induction we choose the next φ-subset, with the help of first type transformations we eliminate subwords (looking through all leading terms of all previous φ-subsets and of all leading terms of the fixed subset). Using the second type transformations we join all compositions thus obtaining higher order φ-subsets.

It is clear that the set S so constructed is stable and $\mathrm{id}(R) = \mathrm{id}(S)$. Since $|X| < \infty$, the number of pairwise different monomials with fixed weight φ_0 is finite. We see now that we used only a finite number of steps in our algorithm to construct φ-subsets with $\varphi \leq n \in \mathbb{N}$ because the set R is φ-homogeneous. From Theorem 3.10 it now follows that if, for $a \in L(X)$, the word $\hat{a}$ does not contain any subword $\hat{b}$, $b \in S$, then $a \notin \mathrm{id}(S) = \mathrm{id}(R)$ (by stability of S, the checking procedure can be done in only a finite number of steps). If $\hat{a}$ contains a subword $\hat{b}$, $b \in S$, then $\hat{a} = \alpha A \hat{b} B$, $\alpha \in K$, $A, B \in S(X)$, and, as above, by Lemmas 3.6 and 3.2 we observe that there is an element $[A(b)B]$ such that $\widehat{[A(b)B]} = \beta A \hat{b} B$, $\beta \in K$, $A, B \in S(X)$. (For $L^p(X)$ it is possible that $[A(b)B] = [C(b)D]^{p^k}$, $C, D \in S(X)$.) Now we consider $a' = a - (\alpha/\beta)[A(b)B]$ and observe that $\hat{a}' < \hat{a}$. Therefore it is sufficient to solve our problem for a'. Thus in finitely many steps we find out whether a is in $\mathrm{id}(R)$ or not. This completes the proof of the theorem. $\square$

3.12. Proposition. *Let $z \in X$, $a \in L(X)(a \in L^p(X))$, $l_z(a) \neq 0$, a be a G-homogeneous element (we may assume that the weight φ is such that $l_z(\hat{a}) \neq 0$). If a is regular word (in particular, if a is odd: $d(a) \in G_-$) then the leading term of any element of $\mathrm{id}(a)$ contains the subword $\hat{a}$ and therefore the Freiheitssatz is valid and there exists an algorithm for solving the equality problem for colour Lie superalgebras and for colour Lie p-superalgebras with one defining relation $a = 0$.*

Proof. Recall that the Freiheitssatz means that $\mathrm{id}(a) \cap L(X \setminus z) = 0$. Using Lemma 3.4, Theorem 3.10 and the final part of the proof of Theorem 3.11 (in which we did not use the φ-homogeneity assumption) we receive the assertion of our theorem. $\square$

3.13. Remark. Consider the case of $L(X)$. Let $a \in L(X)$, $\hat{a} = 2uu$, where u is a regular word, $d(u) \in G_-$, $a = [u, u] + \sum \alpha_i v_i$ where $\alpha_i \in K$, v_i is an s-regular monomial, $\bar{v}_i < uu$, $\bar{v}_1 > \bar{v}_i$ for $i \neq 1$ (first by the weight φ and then lexicographically), $\bar{v}_1 \neq au$ for $\bar{v}_1 > u$, $\bar{v}_1 \neq ua$ for $u > \bar{v}_1$, $a \in S(X)$, $z \in X$, $l_z(\hat{a}) \neq 0$. Then $\mathrm{id}(a) = \mathrm{id}(S)$, $S = \{a, (a, a)\}$ where (a, a) is the composition, S is a stable set. Therefore, for $\mathrm{id}(a)$, we have an algorithmic solution of the equality problem. Moreover, from Theorem 3.10, the resulting form of (a, a) and the condition $l_z(\hat{a}) \neq 0$ we immediately derive the Freiheitssatz.

§4. Free products with amalgamated subalgebra

In this section we construct bases of free products of colour Lie superalgebras (Theorem 4.4) and colour Lie p-superalgebras (Theorem 4.7) with amalgamated subalgebra.

4.1. Definition. Let K be a field, T a set, and H^0, H_α, $\alpha \in T$, colour Lie K-superalgebras (with the same group G and skew symmetric bilinear form ε), H_α^0 be a G-homogeneous subalgebra of H_α, $\delta_\alpha: H^0 \to H_\alpha^0$ an isomorphism of colour Lie superalgebras. We say that a colour Lie K-superalgebra Q (with the group G and the form ε) is the *free product* of colour Lie superalgebras H_α, $\alpha \in T$, with amalgamated homogeneous subalgebra H^0 (notation: $Q = \prod_{\alpha \in T}^{*}{}_{H^0} H_\alpha$) if the following conditions are satisfied.
1) There are homogeneous subalgebras H_α' in Q, isomorphisms $\sigma_\alpha: H_\alpha \to H_\alpha'$ of colour Lie superalgebras such that $\sigma_\alpha \delta_\alpha(h) = \sigma_\beta \delta_\beta(h)$ for all $\alpha, \beta \in T$ and $h \in H^0$.
2) If M is a colour Lie K-superalgebra (with the same group G and form ε), $\gamma_\alpha: H_\alpha \to M$ a homomorphism of colour Lie superalgebras, $\gamma_\alpha \delta_\alpha(h) = \gamma_\beta \delta_\beta(h)$ for all $\alpha, \beta \in T$ and $h \in H^0$, then there is one and only one homomorphism ψ: $\prod_{\eta \in T}^{*}{}_{H^0} H_\eta \to M$ of colour Lie superalgebras such that $\psi \sigma_\alpha = \gamma_\alpha$ for all $\alpha \in T$, i.e. the following diagram is commutative.

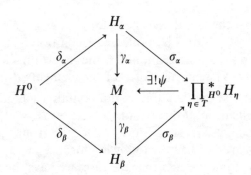

It is clear that the colour Lie superalgebra $\prod_{\alpha \in T}^{*}{}_{H^0} H_\alpha$ is defined uniquely (up to an isomorphism of colour Lie superalgebras).

4.2. The construction of the free product. Let $B^0 = \{f_\beta, \beta \in T^0\}$ be a G-homogeneous basis in H^0, $B_\alpha^0 = \{e_{\alpha\beta}, e_{\alpha\beta} = \delta_\alpha(f_\beta), \alpha \in T, \beta \in T^0\}$, and let B_α be a homogeneous basis in H_α, $B_\alpha^0 \subset B_\alpha$ and $B_\alpha = \{e_{\alpha\beta}, \beta \in T_\alpha\}$ for all $\alpha \in T$. Then the multiplication in H_α for all $\alpha \in T$ can be given using the structure constants $p_{\alpha\beta\gamma}^\tau \in K$: $[e_{\alpha\beta}, e_{\alpha\gamma}] = \sum_\tau p_{\alpha\beta\gamma}^\tau e_{\alpha\tau}$, $\beta, \gamma \in T_\alpha$. For any $\alpha \in T$ consider a G-graded set $X_\alpha = \{x_{\alpha\beta}, \beta \in T_\alpha, d(x_{\alpha\beta}) = d(e_{\alpha\beta})\}$.

Let X be the set obtained from $\bigcup_{\alpha \in T} X_\alpha$ by the identification $x_{\alpha\gamma} = x_{\beta\gamma}$ for $\gamma \in T^0$. Consider T^0 with a total order and extend it to a total order of T_α for all $\alpha \in T$ such that $\beta > \gamma$ for $\beta, \gamma \in T_\alpha$, $\beta \notin T^0$, $\gamma \in T^0$. Let J be an ideal of the free colour Lie superalgebra $L(X)$ generated by the following subset of G-homogeneous elements $D = \{d_{\alpha\beta\gamma}\} = \{[x_{\alpha\beta}, x_{\alpha\gamma}] - \sum_\tau p^\tau_{\alpha\beta\gamma} x_{\alpha\tau}, \alpha \in T, \beta, \gamma \in T_\alpha$, where $\beta > \gamma$ if $x_{\alpha\beta} \in X_+$ and $\beta \geq \gamma$ if $x_{\alpha\beta} \in X_-\}$.

It is easy to see that the quotient-algebra $L(X)/J$ is isomorphic to the algebra $\prod^*_{H^0} H_\alpha$. (For $a = \sum q_{\alpha\beta} e_{\alpha\beta} \in H_\alpha$, $q_{\alpha\beta} \in K$, we set $\sigma_\alpha(a) = \sum q_{\alpha\beta} \tilde{x}_{\alpha\beta} \in L(X)/J$. Then using the definition of $L(X)$ and the Homomorphism Theorem we get Parts 1 and 2 of Definition 4.1.)

Now we order X in the following way: for $x_{\alpha\beta} \neq x_{\gamma\delta}$ we set $x_{\alpha\beta} > x_{\gamma\delta}$ if $\alpha > \gamma$ or $\alpha = \gamma$ and $\beta > \delta$.

4.3. Definition. We say that a regular monomial $w \in \Gamma(X)$ is *special* if the word $\bar{w}$ does not contain subwords $x_{\alpha\beta} x_{\alpha\gamma}$, $\beta > \gamma$, and also no subwords $x_{\alpha\beta} x_{\alpha\beta}$, where $\beta \geq \gamma$ for $x_{\alpha\beta} \in X_-$. We say that a monomial $w \in \Gamma(X)$ is *s-special* if either w is a special monomial or $w = [v, v]$, where v is a special monomial, $d(v) \in G_-$, $l_X(v) > 1$.

4.4. Theorem. *Let K be a field, char $K \neq 2, 3$. Then the cosets the representatives of which are s-special monomials form a linear basis of the algebra $L(X)/J = \prod^*_{H^0} H_\alpha$.*

Proof. Let φ be just the length ($\varphi(x) = 1$ for all $x \in X$). We show that the set of generators $D = \{d_{\alpha\beta\gamma}\}$ for J is stable (see Definition 3.9). Then from Theorem 3.10, Definition 4.3 and Lemma 3.6 the assertion of our theorem will follow immediately. Viewing the form of the elements $d_{\alpha\beta\gamma}$, we see that the words $\widehat{d_{\alpha\beta\gamma}}$ are pairwise distinct and do not contain each other as a subword. Therefore, for verification of the stability of the set $D = \{d_{\alpha\beta\gamma}\}$, it is sufficient to consider the case of overlaps. Since for $\beta \in T^0$ from $x_{\alpha\beta} > x_{\alpha\gamma}$ it follows that $x_{\alpha\gamma} \in H^0_\alpha$, any overlap has the following form: $x_{\alpha\beta} x_{\alpha\gamma} x_{\alpha\delta}$, where $\beta > \gamma$ for $x_{\alpha\beta} \in X_+$ and $\beta \geq \gamma$ for $x_{\alpha\beta} \in X_-$. In these cases we have the following compositions: $a = [[x_{\alpha\beta}, x_{\alpha\gamma}] - \sum_\tau p^\tau_{\alpha\beta\gamma} x_{\alpha\tau}, x_{\alpha\delta}] - [x_{\alpha\beta}, [x_{\alpha\gamma}, x_{\alpha\delta}] - \sum_\tau p^\tau_{\alpha\beta\delta} x_{\alpha\tau}]$ or $b = [[x_{\alpha\beta}, x_{\alpha\beta}] - \sum_\tau p^\tau_{\alpha\beta\beta} x_{\alpha\tau}, x_{\alpha\beta}]$. Thus we have $a = [[x_{\alpha\beta}, x_{\alpha\gamma}], x_{\alpha\delta}] - [x_{\alpha\beta}, [x_{\alpha\gamma}, x_{\alpha\delta}]] + \sum c_\tau x_{\alpha\tau}$, $c_\tau \in K$. Furthermore, by the Jacobi identity, $a = \varepsilon(x_{\alpha\beta}, x_{\alpha\gamma})[[x_{\alpha\beta}, x_{\alpha\delta}], x_{\alpha\gamma}] + \sum c_\tau x_{\alpha\tau} = \varepsilon(x_{\alpha\beta}, x_{\alpha\gamma})[[x_{\alpha\beta}, x_{\alpha\delta}] - \sum p^\tau_{\alpha\beta\gamma} x_{\alpha\tau}, x_{\alpha\gamma}] + \varepsilon(x_{\alpha\beta}, x_{\alpha\gamma}) \sum p^\tau_{\alpha\beta\delta} [x_{\alpha\tau}, x_{\alpha\beta}] + \sum c_\tau x_{\alpha\tau} = \varepsilon(x_{\alpha\beta}, x_{\alpha\gamma})[d_{\alpha\beta\gamma}, x_{\alpha\gamma}] + \sum q_{\alpha\psi\theta} d_{\alpha\psi\theta} + \sum d_\tau x_{\alpha\tau}$, where $q_{\alpha\psi\theta} \in K$, $\psi = \begin{cases} \tau & \text{for } \tau \geq \beta \text{ (then } \theta = \beta) \\ \beta & \text{in other cases (then } \theta = \tau) \end{cases}$. Moreover, by the Jacobi identity in H_α we have $\sum d_\tau x_{\alpha\tau} = 0$. Thus we obtain $a = \varepsilon(x_{\alpha\beta}, x_{\alpha\gamma})[d_{\alpha\beta\gamma}, x_{\alpha\gamma}] + \sum q_{\alpha\psi\theta} d_{\alpha\psi\theta}, [\widehat{d_{\alpha\beta\delta}}, x_{\alpha\gamma}] = x_{\alpha\beta} x_{\alpha\delta} x_{\alpha\gamma} < x_{\alpha\beta} x_{\alpha\gamma} x_{\alpha\delta}, \widehat{d_{\alpha\psi\theta}}$

$< x_{\alpha\beta} x_{\alpha\gamma} x_{\alpha\delta}$ because of $l(d_{\alpha\psi\theta}) = 2 < 3$. For the composition b we have $b = -\sum_\tau p^\tau_{\alpha\beta\beta}[x_{\alpha\tau}, x_{\alpha\beta}] = \sum q_{\alpha\psi\theta} d_{\alpha\psi\theta} + \sum d_\tau x_{\alpha\tau}$, where $q_{\alpha\psi\theta}, d_\tau \in K$, and ψ is defined as above. Moreover, taking into account the definitions of the constants $p^\tau_{\alpha\beta\gamma}$ we see that $\sum d_\tau x_{\alpha\tau} = 0$. Therefore we obtain $b = \sum q_{\alpha\psi\theta} d_{\alpha\psi\theta}$. Thus, the set $D = \{d_{\alpha\beta\gamma}\}$ is stable. This completes the proof of the theorem. □

If we consider G-graded associative K-algebras (without identity element) in place of colour Lie superalgebras in Definition 4.1 then we have the definition of the free product of G-graded associative K-algebras with an amalgamated homogeneous subalgebra (which is determined uniquely up to isomorphism).

Recall that for a colour Lie superalgebra L the universal enveloping K-algebra (without identity element) $U^+(L)$ was defined in Subsection 1.9 of Chapter 1 (see also Section 2 of Chapter 3).

4.5. Proposition.

$$U^+\left(\prod_{\alpha \in T}^* {}_{H^0} H_\alpha\right) \cong \prod_{\alpha \in T}^* {}_{U^+(H^0)} U^+(H_\alpha).$$

Proof. Consider the following commutative diagram.

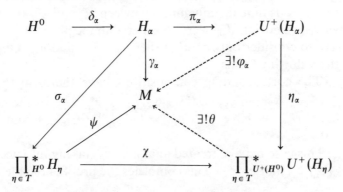

Here M is a G-graded associative K-algebra, π_α, σ_α, η_α are canonical embeddings, γ_α, φ_α, ψ, θ are homomorphisms and $\gamma_\alpha \delta_\alpha = \gamma_\beta \delta_\beta$ for all $\alpha, \beta \in T$. Now the proposition follows from the universal properties of free products and universal enveloping algebras. □

If, in Definition 4.1, in place of colour Lie superalgebras we consider colour Lie p-superalgebras then we have the definition of the free product of colour Lie p-superalgebras with amalgamated homogeneous subalgebra. It is clear that such a product is determined uniquely (up to an isomorphism of colour Lie p-superalgebras). Let J be an ideal of the free colour Lie p-superalgebra $L^p(X)$ generated by elements $d_{\alpha\beta\gamma}$ and $h_{\alpha\beta} = x^p_{\alpha\beta} - \sum_\tau q^\tau_{\alpha\beta} x_{\alpha\tau}$, where $\alpha \in T$, β,

$\tau \in T_\alpha$, $q_{\alpha\beta}^\tau \in K$, $x_{\alpha\beta} \in X_+$ and $e_{\alpha\beta}^p = \sum_\tau q_{\alpha\beta}^\tau e_{\alpha\tau}$ in H_α (see 4.2). It is clear that the algebra $L^p(X)/J$ is the free product of colour Lie p-superalgebras H_α with amalgamated homogeneous subalgebra H^0.

4.6. Definition. We say that a regular monomial $w \in \Gamma(X)$ is *p-special* if the word $\bar{w}$ does not contain subwords of the form $x_{\alpha\beta}x_{\alpha\gamma}$, $\beta > \gamma$, $x_{\alpha\beta}x_{\alpha\gamma}$, $\beta \geq \gamma$ with $x_{\alpha\beta} \in X_-$, or $x_{\alpha\beta}^p$ with $x_{\alpha\beta} \in X_+$. A monomial $w \in \Gamma(X)$ is called *ps-special* if either w is a p-special monomial or $w = [u, u]$ where u is a p-special monomial, $d(u) \in G_-$, $l_X(u) > 1$, or $w = v^{p^t}$ where $t \in \mathbb{N}$, v is a p-special monomial, $d(v) \in G_+$, $l_X(v) > 1$.

4.7. Theorem. *Let K be a field, char $K = p > 3$. Then the cosets, the representatives of which are ps-special monomials, form a linear basis of $L^p(X)/J$.*

Proof. Consider φ as just the usual length. We show that the subset $\{d_{\alpha\beta\gamma}\} \cup \{h_{\alpha\beta}\}$ of the ideal J is stable (in the sense of Definition 3.9). Then the assertion of our theorem will follow from Theorem 3.10, Definition 4.7 and Lemma 3.6. Looking at the form of the elements $d_{\alpha\beta\gamma}$, $h_{\alpha\beta}$ we see that the words $\widehat{d_{\alpha\beta\gamma}}$, $\widehat{h_{\alpha\beta}}$ are distinct and do not contain each other as a subword. Therefore, to verify the stability of the set $\{d_{\alpha\beta\gamma}\} \cup \{h_{\alpha\beta}\}$ it is sufficient to consider the case of overlaps. It is obvious that all ambiguities between $x_{\alpha\beta}^p$ and $x_{\alpha\beta}^p$ are resolvable. Overlaps between elements $d_{\alpha\beta\gamma}$ were considered in the proof of Theorem 4.4. Thus we have to consider only overlaps between $d_{\alpha\beta\gamma}$ and $h_{\alpha\beta}$. Observe that they have the following forms:

a) $x_{\alpha\beta}^p x_{\alpha\gamma}$. The corresponding compositions have the form $(h_{\alpha\beta}, d_{\alpha\beta\gamma}) = [x_{\alpha\beta}^p - \sum q_{\alpha\beta}^\tau x_{\alpha\tau}, x_{\alpha\gamma}] - (\text{ad } x_{\alpha\beta})^{p-1}([x_{\alpha\beta}, x_{\alpha\gamma}] - \sum p_{\alpha\beta\gamma}^\tau x_{\alpha\tau})$. In the colour Lie p-superalgebra H_α we have $[e_{\alpha\beta}^p, e_{\alpha\gamma}] = (\text{ad } e_{\alpha\beta}^{p-1})[e_{\alpha\beta}, e_{\alpha\gamma}]$ and therefore $(h_{\alpha\beta}, d_{\alpha\beta\gamma}) = 0$.

b) $x_{\alpha\beta}x_{\alpha\gamma}^p$. This case may be treated analogously, taking into account that $[e_{\alpha\beta}, e_{\alpha\gamma}^p] = [e_{\alpha\beta}, e_{\alpha\gamma}](\text{ad}' e_{\alpha\gamma})^{p-1}$. This completes the proof of the theorem. □

Comments to Chapter 3

A.I. Shirshov invented the composition method for Lie algebras which became an extremely useful tool in Ring Theory for solving quite a few algorithmic problems (see surveys [Bahturin, Slin'ko, Shestakov, 1981], [Bokut', Kukin, 1987], [Bokut', L'vov, Kharchenko, 1988]). Observe that the Composition Lemma is a key stone of the theory of canonical bases (so called Gröbner bases) from the point of view of Computer Algebra (see [Computer Algebra, 1983], [Latyshev, 1988], [Ufnarovsky, 1990]).

A.I. Shirshov [Shirshov, 1962a, 1984] used the Composition Lemma to prove the Freiheitssatz for Lie algebras with one defining relation (the corresponding statement for groups had been proved earlier by W. Magnus [Magnus, 1930]). We mention also the papers [Gerasimov, 1976], [Lewin J., Lewin T., 1968], [Dicks, 1972], [Makar-Limanov, 1985], [Hedges, 1987] devoted to the study of associative algebras with one defining relation. Free products of Lie algebras with amalgamated subalgebra have been considered by A.I. Shirshov in [Shirshov, 1962b], subalgebras of these products have been studied by G.P. Kukin in [Kukin, 1972a, 1974].

Our exposition of the Diamond Lemma in Section 1 is essentially based on the paper of G.M. Bergman [Bergman, 1978]. The PBW-theorem for colour Lie superalgebras over fields can be found in [Scheunert, 1979a]. In the proof of Theorem 2.12 we have used an idea from [Umirbayev, 1990]. S.A. Agalakov remarks in his paper of 1984 that a free associative algebra of rank at least two possesses subalgebras which are not finitely separable. He also gives in [Agalakov, 1989] an example of a Lie algebra with one defining relation over a field of characteristic zero which is not residually finite. The results of §3 and §4 have been obtained by A.A. Mikhalev in [Mikhalev, 1986, 1989].

Chapter 4

Identities in enveloping algebras

§1. Main results

In this section we present the main results of this chapter on identities in enveloping algebras. Our goal is to find necessary and sufficient conditions for a colour Lie superalgebra under which its restricted or universal enveloping algebra satisfies a non-trivial associative identity.

1.1. Theorem. *Let G be a finite group and let $L = \bigoplus_{g \in G} L_g$ be a restricted colour Lie superalgebra over a field of positive characteristic $p > 2$. Then the restricted enveloping algebra $u(L)$ satisfies a non-trivial identity if, and only if, there exist restricted homogeneous ideals $Q \subset R \subset L$ such that*
1) $\dim L/R < \infty$, $\dim Q < \infty$;
2) $R^2 \subset Q$, $Q^2 = 0$;
3) *Q has a nilpotent p-map (i.e. for each even homogeneous $x \in G_g$, $g \in G_+$ we have $x^{[p^n]} = 0$ for some integer n).*

This theorem will also be proved in the following modification. We may assume that Q, R are not ideals, but subalgebras. In this case bounds on the dimensions in Claim 1 can be established, provided that $u(L)$ satisfies a polynomial identity of given degree. Sufficiency will also be proved without the assumption that Q is abelian; and some bounds on the degree of the identity will be given as functions of the dimensions in Claim 1. As a corollary we will get a theorem on identities in the *universal enveloping algebra $U(L)$* in the positive characteristic case, where the colour Lie superalgebra L need not, in general, be restricted.

In the case of characteristic zero we will prove the following result.

1.2. Theorem. *Let G be a finite group and let $L = \bigoplus_{g \in G} L_g$ be a colour Lie superalgebra over a field of characteristic zero. Then a universal enveloping algebra $U(L)$ satisfies a non-trivial identity if, and only if, there exists a homogeneous L_+-submodule $M \subset L_-$ such that*
1) $\dim L_-/M < \infty$, $\dim[L_+, M] < \infty$;
2) *L_+ is abelian and $[M, M] = 0$.*

We will find some bounds on the dimensions in Claim 1 as functions of the degree of the identity in $U(L)$ and vice versa. The proof avoids tedious computations connected with the p-map which are needed in the proof of the previous theorem, but some additional computations concerning the action of L_+ on L_- are necessary.

One can easily find an analogy between these two theorems. Indeed, R, Q are analogous to $L_+ \oplus M$ and $[L_+, M]$.

1.3. The question of the existence of a non-trivial identity in enveloping algebras is connected to the question of the existence of a bound for the dimensions of all irreducible representations. This second important problem is treated in Chapter 5.

Lemma. *Let A be an associative PI-algebra. Then the dimensions of all irreducible A-representations over their endomorphism fields are bounded by some constant.*

Proof. Let V be an irreducible A-module, $D = \text{End}_A V$ and set $I = \text{Ann}_A V$. Then, by the Density Theorem [Jacobson, 1956], either $\dim_D V = n < \infty$ and $A/I \cong M_n(D)$ or $\dim_D V = \infty$ and for any integer n there exists a subalgebra $B \subset A/I$ which maps homomorphically onto $M_n(D)$. By Kaplansky's Theorem [Herstein, 1968], $M_n(D)$ does not satisfy any identity of degree $2n - 1$. Now the result is obvious. □

Example. Let Γ_n be the *Heisenberg algebra*

$$\Gamma_n = \langle x_1, \ldots, x_n, y_1, \ldots, y_n, z | [x_i, y_j] = \delta_{i,j} z \rangle$$

where all other commutators are trivial. Define the p-map as follows: $x_i^{[p]} = y_i^{[p]} = 0$, $i = 1, \ldots, n$; $z^{[p]} = z$. Let Γ stand for the Heisenberg algebra with countable basis. Then $u(\Gamma)$ does not satisfy any identity. To verify this assertion we may assume that the ground field K is algebraically closed because all multilinear identities remain valid after an extension of the ground field.

One has a Γ_n-module of dimension p^n:

$$V = K[t_1, \ldots, t_n]/(t_1^p, \ldots, t_n^p);$$

$$x_i \circ f(t) = \frac{\partial}{\partial t_i} f(t); \qquad y_i \circ f(t) = t_i f(t); \qquad z \circ f(t) = f(t); \qquad f(t) \in V.$$

It is not difficult to verify that V is irreducible. Evidently, $\text{End}_{\Gamma_n} V = K$. As above, $u(\Gamma_n)$ does not satisfy any identity of degree less then $2p^n$. Thus, $u(\Gamma)$ is not a *PI*-algebra.

This example gives a negative answer to a question raised by V.K. Kharchenko [Dnestrovskaia tetrad, Problem 2.118]: is it true that the existence of a non-trivial Lie identity in a p-algebra L implies the existence of a non-trivial associative identity in $u(L)$? In the example Γ is a 2-step nilpotent Lie p-algebra.

Remark. In the case of an algebraically closed field, there are, up to the choice of bases, two options for the p-map: $z^{[p]} = z$ and $z^{[p]} = 0$. Let us remark that in the second case $u(\Gamma)$ does satisfy an identity $[X, Y]^p \equiv 0$. To see this, one should only remark that all non-trivial terms in the commutator of two elements $a, b \in u(\Gamma)$ contain, as a factor, a central element z which is nilpotent, i.e. $z^{[p]} = 0$.

1.4. Theorem. *Let A, B be associative algebras over a field K satisfying non-trivial identities of degrees d_1, d_2, respectively. Then $A \otimes_K B$ also is a PI-algebra satisfying a non-trivial identity of degree $n = 3(d_1 - 1)^2(d_2 - 1)^2$.*

Proof. See [Bahturin, 1987a]; the bound for the degree of the identity may be derived easily by modifying arguments there. □

1.5. Theorem [Procesi, Small, 1968]. *Let A satisfy some power of a standard identity $(S_{2m})^l \equiv 0$. Then $M_n(A)$ satisfies also some power of a standard identity $(S_{2nm})^t \equiv 0$ for some $t \in \mathbb{N}$.* □

1.6. Abelian colour Lie superalgebras. If L is an abelian (restricted) Lie algebra then $U(L)$ ($u(L)$) is commutative, i.e. it is a PI-algebra. We will prove a similar result for the colour case.

Proposition. *Let G be a finite group and let $L = \bigoplus_{g \in G} L_g$ be an abelian colour Lie superalgebra over an arbitrary field K. Then the following holds.*

$$U(L) \subset M_{|G|}(R) \otimes_k \Lambda(Y)$$

for some extension $K \subset k$, some commutative k-algebra R, and the Grassmann algebra $\Lambda(Y)$ with a sufficiently large number of variables Y.

Proof. Let us fix $n = |G|$. Without loss of generality we may assume that $\dim L_g \geq 1$ for all $g \in G$. Take the algebra of colour polynomials in n variables $A = K^{\varepsilon'}[X_g | g \in G]$, defined by the relations

$$X_a \cdot X_b = \varepsilon'(a, b) X_b \cdot X_a; \quad \varepsilon'(a, b) = \varepsilon(a, b)\tilde{\varepsilon}(a, b); \quad \tilde{\varepsilon}(a, b) = \pm 1; \quad a, b \in G,$$

where $\tilde{\varepsilon}(a,b) = -1$ only in the case $a, b \in G_-$. The algebra A is in fact the algebra of colour polynomials but not superpolynomials since $\varepsilon'(a,a) \neq -1$, $a \in G$, by definition. Let us consider the following purely transcendental extension of K and the Grassmann algebra $\Lambda(Y)$:

$$k' = K[t_g^\alpha | g \in G_+, \alpha \in \Xi_g], \qquad \Lambda(Y) = \Lambda(y_g^\alpha | g \in G_-, \alpha \in \Xi_g),$$

where in both cases $|\Xi_g| = \dim L_g$, $g \in G$. Let $\bar{A} = A \otimes_K k'$ denote the algebra resulting from A by an extension of the ground field.

One has an embedding

$$L \cong L' = \bigoplus_{g \in G_+} \langle X_g t_g^\alpha | \alpha \in \Xi_g \rangle_K \oplus \bigoplus_{g \in G_-} \langle X_g y_g^\alpha | \alpha \in \Xi_g \rangle_K \subset \bar{A} \otimes_{k'} \bar{\Lambda}(Y),$$

which can be naturally extended as follows:

$$U(L) \subset \bar{A} \otimes_{k'} \bar{\Lambda}(Y) \subset B_{\bar{k}'} \overline{\bar{\Lambda}(Y)}, \qquad (1)$$

where $k' \subset k$ is an algebraic closure and two bars denote the extension of the ground field to k, and $B = \bar{\bar{A}} = k^{\varepsilon'}[X_g | g \in G]$. Let us prove that the following holds:

$$B \subset \prod_\rho B/\mathrm{Ker}\,\rho \cong \prod_\rho M_{\dim \rho}(k), \qquad (2)$$

where the Cartesian product is taken over all irreducible representations ρ of B. The first embedding follows from the triviality of the Jacobson radical of an arbitrary colour polynomial ring (Subsection 2.8 of Chapter 3). The elements $X_g^n, g \in G, n = |G|$ are central in B, hence by Schur's Lemma they act on a given representation ρ as scalars α_g, $g \in G$, which are algebraic over k (this will be proved in Subsection 1.4 of Chapter 5). Hence $\alpha_g \in k$ and one has in fact a representation of the n^2-dimensional algebra $B/((X_g^n - \alpha_g \cdot 1)|g \in G)$. By the Density Theorem we get (2) and the bound $\dim \rho \leq n$.

Finally, for the ring $R = \prod_\rho k$ the embedding (2) is extended to the following: $B \subset M_n(R)$. Together with (1) the proof of the proposition is complete. $\square$

Corollary. *Let G be a finite group and $L = \bigoplus_{g \in G} L_g$ a (restricted) abelian colour Lie superalgebra over an arbitrary field. Then $U(L)$ ($u(L)$) is a PI-algebra. Moreover it satisfies the following identities:*
1) *an identity of degree $48|G|^2$,*
2) *some power of the standard identity $(S_{2|G|})^t \equiv 0$, $t \in \mathbb{N}$,*
3) *the standard identity $S_{2|G|} \equiv 0$ if $G = G_+$.*

§1. Main results

Proof. 1. It is well-known that the Grassmann algebra satisfies an identity $[X,[Y,Z]] \equiv 0$ and the ring $M_n(R)$, where R is commutative, satisfies the identity $S_{2n} \equiv 0$. The desired conclusion now follows from Theorem 1.4.

2. The Grassmann algebra satisfies $S_2^2 = [X, Y]^2 \equiv 0$ since

$$[X,[X \cdot Y, Y]] = [X,[X, Y] \cdot Y] = [X,[X, Y]] \cdot Y + [X, Y] \cdot [X, Y].$$

Using Theorem 1.5 we are done.

3. Follows from the proof of the proposition because the Grassmann algebra does not appear in this case. □

Let us explain why in Theorem 1.1 and the corollary in 1.9 (see below) the condition of G being finite is essential. Indeed, we shall give examples of abelian colour Lie superalgebras with infinite group G whose enveloping algebras do not satisfy any identity.

Suppose that $\lambda \in K$ is not a root of unity. Set $G = \mathbb{Z} \times \mathbb{Z}$ and $\alpha = (1,0)$, $\beta = (0,1)$. Then the conditions $\varepsilon(\alpha, \beta) = \lambda$, $\varepsilon(\alpha, \alpha) = \varepsilon(\beta, \beta) = 1$ uniquely determine a bilinear alternating form $\varepsilon \colon G \times G \to K^*$. Obviously, one has $G = G_+$.

Examples. 1) The universal enveloping algebra $U(L)$ for the two-dimensional abelian colour Lie superalgebra $L = \langle x \rangle \oplus \langle y \rangle$ with $d(x) = \alpha \in G$, $d(y) = \beta \in G$ does not satisfy any identity.

2) If char $K = p > 0$ then the subspace

$$\tilde{L} = \bigoplus_{i=0}^{\infty} (\langle x^{p^i} \rangle \oplus \langle y^{p^i} \rangle) \subset U(L)$$

is naturally equipped with the structure of an abelian colour Lie p-superalgebra such that $u(\tilde{L}) \cong U(L)$ does not satisfy any identity.

Proof. 1. It is obvious that $U(L)$ is a colour polynomial ring in two variables x, y with the relation $xy = \lambda yx$. Let us consider the following infinite-dimensional L-module:

$$V = \langle v_i | i \in \mathbb{Z} \rangle; \qquad x \circ v_i = v_{i-1}, \qquad y \circ v_i = \lambda^i v_{i+1}, \qquad i \in \mathbb{Z}.$$

We choose an element

$$0 \neq v = \sum_{i=0}^{m} \mu_i v_{t_i}; \qquad 0 \neq \mu_i \in K, \qquad m \geq 0, \qquad t_i \neq t_j, \qquad i \neq j.$$

Then

$$(xy)^s \circ v = \sum_{i=0}^{m} \lambda^{(st_i)} \mu_i v_{t_i}; \qquad 0 \leq s \leq m, \qquad \mu_i \in K.$$

In order to express $v_{t_0}, \ldots, v_{t_m}$ via $(xy)^s \circ v$, $0 \le s \le m$, it is sufficient to show that the determinant $\|\mu_i \lambda^{st_i}\|$, $0 \le s, i \le m$, is non-trivial. Evidently, this is, up to a scalar $\mu_0 \cdot \ldots \cdot \mu_m \ne 0$, the Vandermonde determinant $\|\lambda^{st_i}\|$, $0 \le s, i \le m$. This determinant is trivial only in the case where $\lambda^{t_i} = \lambda^{t_j}$ for some pair $0 \le i < j \le m$. Since λ is not a root of unity V is an absolutely irreducible module. Therefore, by the Density Theorem, $U(L)$ is not a PI-algebra.

2. The p-structure on $\tilde{L}$ is obtained by restricting the p-map of $U(L)$. □

1.7. Sufficiency for restricted Lie superalgebras. Let us prove the sufficiency of the conditions in Theorem 1.1 without supposing that Q, R are ideals and that Q is abelian.

Proposition. *Let G be a finite group and let $L = \bigoplus_{g \in G} L_g$ be a restricted colour Lie superalgebra over a field of positive characteristic $p > 2$. Suppose that there exist restricted homogeneous subalgebras $Q \subset R \subset L$ such that*
1) $\dim L/R = t < \infty$, $\dim Q = m < \infty$;
2) $R^2 \subset Q$;
3) Q *has a nilpotent p-map*.

Then $u(L)$ satisfies a non-trivial identity of degree $C|G|^4 p^{2(m+t)}$ for some constant C.

Proof. We consider canonical embeddings $u(Q) \subset u(R) \subset u(L)$. We show that the augmentation ideal $I = u^+(Q)$ of $u(Q)$ is nilpotent. Consider the set of linear transformations $A = \{\text{ad } x|_{u(Q)} | x \in Q_g, g \in G\}$ defined on the finite-dimensional space $u(Q)$. The set A is weakly closed (in the sense of [Jacobson, 1962]). Now let $x \in Q_g$: if $g \in G_+$, then $(\text{ad } x)^{p^n} = \text{ad}(x^{[p^n]}) = 0$ for some n; if $g \in G_-$, then $(\text{ad } x)^2 = \frac{1}{2}\text{ad}[x, x]$ with $[x, x] \in G_+$, and we return to the first case. Therefore all elements in A are nilpotent, and by Engel's Theorem on weakly closed nil-sets [Jacobson, 1962] the associative hull of A is nilpotent. On the other hand this associative hull is isomorphic to $u^+(Q)$. It follows that $I^{\dim u(Q)} = 0$, $\dim u(Q) = p^{\dim Q_+} 2^{\dim Q_-} \le p^m$.

By hypothesis Q is an ideal in R. The ideal J in $u(R)$ generated by Q is also nilpotent: $J = u(R) \cdot Q \cdot u(R) = u(R) \cdot Q$, $J^{p^m} = u(R)Q^{p^m} = 0$.

We have $u(R)/J \cong u(R/Q)$. By the hypothesis R/Q is abelian, hence, by Corollary 1.6, $u(R/Q)$ satisfies a non-trivial identity of degree $48|G|^2$. We conclude that $u(R)$ satisfies some non-trivial identity of degree $d_1 = 48|G|^2 p^m$.

By identifying any $x \in u(L)$ with a right multiplication we obtain an embedding of $u(L)$ into the endomorphism ring of a free module $_{u(R)}u(L)$ of finite rank $p^{\dim L_+/R_+} 2^{\dim L_-/R_-} \le p^t$. Thus we have shown that

$$u(L) \subset \text{End}_{u(R)} u(L) \subset M_{p^t}(u(R)) \cong M_{p^t}(K) \otimes_K u(R).$$

By Kaplansky's Theorem $M_{p^t}(K)$ satisfies the non-trivial identity $S_{2p^t} \equiv 0$ of degree $d_2 = 2p^t$. Finally, using 1.3 we obtain a non-trivial identity for $u(L)$ of degree $n = 2^{10}3^3|G|^4p^{2(m+t)}$. □

Remark. Let $Q \subset R \subset L$ be ideals satisfying the hypotheses of the proposition. Then we can consider L/Q and its restricted enveloping algebra $u(R/Q)$ which as above can be embedded in $M_{p^t}(u(R/Q))$. Using 1.5 we get

$$u(L) \subset M_{|G|p^t}(A) \otimes_K \Lambda(Y),$$

for some extension $K \subset k$, some commutative k-algebra A, and the Grassmann algebra $\Lambda(Y)$ with a sufficiently large number of variables Y.

As above we derive that the ideal $J = u(L)Q$ of $u(L)$, $u(L)/J \cong u(L/Q)$, is nilpotent: $J^{p^m} = 0$. By the same argument as in Corollary 1.6, $u(L)$ satisfies some power of the standard identity $(S_{2|G|p^t})^l \equiv 0$, $l \in \mathbb{N}$.

1.8. Sufficiency in the case of characteristic zero.

Proposition. *Let G be a finite group and let $L = \bigoplus_{g \in G} L_g$ be a colour Lie superalgebra over a field of characteristic zero. We assume that there exists a homogeneous L_+-submodule $M \subset L_-$ such that*
1) $\dim L_-/M = t < \infty$, $\dim[L_+, M] = m < \infty$;
2) L_+ *is abelian and* $[M, M] = 0$.
Then $U(L)$ satisfies a non-trivial identity of degree $C|G|^2 4^t m$ for some constant C.

Proof. We set $H = L_+ \oplus M$ and $N = [L_+, M]$. Then we get a chain of ideals $L \supset H \supset N$ and the corresponding chain of enveloping algebras $u(L) \supset u(H) \supset u(N)$. Now N is an abelian ideal, finite dimensional, and $N \subset L_-$. By direct computation we get $I^{m+1} = 0$ for $I = U^+(N)$. For the ideal $J = NU(L) = U(L)N$ of $U(L)$ this implies that $J^{m+1} \subset U(L)N^{m+1} = 0$.
Set $\bar{L} = L/N$ and $\bar{H} = H/N$. Identifying any $x \in U(\bar{L})$ with the right multiplication by x we obtain an embedding of $U(\bar{L})$ into the endomorphism ring of the free module $_{U(\bar{H})}U(\bar{L})$ of rank 2^t:

$$U(\bar{L}) \subset \mathrm{End}_{U(\bar{H})} U(\bar{L}) \subset M_{2^t}(U(\bar{H})) \cong M_{2^t}(K) \otimes_K U(\bar{H}).$$

Using Proposition 1.6 for the abelian algebra $\bar{H}$ we get

$$U(\bar{L}) \subset M_{2^t}(K) \otimes_K M_{|G|}(R) \otimes_k \Lambda(Y) \subset M_{2^t|G|}(R) \otimes_k \Lambda(Y)$$

for some extension $K \subset k$, some commutative k-algebra R, and the Grassmann algebra $\Lambda(Y)$. Since the Grassmann algebra satisfies the identity $[X, [Y, Z]] \equiv 0$ of the degree 3 using 1.4 we conclude that $U(\bar{L})$ satisfies some non-trivial identity of degree $48|G|^2 4^t$.

The conclusion follows if we recall that $J^{m+1} = 0$ and $U(L)/J \cong U(L/N) = U(\tilde{L})$. □

Remark. The same argument as in the preceding subsection yields that $U(L)$ satisfies a power of the standard identity $(S_{|G|2^{t+1}})^l \equiv 0$, $l \in \mathbb{N}$.

1.9. Identities in universal envelopes in positive characteristic.

Proposition. *Let* char $K = p > 0$ *and let* $L = \bigoplus_{g \in G} L_g$ *be a (restricted) colour Lie superalgebra. Suppose that* $P \subset G$ *is a primary p-subgroup;* $\tilde{G} = G/P$. *Then L can be endowed with the structure of a (restricted) colour Lie superalgebra $\tilde{L}$ with group $\tilde{G}$. Moreover,* $U(L) \cong U(\tilde{L})$ ($u(L) \cong u(\tilde{L})$).

Proof. If $A = \bigoplus_{g \in G} A_g$ is any graded algebra and $G_0 \subset G$ is a subgroup then it is obvious that

$$A_{\alpha+a} \cdot A_{\beta+b} \subset A_{(\alpha+\beta)+(a+b)}; \qquad \alpha, \beta \in G, \qquad a, b \in G_0.$$

For $f \in \tilde{G} = G/G_0$ we set $A_f = \bigoplus_{g \in G, g \in f} A_g$. Then A becomes a $\tilde{G}$-graded algebra of the form $A = \bigoplus_{f \in \tilde{G}} A_f$.

Now let $L = \bigoplus_{g \in G} L_g$ be a colour Lie superalgebra with a form $\varepsilon \colon G \times G \to K^*$ and $G_0 \subset G$ a subgroup with $\varepsilon(G_0, G) = 1$. Then ε induces a new form $\tilde{\varepsilon} \colon \tilde{G} \times \tilde{G} \to K^*$, $\tilde{G} = G/G_0$, given by $\varepsilon(\alpha + a, \beta + b) = \varepsilon(\alpha, \beta)$, $\alpha, \beta \in G$, $a, b \in G_0$. So, $\tilde{L} = \bigoplus_{f \in \tilde{G}} L_f$ is a colour Lie superalgebra with the group $\tilde{G}$ and the form $\tilde{\varepsilon}$. The isomorphism is obvious if we view an enveloping algebra determined by certain relations.

Now let $P = G_0$ be as in the hypothesis. Then for $a \in P$ one has $p^n a = 0$ for some integer n and

$$1 = \varepsilon(0, \alpha) = \varepsilon(p^n a, \alpha) = \varepsilon(a, \alpha)^{p^n}, \qquad \alpha \in G.$$

This implies $\varepsilon(P, G_0) = 1$, and we apply the above argument for $G_0 = P$. If L is restricted then the p-map on $\tilde{G}$-components is defined as a restriction of the p-map in $u(L)$. □

Definition. A polynomial of the type $f(X) = \sum_{i \geq 0} \lambda_i X^{p^{ni}}$, $\lambda_i \in K$ will be called a p^n-*polynomial*.

Lemma. *Any polynomial divides some p^n-polynomial.*

Proof. Suppose that $g(X)$ is a polynomial of degree m. We represent $X^{p^{ni}}$ in the form

$$X^{p^{ni}} = g(X) \cdot q_i(X) + r_i(X), \qquad \deg r_i(X) < m, \qquad i = 0, \ldots, m.$$

One derives that $r_0(X), \ldots, r_m(X)$ are linearly dependent and the result follows. □

The following is a consequence of Theorem 1.1.

Corollary. *Let G be a finite group and let $L = \bigoplus_{g \in G} L_g$ be a colour Lie superalgebra over a field of positive characteristic $p > 2$. Then its universal enveloping algebra $U(L)$ satisfies a non-trivial identity if, and only if, there exist homogeneous ideals $B \subset A \subset L$ such that*
1) $\dim L/A < \infty$, $\dim B < \infty$;
2) $A^2 \subset B$;
3) $B = B_-$ (i.e. $B_+ = 0$);
4) *All inner derivations* $\operatorname{ad} x$, $x \in L_g$, $g \in G_+$, *defined on the whole of the superalgebra, are algebraic and their degrees are bounded by some constant.*

Proof. We know that $U(L)$ carries the natural structure of a colour Lie p-superalgebra. We define the p-map on the homogeneous components $[U(L)]_g$, $g \in G_+$, as the conventional p-power. Then for the p-hull $L_{[p]} = \bigoplus_{g \in G_+, i \geq 0} L_g^{p^i} \bigoplus_{g \in G_-} L_g$ we easily have $U(L) \cong u(L_{[p]})$. Suppose that $U(L)$ is a PI-algebra. Then there exist restricted homogeneous ideals $Q \subset R \subset L_{[p]}$ satisfying 1), 2) and 3) of Theorem 1.1. Note that $Q_+ = 0$, hence $A = R \cap L$, $B = Q \cap L$ are ideals as required.

By the Proposition we may assume that $(|G|, p) = 1$. Take $n \in \mathbb{N}$ such that $p^n \equiv 1 \pmod{|G|}$. Next let $x \in L_g$, $g \in G_+$, and consider the set $\{x^{[p^{ni}]} | i \geq 0\}$ which is contained in L_g. Since $\dim L_{[p]}/R < \infty$, using the Lemma we find that there exists a p^n-polynomial $f(x) = \sum_{i \geq 0} \lambda_i X^{p^{(ni)}}$ such that $y = f(x) \in R$, hence $\operatorname{ad} yL \subset R$, $(\operatorname{ad} y)^p L \subset R^2 \subset Q$. Among the p-powers of the operator

$$(\operatorname{ad} y)^p = \tilde{f}((\operatorname{ad} x)^p), \qquad \tilde{f}(X) = \sum_{i \geq 0} \lambda_i^p X^{p^{ni}},$$

considered as acting on the finite-dimensional space Q, we find some linearly dependent. If we multiply the relation thus obtained on the right by $(\operatorname{ad} y)^p$ then we find that the operator $\operatorname{ad} x$ is algebraic.

Let us prove the converse implication directly. By the hypotheses $U(B)$ is a colour Grassmann algebra, hence for $I = U^+(B)$ one has $I^{\dim B + 1} = 0$. Since A/B is abelian, Corollary 1.6 shows that $U(A/B)$ is a PI-algebra. As in 1.8 for the ideal $J = BU(A) = U(A)B$ of $U(A)$ one has $J^{\dim B + 1} = 0$. The isomorphism $U(A)/J \cong U(A/B)$ implies the existence of a non-trivial identity for $U(A)$.

Recall that we have $(|G|, p) = 1$ without loss of generality. Let n be such that $p^n \equiv 1 \pmod{|G|}$. For $x \in L_g$, $g \in G_+$, this implies that $x^{[p^n]} \in L_g$.

Choose homogeneous bases $L_+ = \langle a_1, \ldots, a_t \rangle \oplus A_+$, $L_- = \langle b_1, \ldots, b_m \rangle \oplus A_-$. By our assumption there exist p^n-polynomials $f_1(X), \ldots, f_t(X)$ with $f_i(\operatorname{ad} a_i) = 0$, $i = 1, \ldots, t$, because each polynomial divides some p^n-polynomial. Thus $\operatorname{ad}(f_i(a_i)) = f_i(\operatorname{ad} a_i) = 0$. Therefore $v_i = f_i(a_i)$, $i = 1, \ldots, t$, are central in the colour Lie superalgebra $[U(L)]$, i.e. $v_i w = \varepsilon(\alpha, \beta) w v_i$ provided that $w \in U(L)_\beta$ and $v_i \in U(L)_\alpha$. Remark that by construction the v_i, $i = 1, \ldots, t$, are homogeneous. Then the $z_i = v_i^{|G|}$, $i = 1, \ldots, t$, are central in the associative algebra $U(L)$. By analogy with [Jacobson, 1962] $U(L)$ has the following basis:

$$a_1^{\alpha_1}, \ldots, a_t^{\alpha_t} b_1^{\beta_1}, \ldots, b_m^{\beta_m} z_1^{\gamma_1}, \ldots, z_t^{\gamma_t} \cdot h_i,$$

$$0 \leq \alpha_j < \deg f_j |G|; \qquad \beta_j \in \{0, 1\}, \qquad \gamma_j \in \mathbb{N},$$

where the h_i are elements in the standard basis for $U(A)$. Now D, the subalgebra generated by $U(A)$ and $z_1, \ldots, z_t$, satisfies all multilinear identities of $U(A)$. Since ${}_D U(L)$ is a free module of finite rank, by analogy with 1.8, $U(L)$ is embedded in some matrix ring over D. Therefore $U(L)$ is a PI-algebra. □

Remark. In this Corollary sufficiency has in fact been proved without supposing that A, B are ideals and that there exists a unique constant which bounds the degrees of the inner derivations.

1.10. Necessity for restricted envelopes. We will reduce the proof of Theorem 1.1 to the proof of the following theorem where we obtain *subalgebras* $Q \subset R \subset L$ satisfying Conditions 1), 2) and 3). Our goal is also to derive some bounds.

Lemma. *Let L be a restricted colour Lie superalgebra and let $M \subset L$ be a homogeneous restricted subalgebra of finite codimension. Then there exists also a homogeneous restricted ideal $\tilde{M} \subset L$ of finite codimension in L which is contained in M.*

Proof. We define a chain of restricted subalgebras M^i, $i \in \mathbb{N}$. Let $M^0 = M$; suppose that we have defined M^j, $0 \leq j \leq i - 1$. Now let M^i be the kernel of the adjoint action of M^{i-1} on L/M^{i-1}. Using induction it is easy to see that $\dim L/M^i < \infty$, $i \in \mathbb{N}$. Again by induction one has

$$M^i = \{x \in L | x \in M, [L, x] \subset M, \ldots, [L, \ldots, L, x] \subset M\},$$

where the latter commutator is i-fold. Hence

$$M^i = \{x \in L | x \in M, [y_1, x] \in M, \ldots, [y_1, \ldots, y_i, x] \in M, \forall y_j \in L\}.$$

Let $v_1, \ldots, v_n \in L_+$ and $v_{n+1}, \ldots, v_m \in L_-$ be the homogeneous elements which form a basis for L modulo M.

§1. Main results

As in the Poincaré-Birkhoff-Witt Theorem, it is not difficult to show that M^i is the set of x such that

$$x \in M, \quad [v_{\alpha_1}, x] \in M, \quad [v_{\alpha_1}, v_{\alpha_2}, x] \in M, \ldots, \quad (3)$$
$$[v_{\alpha_1}, v_{\alpha_2}, \ldots, v_{\alpha_i}, x] \in M, \quad \alpha_1 \leq \alpha_2 \leq \cdots \leq \alpha_i.$$

If v is an even basis element then using the decomposition $v^{[p]} = \sum_{j=1}^n \lambda_j v_j + z$, $z \in M$, we can cancel out $(\operatorname{ad} v)^p = \operatorname{ad}(v^{[p]})$ from (3). If v is an odd basis element then we use the relation $(\operatorname{ad} w)^2 = \frac{1}{2}\operatorname{ad}([w,w])$ and a similar decomposition for $[w,w]$.

Thus, we have shown that

$$M^i = \{x \in L \mid [v_1^{\alpha_1}, \ldots, v_n^{\alpha_n}, v_{n+1}^{\beta_{n+1}}, \ldots, v_m^{\beta_m}, x] \in M;$$
$$0 \leq \alpha_j < p, \beta_j \in \{0, 1\}, \alpha_1 + \cdots + \alpha_n + \beta_{n+1} + \cdots + \beta_m = i\}.$$

It is clear now that the above chain must stabilize, i.e. $M^l = M^{l+1}$ with $l = (p-1)n + (m-n)$. It follows that M^l is the required ideal in L. □

Theorem. *Let G be a finite group and let $L = \bigoplus_{g \in G} L_g$ be a restricted colour Lie superalgebra over a field K of positive characteristic $p > 2$. Suppose that $u(L)$ satisfies a non-trivial identity of degree d. Then there exist restricted homogeneous subalgebras $Q \subset R \subset L$ such that:*
1) $\dim L/R < \infty$, $\dim Q < \infty$;
2) $R^2 \subset Q$, $Q^2 = 0$;
3) Q has a nilpotent p-map;
4) Moreover, if K is perfect then

$$\dim L/R < 4^{4|G|d^4}, \quad \dim Q < 2^{4|G|d^4}.$$

This theorem will be proved in Sections 2, 3.

Proof of the necessity of the conditions in Theorem 1.1. The assertion is an immediate consequence of Theorem 1.10. Indeed, we obtain subalgebras $Q \subset R \subset L$ satisfying Conditions 1, 2, 3 and we must construct a chain of ideals. By the Lemma we get an ideal $\tilde{R} \subset L$ with $\tilde{R} \subset R$, $\dim L/\tilde{R} < \infty$. Consider $\tilde{Q} = (\tilde{R}^2)_{[p]} \subset Q$. Then we arrive at the chain $\tilde{Q} \subset \tilde{R} \subset L$ which completes the proof. □

1.11. θ-identities. Fix $\theta \in K$. We introduce the notation $[a,b]_\theta = ab - \theta ba$ for elements a, b in an associative algebra.

Let $A = A(X_1,\ldots,X_n,\ldots,Y_1,\ldots,Y_n,\ldots)$ be the free associative algebra in countably many variables. For any permutation $\pi \in \mathrm{Sym}(n)$ we define $f_\pi^\theta \in A$ by

$$f_\pi^\theta = [X_1, Y_{\pi(1)}]_\theta \cdot \ldots \cdot [X_i, Y_{\pi(i)}]_\theta \cdot \ldots \cdot [X_n, Y_{\pi(n)}]_\theta.$$

The following elements in A will be called θ-*polynomials*:

$$f^\theta(X_1,\ldots,X_n,Y_1,\ldots,Y_n) = \sum_{\pi \in \mathrm{Sym}(n)} \lambda_\pi f_\pi^\theta, \qquad \lambda_\pi \in K. \tag{4}$$

Proposition. *Suppose that an associative algebra B over a field K satisfies some non-trivial identity of degree d. Then for any $\theta \in K$ it also satisfies a non-trivial identity of the special type*

$$f^\theta(X_1,\ldots,X_n,Y_1,\ldots,Y_n) \equiv 0, \tag{5}$$

where $n = 3d^4$ and f^θ is of type (4). *Moreover, we may take $\lambda_1 = 1$.*

Proof. We denote by $P_m(Z_1,\ldots,Z_m)$ the subspace of all multilinear polynomials in m variables $Z_1, \ldots, Z_m$ in the free associative algebra $\tilde{A} = \tilde{A}(Z_1,\ldots,Z_i,\ldots)$ on a countable set of variables. By $T_m(Z_1,\ldots,Z_m)$ we denote the subspace of elements in $P_m(Z_1,\ldots,Z_m)$ which are identities in the PI-algebra B. It is known that in this case the following estimate holds (see e.g. [Bahturin, 1987a]):

$$\dim P_m(Z_1,\ldots,Z_m)/T_m(Z_1,\ldots,Z_m) < d^{2m}, \qquad m \in \mathbb{N}. \tag{6}$$

Next we apply this inequality to A. Let $P_{2n}(X_1,\ldots,X_n,Y_1,\ldots,Y_n) \subset A$ be the subspace of multilinear polynomials of degree $2n$ depending on the variables $X_1, \ldots, X_n, Y_1, \ldots, Y_n$. In this subspace there exist $n!$ polynomials of type f_π^θ, $\pi \in \mathrm{Sym}(n)$ which are linearly independent. The latter fact is clear from the form of a standard basis of a free associative algebra. Applying (6), we get

$$\dim P_{2n}(X_1,\ldots,X_n,Y_1,\ldots,Y_n)/T_{2n}(X_1,\ldots,X_n,Y_1,\ldots,Y_n) < d^{4n}, \qquad n \in \mathbb{N}.$$

If $n! > d^{4n}$ then one immediately has the desired identities (5). Since $n! > (n/e)^n > (n/3)^n$ the number $n = 3d^4$ is sufficiently large and the result follows. The coefficient λ_1 may be taken non-zero by a renumbering of the variables $Y_1, \ldots, Y_n$. $\square$

Lemma. *The non-trivial element $f^\theta \in A$ of type* (4) *may be written in the form*

$$f^\theta = f^\theta(X_1,\ldots,X_n,Y_1,\ldots,Y_n) = \sum_{i=1}^n [X_1, Y_i]_\theta \cdot g_i^\theta, \qquad g_i^\theta \in A, \tag{7}$$

where

$$g_i^\theta = g_i^\theta(X_2, \ldots, X_n, Y_1, \ldots, \hat{Y}_i, \ldots, Y_n)$$

$$= \sum_{\pi(1)=i,\, \pi \in \mathrm{Sym}(n)} \lambda_\pi [X_2, Y_{\pi(2)}]_\theta \cdot \ldots \cdot [X_n, Y_{\pi(n)}]_\theta$$

and at least one of the g_i^θ is non-zero. The hat here denotes the argument which is omitted.

Proof. The decomposition is obvious. One of the g_i^θ is non-zero because f^θ is non-zero. In fact, all g_i^θ are θ-polynomials of type (4), it suffices only to change indexes. □

For convenience, the index θ will be sometimes omitted. Let us show how θ-identities can be used for the study of a generalization of Example 1.2.

Example. Let G be a finite group and $\varepsilon: G \times G \to K^*$ a bilinear alternating form. Suppose that $\alpha, \beta \in G$ such that $\alpha + \beta \in G_+$. Consider the infinite-dimensional *colour Heisenberg algebra*:

$$\Gamma(\alpha, \beta) = \langle a_1, a_2, \ldots, b_1, b_2, \ldots, z | [a_i, b_j] = \delta_{i,j} z \rangle,$$

where all other commutators are trivial and $d(a_i) = \alpha$, $d(b_i) = \beta$, $i \geq 1$; $d(z) = \alpha + \beta$. Then $U(\Gamma(\alpha, \beta))$ (or $u(\Gamma(\alpha, \beta))$ if $\mathrm{char}\, K = p > 0$ and the p-map is defined such that $z^{[p]} = \lambda z$, $0 \neq \lambda \in K$) does not satisfy any identity.

Proof. Otherwise there exists a θ-identity of type (5) for the enveloping algebra where $\theta = \varepsilon(\alpha, \beta)$. Let us substitute in it the elements $X_1 = a_1, \ldots, X_n = a_n$; $Y_1 = b_1, \ldots, Y_n = b_n$:

$$\sum_{\pi \in \mathrm{Sym}(n)} \lambda_\pi [a_1, b_{\pi(1)}]_\theta \cdot \ldots \cdot [a_n, b_{\pi(n)}]_\theta = 0.$$

Here all brackets coincide with the operation in $\Gamma(\alpha, \beta)$, i.e. $[a_i, b_j] = \delta_{i,j} z$. The only non-trivial term in (8) corresponds to the trivial permutation. Thus $z^n = 0$, a contradiction. □

§2. Delta-sets

Next we are going to use a theorem of P.M. Neumann, see e.g. [Bahturin, 1987a].

2.1. Theorem. *Let $\phi: U \times V \to W$ be a bilinear map, where U, V, W are vector spaces over a field K. Suppose that for each $u \in U$ the codimension of its annihilator in V is bounded by m, and for each $v \in V$ the codimension of its annihilator in U is bounded by l. Then $\dim \langle \phi(U,V) \rangle_K \leq ml$.* □

2.2. We consider the colour Lie superalgebra $L = \bigoplus_{g \in G} L_g$ over an arbitrary field K and introduce some useful sets.

Definition. For any $\alpha, \beta \in G$, $m \in \mathbb{N}$ we set
1) $\delta^m_{\alpha,\beta}(L) = \{a \in L_\alpha \mid \dim[a, L_\beta] \leq m\}$;
2) $\Delta^m_\alpha(L) = \bigcap_{\beta \in G} \delta^m_{\alpha,\beta}(L)$;
3) $\Delta_\alpha(L) = \bigcup_{m \in \mathbb{N}} \Delta^m_\alpha(L)$;
4) $\Delta(L) = \bigoplus_{\alpha \in G} \Delta_\alpha(L)$.

If there is no ambiguity, the argument L in the delta-sets will be omitted. By $\delta_{i,j}$ we shall denote Kronecker's delta.

We consider a natural bilinear mapping $\phi: L_\alpha \times L_\beta \to L_{\alpha+\beta}$ which is induced by the operator in the superalgebra. Then $\delta^m_{\alpha,\beta}(L)$ is the set of all $a \in L_\alpha$ such that the codimension of its annihilator in L_β is bounded by m. Also $\Delta^m_\alpha(L)$ is the set of all $a \in L_\alpha$ such that $\dim[a, L_\beta] \leq m$ for all $\beta \in G$; this, in particular, implies $\dim[a, L] \leq m|G|$.

Lemma. 1) *If $x_1, \ldots, x_n \in \delta^m_{\alpha,\beta}(L)$ (or $\Delta^m_\alpha(L)$) then $\lambda_1 x_1 + \cdots + \lambda_n x_n \in \delta^{nm}_{\alpha,\beta}(L)$ ($\Delta^{nm}_\alpha(L)$ respectively) for any $\lambda_1, \ldots, \lambda_n \in K$.*
2) *If $x \in \Delta^m_\alpha(L)$, $y \in L_\gamma$ then $[x, y] \in \Delta^{2m}_{\alpha+\gamma}(L)$.*
3) *If L is restricted and $x \in \Delta^m_\alpha$, $\alpha \in G_+$ then $x^{[p]} \in \Delta^m_{p\alpha}$.*
4) $\Delta^m_\alpha(L) \subset \Delta^{m+1}_\alpha(L)$.

Proof. The result follows from the inclusions

$$[\lambda_1 x_1 + \cdots + \lambda_n x_n, L_\beta] \subset [x_1, L_\beta] + \cdots + [x_n, L_\beta],$$

$$[[x,y], L_\beta] \subset [y, [x, L_\beta]] + [x, L_{\beta+\gamma}],$$

$$[x^{[p]}, L_\beta] \subset [x, \ldots, x, L_\beta] \subset [x, L_\beta]. \qquad \square$$

In the general case $\delta^m_{\alpha,\beta}(L), \Delta^m_\alpha(L) \subset L_\alpha$ are not subspaces. But according to the lemma $\Delta_\alpha(L) \subset L_\alpha$ are vector subspaces. Hence the plus sign is justified in our definition. Moreover, this lemma implies the following

Corollary. $\Delta = \bigoplus_{g \in G} \Delta_g$ *is a homogeneous ideal in any colour Lie superalgebra L. This ideal is restricted if the superalgebra is restricted.* □

§2. Delta-sets

2.3. The crucial point in this chapter is the following theorem on "weak" identities in enveloping algebras. We make use of it also in the next chapter.

Theorem. *Let $L = \bigoplus_{g \in G} L_g$ be a colour (restricted) Lie superalgebra over an arbitrary field K and fix $\alpha, \beta \in G$. Suppose that in $U(L)$ ($u(L)$) some non-trivial θ-polynomial $f(X_1, \ldots, X_n, Y_1, \ldots, Y_n)$, where $\theta = \varepsilon(\beta, \alpha)$, satisfies*

$$f(x_1, \ldots, x_n, y_1, \ldots, y_n) = \sum_{\pi \in \text{Sym}(n)} \lambda_\pi [x_1, y_{\pi(1)}]_\theta \cdot \ldots \cdot [x_n, y_{\pi(n)}]_\theta = 0, \quad \lambda_\pi \in K$$

for any $x_1, \ldots, x_n \in L_\beta$, $y_1, \ldots, y_n \in L_\alpha$. Then any n elements in L_α are linearly dependent modulo $\delta_{\alpha,\beta}^{n^2}$.

Proof. Set $\gamma = \alpha + \beta$. Our arguments hold for both restricted and universal enveloping algebras. We consider the case of restricted enveloping algebras.

Pick arbitrary $a_1, \ldots, a_n \in L_\alpha$ and substitute in f the following elements:

$$Y_1 = a_1, \ldots, Y_n = a_n; \quad X_1 = x_1, \ldots, X_n = x_n,$$

where $x_1, \ldots, x_n$ are arbitrary elements in L_β. From now on, if symbols $x_1, \ldots, x_n$ (or simply x) occur in some relation, this means that the relation holds for any $x_1, \ldots, x_n \in L_\beta$ ($x \in L_\beta$).

We remark that $[x, y]_\theta = xy - \theta yx$, $x \in L_\beta$, $y \in L_\alpha$, coincides with the operation in L and that the index θ will be omitted henceforth.

Using the decomposition in Lemma 1.11 one has

$$f(x_1, \ldots, x_n, a_1, \ldots, a_n) = \sum_{\pi \in \text{Sym}(n)} \lambda_\pi [x_1, a_{\pi(1)}] \cdot \ldots \cdot [x_n, a_{\pi(n)}]$$

$$= \sum_{i=1}^{n} [x_1, a_i] \cdot g_i(x_2, \ldots, x_n, a_1, \ldots, \hat{a}_i, \ldots, a_n) \equiv 0,$$

$$x_1, \ldots, x_n \in L_\beta. \tag{1}$$

Note that the left hand side is the sum of products, each consisting of n factors in L, hence the left hand side is always an element in $u^n(L) = u^n$.

Since (1) holds, it suffices to prove that under the condition that for any fixed elements $a'_1, \ldots, a'_m \in L_\alpha$ we have

$$f'(x_1, \ldots, x_m, a'_1, \ldots, a'_m) \in u^{m-1}; \quad x_1, \ldots, x_m \in L_\beta,$$

where $f'(X_1, \ldots, X_m, Y_1, \ldots, Y_m) \in A$ is a non-trivial θ-polynomial, the elements $a'_1, \ldots, a'_m$ are linearly dependent modulo $\delta_{\alpha,\beta}^{m^2}$. We omit, for convenience, the dashes and proceed by induction on m.

If $m = 1$ then $f(x_1, a_1) = [x_1, a_1] = 0$, for any $x_1 \in L_\beta$, hence $a_1 \in \delta^0_{\alpha,\beta}$.

Now suppose that our result is true for $m - 1$, $m > 1$. Without loss of generality we assume that $0 \neq g_1(X_2, \ldots, X_m, Y_2, \ldots, Y_m) \in A$. There are two cases to consider. In the first case, for any $x_2, \ldots, x_m \in L_\beta$ we have $g_1(x_2, \ldots, x_m, a_2, \ldots, a_m) \in u^{m-2}$, hence by the induction hypothesis $a_2, \ldots, a_m$ are linearly dependent modulo $\delta^{(m-1)^2}_{\alpha,\beta} \subset \delta^{m^2}_{\alpha,\beta}$.

In the remaining case there exist $b_2, \ldots, b_m \in L_\beta$ such that we have $\deg(g_1(b_2, \ldots, b_m, a_2, \ldots, a_m)) = m - 1$. In the decomposition in the lemma in 1.11 we fix $x_2 = b_2, \ldots, x_m = b_m$:

$$\sum_{i=1}^{m} [x_1, a_i] \cdot g_i(b_2, \ldots, b_m, a_1, \ldots, \hat{a}_i, \ldots, a_m) \in u^{m-1}, \quad x_1 \in L_\beta.$$

If we write x in place of x_1 and translate the summands where $g_i = g_i(b_2, \ldots, b_m, a_1, \ldots, \hat{a}_i, \ldots, a_m) \in u^{m-2}$ to the right hand side then we obtain

$$\sum_{i=1}^{r} [x, a_i] \cdot g_i \in u^{m-1}; \quad x \in L_\beta; \quad \deg g_i = m - 1; \quad i = 1, \ldots, r. \quad (2)$$

The next step is to prove, by induction on r, that (2) implies that $a_1, \ldots, a_r$ are linearly dependent modulo $\delta^{m^2}_{\alpha,\beta}$. We denote by V the subspace in L_γ spanned by all $[b_i, a_j]$, $2 \leq i \leq m$, $1 \leq j \leq m$. Then $\dim V < m^2$ and $g_1, \ldots, g_r$ are in the subalgebra of $u(L)$ generated by V.

If $r = 1$, then $[x, a_1]g_1 \subset u^{m-1}$ for all $x \in L_\beta$. Let us show that this is possible only in the case where $[L_\beta, a_1] \subset V$, that is, if $a_1 \in \delta^{m^2}_{\alpha,\beta}$. By way of contradiction, suppose that $e = [b, a_1] \notin V$, $b \in L_\beta$. Choose a homogeneous ordered basis for L whose first element is $e \in L_\gamma$ followed by a basis of $V = \langle v_1, \ldots, v_t \rangle \subset L_\gamma$. Now g_1 is the sum of products, each product consisting of $m - 1$ factors of the form $[b_i, a_j] \in V \subset L_\gamma$, which can be expressed as linear combinations of basis elements of V. Since $\deg(g_1) = m - 1$, using the standard basis of the restricted enveloping algebra, we get

$$g_1 = \sum v_{i_1} \ldots v_{i_{m-1}} + v, \quad \deg v < m - 1. \quad (3)$$

Multiplying g_1 by $e = [b, a_1] \notin V$ on the left we obtain an element of degree m. Thus, whenever $x = b$, we have a contradiction to the fact that $[x, a_1]g_1 \in u^{m-1}$ for all $x \in L_\beta$.

Now let $r > 1$. If, in (2), $[L_\beta, a_r] \subset V$ holds then $a_r \in \delta^{m^2}_{\alpha,\beta}$ and the result follows. Thus we assume that $e = [b, a_1] \notin V$ for some $b \in L_\beta$. By analogy with the preceding argument we choose a basis in L. We set $[b, a_i] = \mu_i e + w_i$, $i = 1, \ldots, r - 1$, $\mu_i \in K$, where each w_i is a linear combination of basis elements of L_γ, except e. By setting $x = b$ in (2) we obtain

$$e \cdot (\mu_1 g_1 + \cdots + \mu_{r-1} g_{r-1} + g_r) + w_1 g_1 + \cdots + w_{r-1} g_{r-1} \in u^{m-1}. \quad (4)$$

Set $g = \mu_1 g_1 + \cdots + \mu_{r-1} g_{r-1} + g_r$. Suppose that $\deg g = m - 1$. By analogy with the preceding argument g is of the form (3), and this means that the first summand in (4) has degree m and may be written as

$$e \cdot g = \sum e \cdot v_{i_1} \ldots v_{i_{m-1}} + e \cdot v, \qquad \deg(e \cdot v) < m. \qquad (5)$$

The other summands in (4) either have degree smaller than m or, being written in the form (5), they have basis elements of L distinct from e as first factors. Since we can permute elements in (4) arbitrarily modulo the right hand side, we have a contradiction.

So $g = \mu_1 g_1 + \cdots + \mu_{r-1} g_{r-1} + g_r \in u^{m-2}$. Take this expression for g and substitute in (2):

$$\sum_{i=1}^{r-1} [x, a_i - \mu_i a_r] \cdot g_i \in u^{m-1}, \quad x \in L_\beta; \quad \deg g_i = m - 1, \quad i = 1, \ldots, r - 1.$$

By the inductive assumption $a_1 - \mu_1 a_r, \ldots, a_{r-1} - \mu_{r-1} a_r$ are linearly dependent modulo $\delta_{\alpha,\beta}^{m^2}$ and therefore the set $\{a_1, \ldots, a_r\}$ is linearly dependent. Now the proof of the theorem is complete. □

2.4. Corollary. *Let G be a finite group and let $L = \bigoplus_{g \in G} L_g$ be a (restricted) colour Lie superalgebra over a field K (of positive characteristic $p > 2$). Suppose that $U(L)$ $(u(L))$ satisfies a non-trivial identity of degree d. Then there exist subspaces $W_\alpha \subset \Delta_\alpha^{n^3}, \alpha \in G$, such that $\dim L_\alpha / W_\alpha \leq |G|(n-1)$, where $n = 3d^4$.*

Proof. By Proposition 1.11 the enveloping algebra satisfies some non-trivial θ-identity for $n = 3d^4$ and any $\theta \in K$. Now for each pair $\alpha, \beta \in G$ and $\theta = \varepsilon(\beta, \alpha)$ we apply the theorem. Remark that $d \geq 2$ implies $n = 3d^4 \geq 48$.

By the theorem in order to get the linear span $\langle \delta_{\alpha,\beta}^{n^2} \rangle$ we can restrict ourselves to sums of n elements. Hence, by Lemma 2.2, $\langle \delta_{\alpha,\beta}^{n^2} \rangle \subset \delta_{\alpha,\beta}^{n^3}$. Also by the theorem $\dim L_\alpha / \langle \delta_{\alpha,\beta}^{n^2} \rangle \leq n - 1$. Finally, we conclude that $W_\alpha = \bigcap_{\beta \in G} \langle \delta_{\alpha,\beta}^{n^2} \rangle \subset \Delta_\alpha^{n^3}, \alpha \in G$ are the subspaces required. □

Remark that the existence of a non-trivial identity in $U(L)$ (or $u(L)$) implies that $\Delta(L) = \bigoplus_{\alpha \in G} \Delta_\alpha^N(L)$ for some integer N and $\dim L / \Delta(L) < \infty$.

2.5. Proposition. *Let L satisfy the hypothesis of Corollary 2.4. Then there exists a (restricted) homogeneous subalgebra C such that*
1) *C is nilpotent of rank 2 and its commutator subalgebra is of finite dimension, $\dim C^2 \leq N_1$, $N_1 = |G|^4 n^5 2^{|G|n}$;*
2) *$\dim L/C \leq N_2$, $N_2 = |G|^6 n^9 4^{|G|n}$.*

Proof. Put $A^0 = \bigcup_{\alpha \in G} W_\alpha$, where the W_α are as in Corollary 2.4. Define

$$A^i = \left\{ [y_1, \ldots, y_l, x] \mid x \in A^0, y_j \in \bigcup_{g \in G} L_g, 0 \leq l \leq i \right\}, \qquad i \in \mathbb{N}.$$

These sets are unions of pairwise disjoint subsets $A^i = \bigcup_{g \in G} A^i_g$, $i \in \mathbb{N}$, lying in the respective homogeneous components. Next we consider the linear spans

$$W^i_\alpha = \langle A^i_\alpha \rangle \subset L_\alpha, \qquad \alpha \in G, i \in \mathbb{N}; \ W^i = \bigoplus_{\alpha \in G} W^i_\alpha \subset L.$$

It is obvious that $W^0_\alpha = W_\alpha$, $W^0 = \bigoplus_{\alpha \in G} W_\alpha$. By Corollary 2.4 $\dim L_\alpha / W^0_\alpha \leq |G|(n-1)$. We have a chain of subspaces which stabilizes due to the finite codimension mentioned above:

$$W^0 \subset W^1 \subset W^2 \subset \cdots \subset W^t = W^{t+1} = \cdots \subset L.$$

It is easy to see that we may take $t = |G|(n+1)$. Note that if $W^t = W^{t+1}$ then W^t is an ideal in L. Thus we obtain a homogeneous ideal $H = W^{|G|(n+1)}$.

Using Lemma 2.2, by induction we easily have $A^i_\alpha \subset \Delta^{n^3 2^i}_\alpha$, $\alpha \in G$. In particular,

$$A^{|G|(n+1)}_\alpha \subset \Delta^m_\alpha, \qquad m = n^3 2^{|G|(n+1)}, \qquad \alpha \in G.$$

By Corollary 2.4 all linear spans $W^i_\alpha = \langle A^i_\alpha \rangle$, $\alpha \in G$, $i \in \mathbb{N}$, may be considered to consist of sums of at most $|G|(n+1)$ summands. Now Lemma 2.2 yields

$$H_\alpha = \langle A^{|G|(n-1)}_\alpha \rangle \subset \Delta^N_\alpha, \qquad N = |G| n^4 2^{|G|(n-1)}. \tag{6}$$

Fix $\alpha, \beta \in G$. We now determine the bound for $\dim [H_\alpha, H_\beta]$. Recall that $W_\gamma \subset \Delta^{n^3}_\gamma$, $\gamma \in G$. Applying Theorem 2.1 to the natural bilinear map $W_\alpha \times W_\beta \to W_{\alpha + \beta}$ we conclude that $\dim [W_\alpha, W_\beta] \leq n^6$.

By construction one has bases $H_\alpha = W_\alpha \oplus \langle a_1, \ldots, a_l \rangle$, $H_\beta = W_\beta \oplus \langle b_1, \ldots, b_s \rangle$, where $0 \leq l, s \leq |G|(n-1)$. Since

$$[H_\alpha, H_\beta] \subset [W_\alpha, W_\beta] + \sum_{i=1}^l [a_i, L_\beta] + \sum_{i=1}^s [L_\alpha, b_i],$$

we conclude, using (6), that

$$\dim [H_\alpha, H_\beta] \leq n^6 + (l+s)N \leq n^6 + 2|G|(n-1)N \leq 2|G|nN.$$

The latter inequality follows from $n \geq 48$. Therefore

$$\dim H^2 \leq 2|G|^3 nN \leq |G|^4 n^5 2^{|G|n} = N_1. \tag{7}$$

We set $C = C_H(H^2) = \{x \in H | [x, H^2] = 0\}$. This is a homogeneous restricted subalgebra. It centralizes its commutator, hence C is 2-step nilpotent. Now Claim 1 follows from (7). Notice that C is the intersection of centralizers in H of finitely many basis elements for H^2. By virtue of (6) we obtain

$$\dim H_\alpha / C_\alpha \leq N \dim H^2 \leq NN_1, \qquad \alpha \in G.$$

Finally,

$$\dim L/C = \dim L/H + \dim H/C \leq |G|^2 n + |G| NN_1$$

$$\leq |G|^2 n + |G|^6 n^9 2^{|G|(2n-1)} \leq |G|^6 n^9 4^{|G|n} = N_2,$$

completing the proof. $\square$

§3. Identities in enveloping algebras of nilpotent Lie superalgebras

3.1. Abelian Lie p-algebras. We start with some preliminary remarks on p-algebras. If we are concerned with an abelian Lie p-algebra A then, by definition,

$$(\lambda_1 x_1 + \lambda_2 x_2)^{[p]} = \lambda_1^p x_1^{[p]} + \lambda_2^p x_2^{[p]}; \qquad x_1, x_2 \in A, \qquad \lambda_1, \lambda_2 \in K.$$

On the vector space $\Omega = \langle t^i | i \in \mathbb{N} \rangle_K$ we introduce the following multiplication:

$$(\alpha t^i) \cdot (\beta t^j) = \alpha \beta^{p^i} t^{i+j}, \qquad \alpha, \beta \in K.$$

Then the restricted subalgebras in A are exactly the submodules in the module ${}_\Omega A$, where the action of the ring Ω on A is determined by the action of the generator $t \circ x = x^{[p]}$, $x \in A$. If the main field K is perfect then the ring Ω is a skew domain of principal ideals [Jacobson, 1943]. Hence for a finitely generated abelian Lie p-algebra A we can apply a theorem on the structure of finitely generated modules [Jacobson, 1943]:

Theorem. *Suppose that A is a finitely generated abelian Lie p-algebra over a perfect field K. Then*

$$A = \langle s_1 \rangle_{[p]} \oplus \cdots \oplus \langle s_q \rangle_{[p]},$$

where $\langle s_i \rangle_{[p]} = \langle s_i^{[p^m]} | m \in \mathbb{N} \rangle_K$ is the p-hull for the one-dimensional Lie p-algebra $\langle s_i \rangle_K$. □

3.2. The nil-radical.

Definition. Let V be an abelian colour Lie-p-algebra. Then its *nil-radical* $\text{Rad}(V)$ is defined as the set of all $v \in V$ such that v is nilpotent in $u(L)$.

Since the p-map $[p]: V_g \to V_{pg}$, $g \in G_+$ is homogeneous one concludes that $\text{Rad}(V)$ is a restricted homogeneous subalgebra in V, and it is spanned by all homogeneous elements which are nilpotent with respect to the p-map.

Lemma. *Let V be an abelian colour Lie p-algebra, the ground field K be perfect and let $K \subset k$ be some field extension. Suppose that the bar over a subalgebra denotes an extension of the ground field. Then $\overline{\text{Rad}(V)} = \text{Rad}(\bar{V})$.*

Proof. It is obvious that $\overline{\text{Rad}(V)} \subset \text{Rad}(\bar{V})$. Let $k = \langle f_i | i \in I \rangle_K$ be a basis. Let us show that $\{f_i^p | i \in I\}$ is a linearly independent set. Indeed, if $\sum_{i=1}^m f_i^p \lambda_i = 0$, $\lambda_i \in K$, then there exist $\mu_1, \ldots, \mu_m \in K$ such that $\mu_1^p = \lambda_1, \ldots, \mu_m^p = \lambda_m$, hence $(\sum_{i=1}^m f_i \mu_i)^p = 0$ and therefore $\lambda_1 = \cdots = \lambda_m = 0$.

Consider $v = \sum_i f_i v_i \in \bar{V}_g$, where $v_i \in V_g$, $g \in G$, and suppose that $v^{[p]} = \sum_i f_i^p v_i^{[p]} = 0$. Since $\{f_i^p | i \in I\}$ is linearly independent we conclude that $v_i^{[p]} = 0$, i.e. $v_i \in \text{Rad}(V)_g$. Thus we have proved the inverse embedding $\overline{\text{Rad}(V)} \supset \text{Rad}(\bar{V})$. □

Theorem [Seligman, 1967]. *Let A be a finite-dimensional abelian Lie p-algebra over an algebraically closed field K. Then*

$$A = \langle e_1, \ldots, e_m | e_i^{[p]} = e_i \rangle \oplus \text{Rad}(A).$$ □

Remark. Let L be an arbitrary colour Lie p-algebra. Then we define its nil-radical $\text{Rad}(L) \subset L$ by setting $\text{Rad}(L) = L \cap J(u(L))$, where $J(u(L))$ stands for the Jacobson radical of $u(L)$. If L is finite dimensional one can easily prove that $\text{Rad}(L)$ coincides with the maximal p-nil-ideal.

3.3. *Proof of Theorem* 1.10 *without establishing estimates.* In Proposition 2.5 we have obtained a 2-step nilpotent superalgebra C with finite-dimensional commutator subalgebra and the proof of Theorem 1.10 is now, in fact, reduced to the study of this latter algebra. Indeed, if we find $Q \subset R \subset C$ satisfying conditions 1, 2, 3 they will also be the required subalgebras of L.

Remark that the subalgebra C^2 is not necessarily restricted and the dimension of its p-hull $(C^2)_{[p]}$ may be infinite.

Since $\dim C^2 < \infty$ we can choose a restricted subalgebra $R \subset C$ such that the dimension of its commutator is minimal among all subalgebras of finite codimension in C. In other words, for any subalgebra $H \subset R$, $\dim L/H < \infty$, we have $R^2 = H^2$.

Let P be a prime ideal in $u(R)$. Consider a prime PI-algebra $A = u(R)/P$. Let $Z = Z(A)$ be its centre. Then according to [Jacobson, 1975] AZ^{-1} is finite dimensional over its central field $F = ZZ^{-1}$. In particular, AZ^{-1} is generated over F by the images of finitely many elements in $u(R)$, which can be expressed via finitely many homogeneous elements in R. Since, by construction, $R = \bigoplus_{\alpha \in G} \Delta_\alpha^N(R)$ for some integer N, and these elements are centralized by some subalgebra of finite codimension $H \subset R$. Therefore, an image of H in A is a commutative Lie superalgebra, hence $[R, R] = [H, H] \subset P$. We have shown that R^2 lies in the prime radical of $u(R)$. Hence, the basis elements of R^2, which can be taken homogeneous, are nilpotent with respect to the p-map. Obviously, $Q = (R^2)_{[p]}$ is finite dimensional and the subalgebras $C \supset R \supset Q$ are just as required. $\square$

3.4. Proposition. *Let C be as in 2.5 and let the ground field K be perfect. Then there exists a restricted homogeneous subalgebra D, $D \subset C$, such that*
1) *D is nilpotent of rank 2, $\dim D^2 \leq N_1$;*
2) *$\dim(D^2)_{[p]} < \infty$;*
3) *$\dim C/D \leq N_3$, $N_3 = |G|^7 n^6 2^{|G|n}$.*

Proof. 1. will follow from 2.5.

Since $(C^2)_{[p]}$ is contained in the centre of C, it is abelian. Set $V = ((C^2)_+)_{[p]}$. We consider the mapping $\tau: G_+ \to G_+$, $\tau(g) = pg$, $g \in G_+$. With respect to this mapping the group G_+ is the union of pairwise disjoint components $G_+ = \bigcup_{l=1}^{l_0} X^l$ such that $\alpha, \beta \in X^l$ if, and only if, $\tau^i(\alpha) = \tau^j(\beta)$ for some $i, j \in \mathbb{N}$. Then

$$(\varepsilon(\alpha, \beta))^{p^i} = \varepsilon(p^i\alpha, \beta) = \varepsilon(p^j\beta, \beta) = (\varepsilon(\beta, \beta))^{p^j} = 1.$$

Hence $\varepsilon(\alpha, \beta) = 1$ for any $\alpha, \beta \in X^l$. Therefore, we have a decomposition into the direct sum of restricted Lie p-algebras as follows: $V = \bigoplus_{i=1}^{l_0} W^l$, $W^l = \bigoplus_{\alpha \in X^l} V_\alpha$. Let us fix some l, $1 \leq l \leq l_0$. Set

$$V = W^l \oplus \tilde{W}^l, \quad \tilde{W}^l = \bigoplus_{i \neq l} W^i. \tag{1}$$

Since $\dim C^2 < \infty$, the algebra W^l is finitely generated. Using 4.1 we can write

$$W^l = \langle s_1 \rangle_{[p]} \oplus \cdots \oplus \langle s_q \rangle_{[p]}. \tag{2}$$

Note that, in general, the elements $s_1, \ldots, s_q$ are not homogeneous. A basis of the space $\langle s \rangle_{[p]}$ is formed by either the sequence of all p-powers $s^{[p^j]}, j \in \mathbb{N}$, or an initial segment $s^{[p^j]}, 0 \le j \le j_0$ (transcendental and algebraic cases, respectively). We shall prove that there exists a subalgebra $D \subset C$ without transcendental elements in all decompositions of D similar to (2), (1). This will prove Claim 2.

Let $s = s_1$ be a transcendental element in (2). Set $T = \langle s_2 \rangle_{[p]} \oplus \cdots \oplus \langle s_q \rangle_{[p]} \oplus \tilde{W}^l$ and consider $\rho: V \to K$, the projection on $\langle s \rangle_{[p]}$ with kernel T followed by taking the coefficient of s in the above basis for $\langle s \rangle_{[p]}$. Consider bilinear forms for all $\alpha, \beta \in G$ such that $\alpha + \beta \in X^l$:

$$\phi_{\alpha,\beta}: C_\alpha \times C_\beta \to K; \qquad \phi_{\alpha,\beta}(x,y) = \rho([x,y]); \qquad x \in C_\alpha, \qquad y \in C_\beta.$$

We claim that the following bounds hold for the left and right kernels:

$$\dim C_\alpha / \mathrm{Ker}_l \, \phi_{\alpha,\beta} < n, \qquad \dim C_\beta / \mathrm{Ker}_r \, \phi_{\alpha,\beta} < n.$$

Otherwise there exist $x_1, \ldots, x_n \in C_\alpha$, $y_1, \ldots, y_n \in C_\beta$ such that $\phi(x_i, y_j) = \delta_{i,j}$, $1 \le i, j \le n$. Substitute these into the θ-identity, where $\theta = \varepsilon(\alpha, \beta)$:

$$\sum_{\pi \in \mathrm{Sym}(n)} \lambda_\pi [x_1, y_{\pi(1)}]_\theta \cdot \ldots \cdot [x_n, y_{\pi(n)}]_\theta = 0. \qquad (3)$$

Notice that $u(\langle s \rangle_{[p]}) \cong K[s]$ is a polynomial ring. The brackets in (3) coincide with the Lie superalgebra operation. If we choose the term in (3) corresponding to the identity permutation then we get s^n. Other terms from the identity permutation and all terms from non-identity permutations either contain a factor from T, or all n factors have the form $s^t, t \ge 1$. Furthermore, in at least one case we have $t > 1$ which is a contradiction to (3). Thus we have shown that the desired bounds on the codimensions of the kernels hold. Notice that $\mathrm{Ker}_l \, \phi_{\alpha,\beta} = \mathrm{Ker}_r \, \phi_{\beta,\alpha}$. We define a restricted subalgebra $\tilde{C} \subset C$ as follows: $(\tilde{C})_\alpha = \bigcap_{\alpha + \beta \in X^l} \mathrm{Ker}_l \, \phi_{\alpha,\beta}$. Then we have $\dim(C/\tilde{C})_\alpha \le n|G|^2$ and $\dim C/\tilde{C} \le n|G|^3$. By construction, $\rho((\tilde{C}^2)_+) = 0$. On the other hand (2) implies $\rho((C^2)_+) \ne 0$ because otherwise

$$V = ((C^2)_+)_{[p]} \subset \langle s_1^{[p]} \rangle_{[p]} \oplus T,$$

and this is a contradiction to the transcendency of s.

Therefore, one has a strict inclusion $\tilde{C}^2 \subset C^2$ and after at most $\dim C^2 \le N_1$ steps we are able to get rid of transcendent elements in the decompositions (1), (2), and we obtain the required restricted homogeneous subalgebra D with

$$\dim C/D \le n|G|^3 N_1 = |G|^7 n^6 2^{|G|n} = N_3. \qquad \square$$

3.5. Proposition. *Let D be as in 3.4 and let the main field K be algebraically closed. Then there exists a restricted homogeneous subalgebra F, $F \subset D$, such that*
1) $((F^2)_+)_{[p]}$ *consists of p-nil-elements*;
2) $\dim D/F \le N_3$, $N_3 = |G|^7 n^6 2^{|G|n}$.

Proof. Let $G_+ = \bigcup_{l=1}^{l_0} X^l$ be the disjoint decomposition into the components obtained above. Set $V = ((D^2)_+)_{[p]}$ and fix some component X^l. Then a decomposition into the direct sum of restricted subalgebras $V = W^l \oplus \widetilde{W}^l$ arises, analogous to (1), $A = W^l$ being a finite-dimensional abelian Lie p-algebra. Applying Theorem 3.2, we have

$$A = \langle e_1, \ldots, e_m | e_j^{[p]} = e_j \rangle \oplus \text{Rad}(A), \tag{4}$$

where $e_1, \ldots, e_m$ are not necessarily homogeneous, but $\text{Rad}(A)$ is a homogeneous subalgebra (see 3.2). Denote $T = \langle e_2, \ldots, e_m \rangle \oplus \text{Rad}(A) \oplus \widetilde{W}^l$. Note that T is not a homogeneous subalgebra in general. Let $\rho: V \to \langle e_1 \rangle$ be the projection of V on $\langle e_1 \rangle$ with kernel T. We consider bilinear forms for all pairs $\alpha, \beta \in G$ such that $\alpha + \beta \in X^l$:

$$\phi_{\alpha,\beta}: D_\alpha \times D_\beta \to \langle e_1 \rangle_K; \quad \phi_{\alpha,\beta}(x,y) = \rho([x,y]); \quad x \in D_\alpha, \quad y \in D_\beta.$$

Next we prove that $\dim D_\alpha / \text{Ker}_l \phi_{\alpha,\beta} < n$, and the same bound holds for the right kernel. Otherwise there exist $x_1, \ldots, x_n \in D_\alpha$, $y_1, \ldots, y_n \in D_\beta$ such that $\phi(x_i, y_j) = \delta_{i,j} e_1$, $1 \le i, j \le n$. Substitute these into the θ-identity, where $\theta = \varepsilon(\alpha, \beta)$:

$$\sum_{\pi \in \text{Sym}(n)} \lambda_\pi [x_1, y_{\pi(1)}]_\theta \cdot \ldots \cdot [x_n, y_{\pi(n)}]_\theta = 0. \tag{5}$$

By construction, $[x_i, y_j]_\theta = \delta_{i,j} e_1 + z_{i,j}$, $z_{i,j} \in T$ for $1 \le i, j \le n$. If we choose the term in (5) corresponding to the identity permutation, then we get $0 \ne e_1^n \in u(\langle e_1 \rangle)$. The other terms in (5) are contained in $u(V) u^+(T)$. Therefore, (5) yields a contradiction, so the desired bounds hold true.

We define a subalgebra $\widetilde{D} \subset D$ as follows: $(\widetilde{D})_\alpha = \bigcap_{\alpha+\beta \in X^l} \text{Ker}_l \phi_{\alpha,\beta}$ and, as above, one has $\dim D/\widetilde{D} \le n|G|^3$ and $\rho((\widetilde{D}^2)_+) = 0$. On the other hand $\rho((D^2)_+) \ne 0$ because otherwise $V = ((D^2)_+)_{[p]} \subset T$, a contradiction. Therefore, one has a strict inclusion $\widetilde{D}^2 \subset D^2$ and after at most $\dim D^2 \le N_1$ steps we obtain a required restricted homogeneous subalgebra F, satisfying Claim 1, with

$$\dim D/F \le n|G|^3 N_1 = |G|^7 n^6 2^{|G|n} = N_3;$$

and the result follows. Finally we remark that F_α, $\alpha \in G$, are the kernels of bilinear mappings

$$\Phi_\alpha: D_\alpha \times D_+ \to V/\mathrm{Rad}(V), \qquad \alpha \in G_+,$$

$$\Phi_\alpha: D_\alpha \times D_- \to V/\mathrm{Rad}(V), \qquad \alpha \in G_-, \tag{6}$$

□

3.6. Proposition. *Let D be as in 3.4 and let the ground field K be perfect. Then there exist restricted homogeneous subalgebras $Q \subset R \subset D$ with the following properties.*
1) $\dim D/R \leq 2N_3$, $\dim Q \leq nN_1$.
2) $R^2 \subset Q$, $Q^2 = 0$.
3) *The p-map of Q is nilpotent.*

Proof. Let $K \subset \bar{K}$ be the algebraic closure of the perfect field K, and, given any subalgebra S, we write $\bar{S}$ for $\bar{S} = \bar{K} \otimes_K S$. Then, by Lemma 3.2, for $V = ((D^2)_+)_{[p]}$ one has $\mathrm{Rad}(\bar{V}) = \overline{\mathrm{Rad}(V)}$.

We consider bilinear mappings for $\alpha \in G_+$,

$$\Phi_\alpha: D_\alpha \times D_+ \to V/\mathrm{Rad}(V),$$

$$\Phi'_\alpha: \bar{D}_\alpha \times \bar{D}_+ \to \bar{V}/\mathrm{Rad}(\bar{V}) = \overline{V/\mathrm{Rad}(V)}.$$

By analogy we define bilinear mappings for $\alpha \in G_-$, where D_- and $\bar{D}_-$ appear as second components.

Obviously, $\mathrm{Ker}_l(\Phi'_\alpha) = \overline{\mathrm{Ker}_l(\Phi_\alpha)}$. Using the final remark (6) in the proof of the preceding proposition, we obtain a subalgebra $F = \bigoplus_{\alpha \in G} \mathrm{Ker}_l(\Phi_\alpha)$ with $\dim D/F \leq N_3$ and $(F^2)_+ \subset \mathrm{Rad}(V)$.

By analogy with (1), for $B = ((F^2)_+)_{[p]}$ and some fixed component $X^l \subset G_+$ we obtain a decomposition into the direct sum of restricted subalgebras $B = B^l \oplus \tilde{B}^l$, B^l being a finite-dimensional abelian Lie p-algebra. Using Theorem 3.1, one has

$$B^l = \langle s_1 \rangle_{[p]} \oplus \cdots \oplus \langle s_m \rangle_{[p]}, \tag{7}$$

$s_1, \ldots, s_m$ being nilpotent. Suppose $s = s_1$, and $s^{[p^n]} \neq 0$. As in 3.4, we construct a linear mapping $\rho: B \to K$, and for any pair $\alpha, \beta \in G$ with $\alpha + \beta \in X^l$ we define

$$\phi_{\alpha,\beta}: F_\alpha \times F_\beta \to K; \qquad \phi_{\alpha,\beta}(x, y) = \rho([x, y]); \qquad x \in F_\alpha, y \in F_\beta.$$

We have to prove that $\dim F_\alpha/\mathrm{Ker}_l \phi_{\alpha,\beta} < n$, and that the same bound holds for the right kernel. Otherwise, there exist $x_1, \ldots, x_n \in F_\alpha$, $y_1, \ldots, y_n \in F_\beta$ such

§ 3. Identities in enveloping algebras of nilpotent Lie superalgebras   135

that $\phi_{\alpha,\beta}(x_i, y_j) = \delta_{i,j}$, $1 \leq i, j \leq n$. Substitute these into the θ-identity, where $\theta = \varepsilon(\alpha, \beta)$:

$$\sum_{\pi \in \text{Sym}(n)} \lambda_\pi [x_1, y_{\pi(1)}]_\theta \cdot \ldots \cdot [x_n, y_{\pi(n)}]_\theta = 0.$$

If we choose the term corresponding to the identity permutation, then we get $s^n \neq 0$, as one of the terms. Other terms either have factors in the subalgebra $T = \langle s_2 \rangle_{[p]} \oplus \cdots \oplus \langle s_m \rangle_{[p]} \oplus \tilde{B}^l$, or all their n factors are of the form s^t, $t \geq 1$, and the inequality is strict at least once, a contradiction. By analogy with 3.4, for a restricted subalgebra $\tilde{F} = \bigoplus_{\alpha \in G} (\tilde{F})_\alpha \subset F$, defined by $(\tilde{F})_\alpha = \bigcap_{\alpha+\beta \in X^l} \text{Ker}_l \phi_{\alpha,\beta}$, one has $\dim F/\tilde{F} \leq n|G|^3$ and $\dim \tilde{F}^2 < \tilde{F}^2$.

After at most $\dim F^2 \leq N_1$ steps we obtain a restricted subalgebra R having a decomposition similar to (7) with $s^{[p^n]} = 0$ for all s. This implies $\dim \langle s \rangle_{[p]} \leq n$. Finally, R and $Q = (R^2)_{[p]}$ are as required because

$$\dim D/R = \dim D/F + \dim F/R \leq 2N_3,$$

$$\dim Q \leq \dim R^2 \cdot n \leq N_1 n. \qquad \square$$

3.7. *Proof of Theorem* 1.10. The case where the field is arbitrary has been considered in 3.3.

Now suppose that K is perfect. Propositions 2.5, 3.4, and 3.6 yield the existence of a chain of restricted homogeneous subalgebras $L \supset C \supset D \supset R \supset Q$. The subalgebras R, Q are as required.

We only have to check the estimates

$$\dim L/R = \dim L/C + \dim C/D + \dim D/R \leq N_2 + 3N_3$$

$$= |G|^6 n^9 4^{|G|n} + 3|G|^7 n^6 2^{|G|n} \leq (|G|n)^{10} 4^{|G|n},$$

$$\dim Q \leq nN_1 = |G|^4 n^6 2^{|G|n} \leq (|G|n)^6 2^{|G|n}.$$

It is not difficult to treat the identities of degree $d = 2$ in $u(L)$ directly: L must be abelian. Hence, we may assume that $d > 2$. Recall that $n = 3d^4$ and set $m = 4|G|d^4 \geq 324$. Then

$$\dim L/R < (m^{10} 4^{-m/4}) 4^m < 4^m,$$
$$\dim Q < (m^6 2^{-m/4}) 2^m < 2^m. \qquad (8)$$

For the second inequality we have to show that $m^6 2^{-m/4} < 1$ for $m \geq 324$. If $m_0 = 324$ then the inequality is verified directly. If we pass from m to $m+1$ then the first factor is multiplied at most by $(\frac{325}{324})^6 < 1.02$ whereas the second factor is divided at least by 2^{81}. The first inequality in (8) is treated similarly.

Now the bounds (8) are as required. $\qquad \square$

§ 4. The case of characteristic zero

In this section we prove the necessary conditions in Theorem 1.2 about the existence of a non-trivial identity in $U(L)$, L being a colour Lie superalgebra over a field K, char $K = 0$. We recall that the sufficiency of the conditions has been established in Subsection 1.8.

4.1. Example. Consider the two-dimensional solvable Lie algebra $H = \langle x, y | [x, y] = y \rangle$. It has an infinite-dimensional module $M = K[t]$ with action

$$x \circ f(t) = tf(t), \qquad y \circ f(t) = f(t-1), \qquad f(t) \in M.$$

Since char $K = 0$, it is not difficult to verify that M is absolutely irreducible.

4.2. Example. Let G be a finite group and let $\varepsilon: G \times G \to K^*$ be an alternating bilinear form such that $G = G_+$. Consider a three-dimensional nilpotent colour Lie algebra $H = \langle x, y, z | [x, y] = z \rangle$ with $\langle x \rangle = H_\alpha$, $\langle y \rangle = H_\beta$, $\langle z \rangle = H_{\alpha+\beta}$; $\alpha, \beta \in G$. Set $\lambda = \varepsilon(\alpha, \beta)$. Then H has an infinite-dimensional module $V = \langle v_1, v_2, \ldots \rangle$ with action as follows:

$$x \circ v_i = (i-1)\lambda^{i-1} v_{i-1}, \qquad y \circ v_i = v_{(i+1)}, \qquad z \circ v_i = \lambda^i v_i.$$

Since char $K = 0$, it is easy to verify that V is absolutely irreducible.

In both examples, combining Corollary 1.4 of Chapter 5 and Lemma 1.3, we conclude that $U(H)$ is not a PI-algebra.

4.3. The colour Lie algebra case (i.e. $G = G_+$).

Theorem. *Let L be a colour Lie algebra with $U(L)$ satisfying a non-trivial identity, char $K = 0$. Then L is abelian.*

Proof. To begin with we consider the case where L is finite-dimensional over an algebraically closed field K. Suppose that all transformations $A = \{\operatorname{ad} x | x \in L_g, g \in G\}$ are nilpotent. Then by Engel's Theorem on weakly closed nil-sets [Jacobson, 1962] an associative hull A^* is nilpotent. It follows, for some $m \in \mathbb{N}$, that $A^m L \neq 0$, $A^{m+1} L = 0$. Suppose that $m > 0$. Since A is homogeneous we can choose homogeneous elements $x \in L$, $y \in A^{m+1} L$ with $0 \neq z = [x, y] \in A^m L$. Hence z is central in L. Now we see that $H = \langle x, y, z \rangle$ is a three-dimensional nilpotent colour Lie algebra. By 4.2, $U(H)$ is not a PI-algebra, a contradiction. Consequently, $m = 0$, i.e. L is abelian.

§4. The case of characteristic zero

Now we consider the remaining case, i.e. where not all transformations in A are nilpotent. Let us take $x \in L_\alpha$ with ad x non-nilpotent, let n be the order of $\alpha \in G$ as group element. It follows that $(\operatorname{ad} x)^n$ maps L_g into L_g for any $g \in G$. Since the adjoint action is non-nilpotent, there exists $y \in L_\beta$ with $(\operatorname{ad} x)^n y = \lambda y$, $0 \neq \lambda \in K$. We find $\mu \in K$ with $\mu^n = \lambda$ and set $\xi = \varepsilon(\alpha, \beta)$; it is obvious that ξ is an m-th root of unity, where m divides n. Consider

$$V = \langle (\operatorname{ad} x)^i y | 0 \leq i < n \rangle, \qquad \dim V = n.$$

Since $\varepsilon(\alpha, \beta + i\alpha) = \varepsilon(\alpha, \beta) = \xi$, we have

$$xw = \xi wx + [x, w], \qquad \text{for all } w \in V. \tag{1}$$

Now $X^n - \lambda$ is the minimal polynomial for $\operatorname{ad} x|_V$, so that using (1) yields another basis of $V = \langle v_0, \ldots, v_{n-1} \rangle$ with

$$xv_i = \xi v_i x + v_i v_i, \qquad 0 \leq i < n,$$

where $v_0, \ldots, v_{n-1}$ are the roots of $X^n - \lambda = 0$. Without loss of generality we assume that an initial segment is as follows: $v_i = \mu \xi^i$, $0 \leq i < m$. Then for $v = v_{m-1} \cdot v_{m-2} \cdot \ldots \cdot v_0$ one has

$$xv = \sum_{i=0}^{m-1} \xi^i v_{m-1} \cdot \ldots \cdot v_{m-i} \cdot [x, v_{m-i-1}] \cdot v_{m-i-2} \cdot \ldots \cdot v_0 + \xi^m vx$$

$$= \left(\sum_{i=0}^{m-1} \xi^i \xi^{m-i-1} \mu \right) v + vx = \gamma v + vx,$$

where $\gamma = m\mu\xi^{m-1} \neq 0$. This relation implies that x, v generate a subalgebra in $U(L)$ isomorphic to the universal enveloping algebra for the two-dimensional solvable Lie algebra. Using 4.1, we have a contradiction to the existence of a non-trivial identity in $U(L)$.

Let now K be an arbitrary field. Consider the algebraic closure $\overline{K}$ of K. Then $U(\overline{K} \otimes_K L) \cong \overline{K} \otimes_K U(L)$ satisfies all multilinear identities for $U(L)$. Thus $\overline{K} \otimes_K L$ is abelian, and therefore L is also abelian.

Finally we have to treat the case that L has arbitrary dimension. By Subsection 2.8 of Chapter 3 $U(L)$ is integral and applying Posner's Theorem [Herstein, 1968] one can embed $U(L)$ in a skew field D which is finite-dimensional over its centre Z. Then the colour Lie algebra $\tilde{L} = ZL \subset D$ is finite-dimensional over Z. We consider a basis $e_1, \ldots, e_n$ of $\tilde{L}$ over Z with $e_i \in L$. Then $U(\tilde{L})$ contains a K-subalgebra W, generated by the K-subalgebra L, W being a PI-algebra since it is a homomorphic image of $U(L)$. On the other hand $U(\tilde{L}) = WZ$ because a basis of $U(\tilde{L})$ over Z consists of ordered monomials depending on $e_1, \ldots, e_n$. It follows that $U(\tilde{L})$ is a PI-algebra, hence $\tilde{L}$ is abelian and thus L is abelian. □

138   4. Identities in enveloping algebras

Recall that the converse statement has been proved in Subsection 1.6.

4.4. From now on we assume that L is a colour Lie superalgebra over a field K, char $K = 0$ with $U(L)$ satisfying some non-trivial identity. By 4.3 we have $[L_+, L_+] = 0$.

If we set $\theta = 0$ then Proposition 1.11 implies:

Lemma. *If an associative K-algebra satisfies a non-trivial identity of degree d then it also satisfies some non-trivial identity of type*

$$\sum_{\pi \in \mathrm{Sym}(n)} \lambda_\pi Y_1 X_{\pi(1)} Y_2 X_{\pi(2)} \cdot \ldots \cdot Y_n X_{\pi(n)} \equiv 0, \qquad (2)$$

where $n = 3d^4$, $\lambda_\pi \in K$, $\lambda_1 \neq 0$.

We begin with some "typical" examples of colour Lie superalgebras with universal enveloping algebras not satisfying any identities of sufficiently large degree. In all these examples $[L_-, L_-] = 0$.

4.5. Lemma. *Let L be a Lie superalgebra with*

$$L_0 = \langle x \rangle, \qquad L_1 \supset \{y_0, y_1, \ldots, y_m\}; \qquad [x, y_{i-1}] = y_i, \qquad i = 1, \ldots, m.$$

Then, given n, there exists N_1 such that, for all $m \geq N_1$, $U(L)$ does not satisfy any identity of type (2).

Proof. We include $\{x, y_0, y_1, \ldots, y_m\}$ as initial segment into a totally ordered basis for L. We substitute in (2) the elements of $U(L)$ as follows: $Y_i = y_{(i-1)n}$, $X_i = x^i$, $1 \leq i \leq n$. Then (2) takes the form

$$\sum_{\pi \in \mathrm{Sym}(n)} \lambda_\pi y_0 x^{\pi(1)} y_n x^{\pi(2)} \cdot \ldots \cdot y_{(i-1)n} x^{\pi(i)} \cdot \ldots \cdot y_{(n-1)n} x^{\pi(n)} \equiv 0. \qquad (3)$$

After reducing the left hand side of (3) to the canonical basis we obtain monomials of the type

$$x^s y_{i_1} \ldots y_{i_r}, \qquad s \geq 0, \qquad i_1 < \cdots < i_r. \qquad (4)$$

Consider only those monomials (4) for which $s = 0$. By finitely many elementary transformations $y_i x = xy_i - [x, y_i] = xy_i - y_{i+1}$, followed by rearrangements of the y_i, any monomial in (3) is presented as a linear combination of monomials of type (4). If $\pi = 1$, one has, with $\pm \lambda_1 \neq 0$, the monomial

$$y_1 y_{n+2} y_{2n+3} \cdots y_{(n-1)n+n}. \qquad (5)$$

§4. The case of characteristic zero

We have to prove that this monomial occurs only in the case where $\pi = 1$. By induction, one easily has the formula

$$y_i x^j = \mu_j x^j y_i + \mu_{j-1} x^{j-1} y_{i+1} + \cdots + \mu_0 y_{i+j}, \qquad \mu_r = (-1)^{j-r}\binom{j}{r}. \quad (6)$$

Since y_1 can only appear from the first factor in (3), using $s = 0$ and (6), one has $\pi(1) = 1$. An analogous argument shows that $\pi = 1$. □

Observe that we need the linear independence of $y_1, \ldots, y_m$ in order to verify that the elements in (5) are linearly independent. The maximal index in (3) does not exceed $N_1 = n(n-1) + 1 + 2 + \cdots + n = n(3n-1)/2$.

4.6. Lemma. *Let L be a Lie superalgebra with $L_0 = \langle x \rangle$, $[x, y] = y$ for all $y \in L_1$ and $[L_1, L_1] = 0$. If we fix an identity (2), then there exists N_2 such that $U(L)$ does not satisfy (2) whenever $\dim L_1 = m > N_2$.*

Proof. Let $\{s_1, \ldots, s_n\}$ be a set of non-negative integers. We substitute in (2) the following elements of the enveloping algebra: $X_i = x^i$, $Y_i = y_1^{(i)} \ldots y_{s_i}^{(i)}$, where $\{y_j^{(i)} | j = 1, \ldots, s; i = 1, \ldots, n\}$ is a linearly independent set of L_1. Then we obtain

$$\sum_{\pi \in \mathrm{Sym}(n)} \lambda_\pi y_1^{(1)} \ldots y_{s_1}^{(1)} \cdot x^{\pi(1)} \cdot \ldots \cdot y_1^{(n)} \ldots y_{s_n}^{(n)} \cdot x^{\pi(n)} = 0. \quad (7)$$

As above, we will reduce the monomials in (7), by elementary transformations, to the type analogous to (4). Since $y_1 \ldots y_p x^q = (x - p)^q y_1 \ldots y_p$, setting $\mu_\pi = \pm \lambda_\pi$, we have

$$\left(\sum_{\pi \in \mathrm{Sym}(n)} \mu_\pi (x - s_1)^{\pi(1)} \cdot (x - s_1 - s_2)^{\pi(2)} \ldots (x - s_1 - \cdots - s_n)^{\pi(n)}\right)$$

$$\times y_1^{(1)} \ldots y_{s_1}^{(1)} \cdot \ldots \cdot y_1^{(n)} \ldots y_{s_n}^{(n)} = 0. \quad (8)$$

Using the Poincaré-Birkhoff-Witt Theorem, we conclude that the first factor in (8) is equal to zero for any $s_1, \ldots, s_n$. By well-known properties of polynomials, $s_1, \ldots, s_n$ may be considered as independent variables and the following polynomial is trivial

$$\sum_{\pi \in \mathrm{Sym}(n)} \mu_\pi (x - s_1)^{\pi(1)} \cdot (x - s_1 - s_2)^{\pi(2)} \cdot \ldots \cdot (x - s_1 - s_2 - \cdots - s_n)^{\pi(n)} = 0. \quad (9)$$

It is easy to verify that if $\pi = 1$ then (9) contains a monomial $s_1^1 \cdot s_2^2 \cdot \ldots \cdot s_n^n$ with non-trivial coefficient $\mu_1 = \pm \lambda_1 \neq 0$. Let us show that this monomial appears only if $\pi = 1$. Indeed, its degree with respect to the variable s_n is equal to n, so that we must have $\pi(n) = n$. The same argument should be used for $s_{n-1}, \ldots, s_1$.

Thus we have shown that there exists an n-tuple of non-negative integers $(s_1^0, \ldots, s_n^0)$ such that (7) is not true for this tuple. The bound $N_2 = N_2(n)$ now has not yet been found but its finding is not a difficult problem. □

4.7. Lemma. *Let L be a colour Lie superalgebra over a field of characteristic zero such that*

$$L_+ = L_\alpha = \langle x \rangle, \quad \alpha \in G_+; \quad L_- \supset \langle y_1, \ldots, y_m \rangle \oplus \langle z_1, \ldots, z_m \rangle,$$

$$L_\beta \supset \langle y_1, \ldots, y_m \rangle, \quad \beta \in G_-; \quad [x, y_i] = z_i, \quad i = 1, \ldots, m,$$

($[L_-, L_-] = 0$ and $\{y_i\}$, $\{z_i\}$ are linearly independent). Then $U(L)$ does not satisfy (2), provided that $m > N_3 = n(n+1)/2$.

Proof. We substitute in (2) the following elements

$$Y_1 = y_1, \quad Y_2 = y_2 y_3, \quad Y_3 = y_4 y_5 y_6, \ldots, \quad X_i = x^i, \quad i = 1, \ldots, n,$$

and obtain

$$\sum_{\pi \in \mathrm{Sym}(n)} \lambda_\pi y_1 x^{\pi(1)} \cdot y_2 y_3 x^{\pi(2)} \cdot \ldots \cdot y_{s+1} \ldots y_r x^{\pi(n)} = 0, \tag{10}$$

where $s = n(n-1)/2$, $r = n(n+1)/2$. Set $\mu = \varepsilon(\beta, \alpha)$. Consider an initial segment $\{x, y_1, \ldots, y_r, z_1, \ldots, z_r\}$. We will reduce the monomials in (10) to the canonical ones, analogous to (4), by elementary transformations such as $y_i x = \mu x y_i - z_i$ and $ab = \xi ba$ for odd homogeneous a, b as well as $z_i x = \mu x z_i - [x, z_i]$ and other commutators. We express every element arising via the basis, and continue the procedure.

We are concerned only with those monomials stemming from (10) which have no occurrences of $x, y_1, \ldots, y_r$. Remark that the degree of (10) with respect to x equals r which is the number of our variables $y_1, \ldots, y_r$. On the one hand y_s can be killed only by x. On the other hand $y_{s+1}, \ldots, y_r$ can be killed only by the last factor $x^{\pi(n)}$ and, recalling $r - s = n$, we have $\pi(n) = n$. Finally, this argument shows that the polynomials under consideration appear only if $\pi = 1$. Moreover, one has a unique such polynomial $z_1 z_2 \ldots z_s \ldots z_r$, with coefficient equal to $1! 2! \ldots n! \cdot \alpha_1 \neq 0$. Thus, (10) yields a contradiction. □

4.8. The structure of the L_+-module L_-.

Theorem. *Let L be a colour Lie superalgebra over a field of characteristic zero with $U(L)$ satisfying some identity and $[L_+, L_+] = [L_-, L_-] = 0$. Then there exists an integer N such that $\delta_{\alpha,\beta}^N(L) = L_\alpha$ for all $\alpha \in G_+, \beta \in G_-$.*

Proof. Let us fix $\alpha \in G_+, \beta \in G_-$. By 4.4 $U(L)$ satisfies an identity (2) for some integer n. Pick any $x \in L_\alpha$. First, we consider the case $\alpha \neq 0$. We claim that $\dim[x, L_\alpha] \leq N_3$, $N_3 = n(n+1)/2$. Thus, by Definition 2.2, $x \in \delta_{\alpha,\beta}^{N_3}(L)$.

Assume the contrary, i.e. let $\dim[x, L_\alpha] > N_3$. Then there exists a linearly independent set $y_1, \ldots, y_{N_3+1} \in L_\beta$ with $z_i = [x, y_i] \in L_{\alpha+\beta}, i = 1, \ldots, N_3 + 1$, being also linearly independent. Now Lemma 4.7 yields a contradiction.

Consider the remaining case: $\alpha = 0$. Then $\langle x \rangle \oplus L_\beta$ is an ordinary Lie superalgebra because $\varepsilon(\alpha, \beta) = \varepsilon(0, \beta) = 1$. Set $H = L_\beta$. Without loss of generality we assume that the main field is algebraically closed.

We make H into a $K[t]$-module by setting $f(t) \circ y = f(\mathrm{ad}\, x)y, y \in H$. Then, by Lemma 4.5, this module is periodic. Moreover, each $y \in H$ is annihilated by some polynomial $f_y(t)$ of degree not greater then N_1. Now H decomposes into the direct sum of its weight components. Since the degrees of annihilating polynomials are bounded, the number of these weight components is finite and bounded by N_1. Suppose that some component $H(\lambda)$ has dimension exceeding $N_4 = (N_1 + 1)N_2$, $\lambda \in K$ being the corresponding weight. Let H^λ denote a subspace of eigenvectors with eigenvalue λ. We claim that $\dim H^\lambda > N_2$. Owing to the fact that H is the union of an ascending chain of finite-dimensional $K[t]$-submodules, we may assume that $H(\lambda)$ is finite-dimensional and $m = \dim H(\lambda) > N_4$. We decompose $H(\lambda)$ into the direct sum of its cyclic subspaces. Their dimensions, by Lemma 4.5, do not exceed $N_1 + 1$ and each contains some eigenvector.

First, we assume $\lambda \neq 0$. Consider $Q = \langle x' \rangle \oplus H$, $x' = \lambda^{-1}x$. Then $[x', y] = y$ for all $y \in H$. Besides, an argument as above shows that $\dim H^\lambda > N_2$. Now we can apply Lemma 4.6 and conclude, that $\dim H(\lambda) \leq N_4$ for all $\lambda \neq 0$. If we denote by $\overline{H(0)}$ the sum of all $H(\lambda), \lambda \neq 0$, then $\dim \overline{H(0)} \leq N_5 = N_1 N_4 = N_1(N_1 + 1)N_2$.

Now we want to prove that $\dim H(0)/H^0 \leq N_6 = N_1 N_3$. Suppose the contrary holds. Decompose $H(0)$ into the sum of cyclic modules. Then there exist at least N_3 submodules of dimension greater then one. Thus, we can construct a subalgebra $P = \langle x \rangle \oplus \tilde{P}$ with $\tilde{P}$ being the linear span of $y_1, z_1, \ldots, y_m, z_m$ such that $[x, y_i] = z_i$, $[x, z_i] = 0$, $[\tilde{P}, \tilde{P}] = 0$, $m > N_3$. Now we have got a contradiction to Lemma 4.7.

Thus, we have proved that $\dim H/H^0 \leq N_5 + N_6 = N$, in other words, $\dim[x, L_\beta] \leq N$ for all $x \in L_0$. □

4.9. A filtration. Let $L = L_+ \oplus L_-$ be a colour Lie superalgebra over an arbitrary field K. Then it is natural to consider another colour Lie superalgebra $\tilde{L} = \tilde{L}_+ \oplus \tilde{L}_-$ such that $\tilde{L}_+ \cong L_+$ as colour Lie algebras, $\tilde{L}_- \cong L_-$ as L_+-modules, but $[\tilde{L}_-, \tilde{L}_-] = 0$. It is obvious that the axioms of a colour Lie superalgebra are satisfied by $\tilde{L}$. Let us find a connection between the universal enveloping algebras $U = U(L)$ and $\tilde{U} = U(\tilde{L})$.

We define an ascending filtration in U as follows:

$$U^{-1} = \{0\}, \quad U^0 = U(L_+), \quad U^1 = U^0 + U^0 L_-, \ldots, U^{n+1} = U^n + U^n L_-, \ldots \tag{11}$$

A basis for the space U^n consists of monomials

$$e_{i_1} e_{i_2} \cdots e_{i_s} f_{j_1} f_{j_2} \cdots f_{j_t}, \tag{12}$$

such that $i_1 \leq \cdots \leq i_s$, $j_1 < \cdots < j_t$, $t \leq n$ (here $\{e_i | i \in I\}$, $\{f_j | j \in J\}$ are the homogeneous bases for L_+ and L_-, the sets I, J being linearly ordered). Since the length of the "J-tail" of the product of the two monomials of type (12) does not exceed the sum of the lengths of "J-tails" of the factors, one readily observes that (11) is a filtration. Next we consider the associated graded algebra $B = \mathrm{gr}(U)$. The basis of this algebra consists of all residue classes as follows:

$$e_{i_1} e_{i_2} \cdots e_{i_s} f_{j_1} f_{j_2} \cdots f_{j_t} + U^{t-1}, \qquad t \in \mathbb{N}, \tag{13}$$

where indices are as above. It is not difficult to verify that if U satisfies some multilinear identity $f(X_1, X_2, \ldots, X_m) \equiv 0$ then B also satisfies this identity. Indeed, for any element

$$u_i + U^{t_i - 1} \in B, \qquad u_i \in U^{t_i}, \qquad i = 1, \ldots, m,$$

one has

$$f(u_1 + U^{t_1-1}, \ldots, u_m + U^{t_m-1}) = f(u_1, \ldots, u_m) + U^{t_1 + \cdots + t_m - 1} = 0.$$

Now we want to prove that $\mathrm{gr}(U) \cong U(\tilde{L})$. We remark that $\tilde{U} = U(\tilde{L})$ is an associative algebra with 1, with generators $\{e_i | i \in I\}$, $\{f_j | j \in J\}$, and with defining relations of the form

$$e_i e_j - \lambda_{i,j} e_j e_i = [e_i, e_j] = \sum_l c_{i,j}^l e_l,$$

$$e_i f_j - \mu_{i,j} f_j e_i = [e_i, f_j] = \sum_l d_{i,j}^l f_l, \qquad f_i f_j - \nu_{i,j} f_j f_i = 0,$$

where $c_{i,j}^l \in K$ are the structure constants of L_+, $d_{i,j}^l \in K$ the structure constants of the L_+-module L_-. Elements $\{e_i + U^{-1} | i \in I\}$, $\{f_j + U^0 | j \in J\}$

generate B and satisfy analogous relations:

$$[e_i + U^{-1}, e_j + U^{-1}] = [e_i, e_j] + U^{-1} = \sum_l c^l_{i,j} e_l + U^{-1} = \sum_l c^l_{i,j}(e_l + U^{-1}),$$

$$[e_i + U^{-1}, f_j + U^0] = [e_i, f_j] + U^0 = \sum_l d^l_{i,j} f_l + U^0 = \sum_l d^l_{i,j}(f_l + U^0),$$

$$[f_i + U^0, f_j + U^0] = [f_i, f_j] + U^1 = U^1 = 0,$$

because $[f_i, f_j] \in L_+ \subset U^1$.

Therefore, the mapping $e_i \mapsto e_i + U^{-1}$, $f_j \mapsto f_j + U^0$ extends to a surjective homomorphism ϕ of associative algebras with 1. On the other hand if $u = e_{i_1} \ldots e_{i_s} f_{j_1} \ldots f_{j_t}$ is a basis monomial for $\tilde{U}$ then $\phi(u) = e_{i_1} \ldots e_{i_s} f_{j_1} \ldots f_{j_t} + U^{t-1}$ is a basis monomial (13) for B. Thus, ϕ is an isomorphism and the existence of a non-trivial polynomial identity for $U(L)$ implies the existence of an identity for $U(\tilde{L})$. Using 4.8 one has the following.

Proposition. *Let L be a colour Lie superalgebra over a field of characteristic zero with $U(L)$ being a PI-algebra. Then there exists an integer N such that $\delta^N_{\alpha,\beta}(L) = L_\alpha$ for any pair $\alpha \in G_+$, $\beta \in G_-$.* □

4.10. Let us collect all the information on delta-sets we have obtained provided that $U(L)$ satisfies a non-trivial identity of degree d. By 1.11, for $n = 3d^4$ and any $\theta \in K$, $U(L)$ satisfies some non-trivial θ-identity. By Corollary 2.5, there exist subspaces $W_\gamma \subset \Delta^{n^3}_\gamma$, $\gamma \in G$, such that $\dim L_\gamma/W_\gamma \leq |G|(n-1)$. For even components we have obtained somewhat stronger results. 1. $\delta^0_{\alpha,\alpha'}(L) = L_\alpha$, $\alpha, \alpha' \in G_+$ (L_+ is abelian, see 4.3). 2. There exists an integer N such that $\delta^N_{\alpha,\beta}(L) = L_\alpha$ for all $\alpha \in G_+$, $\beta \in G_-$ (see 4.9). Thus, $\Delta^N_\alpha(L) = L_\alpha$, for all $\alpha \in G_+$.

Lemma. *Let L be as above. Then there exists an L_+-submodule $F \subset L_-$ such that $\dim L_-/F < \infty$, $F \subset \bigoplus_{\beta \in G_-} \Delta^m_\beta(L)$ for some m.*

Proof. Set $W = \bigoplus_{\beta \in G_-} W_\beta \subset L_-$ and consider $\tilde{W} = \bigoplus_{\beta \in G_-} \tilde{W}_\beta$ as follows:

$$\tilde{W} = \langle [x, y] \mid x \in L_\alpha, \alpha \in G_+; y \in W_\beta, \beta \in G_- \rangle.$$

By Lemma 2.2, if we take a commutator then the "width" of elements at most doubles, and if we take a linear span it increases at most $|G|(n-1)$-times since $\dim L_\beta/W_\beta \leq |G|(n-1)$. Thus, $\tilde{W}_\beta \subset \Delta^{2|G|n^4}_\beta$, $\beta \in G_-$. Now either $W = \tilde{W}$ and $F = W$ is the desired L_+-submodule, or $W \neq \tilde{W}$ and we repeat the process. After finitely many steps one has an L_+-submodule $F \subset L_-$, completing the proof. □

Proof of the necessary conditions in Theorem 1.2. By 4.3 L_+ is abelian. We consider bilinear maps:

$$L_\alpha \times F_\beta \to F_{\alpha+\beta}, \qquad \alpha \in G_+, \qquad \beta \in G_-,$$

$$F_\beta \times F_{\beta'} \to L_{\beta+\beta'}, \qquad \beta, \beta' \in G_-.$$

By applying P.M. Neumann's Theorem 2.1, one has

$$\dim[L_+, F] < \infty, \qquad \dim[F, F] < \infty.$$

Now it remains only to get rid of the finite-dimensional space $V = [F, F] \subset L_+$. Choose a homogeneous basis $V = \langle v_1, \ldots, v_t \rangle$. Fix some s, $1 \leq s \leq t$, and let $\rho_s \colon V \to K$ be the operator taking $v \in V$ to the coefficient of v_s. Consider bilinear mappings for all $\beta, \beta' \in G_-$:

$$\psi^s_{\beta,\beta'} \colon F_\beta \times F_{\beta'} \to K; \qquad \psi^s_{\beta,\beta'}(x, y) = \rho_s([x, y]); \qquad x \in F_\beta, \qquad y \in F_{\beta'}.$$

We claim that $\dim F_\beta / \mathrm{Ker}_l \psi^s_{\beta,\beta'} < n$ and the same inequality for the codimension of the right kernel. For, otherwise, there exist $x_1, \ldots, x_n \in F_\beta$, $y_1, \ldots, y_n \in F_{\beta'}$ such that $\psi^s_{\beta,\beta'}(x_i, y_j) = \delta_{i,j}$; $1 \leq i, j \leq n$. Substitute these into the θ-identity, where $\theta = \varepsilon(\beta, \beta')$:

$$\sum_{\pi \in \mathrm{Sym}(n)} \lambda_\pi [x_1, y_{\pi(1)}]_\theta \cdots [x_n, y_{\pi(n)}]_\theta = 0. \tag{14}$$

The identity permutation in (14) as one of the terms yields v_s^n which cannot be cancelled because other non-trivial monomials of degree n in (14) contain $v_j, j \neq s$ as a factor. This contradiction proves the required estimates for the codimensions of the kernels.

Thus, $M = \bigoplus_{\beta \in G_-} M_\beta$, where $M_\beta = \bigcap_{\beta' \in G_-, 1 \leq s \leq t} \mathrm{Ker}_l \psi^s_{\beta,\beta'}$, is as required. Indeed, by our construction $[M, M] = 0$ and $\dim L_-/M = \dim L_-/F + \dim F/M < \infty$. Note that M is the kernel of an L_+-invariant mapping $\Psi \colon F \times F \to V$, hence $M \subset L_-$ is invariant under the action of L_+. Now the proof of Theorem 1.2 is complete. □

Comments to Chapter 4

This is the first work on identities in enveloping algebras for *colour* Lie superalgebras. The approach presented here is due to V.M. Petrogradsky.

Theorem 1.2 for Lie algebras has been established in [Latyshev, 1963]; this is the first result on identities in enveloping algebras. Theorem 1.2 for Lie superalgebras has been proved in [Bahturin, 1985].

The case of universal enveloping algebras of Lie algebras over modular fields is due to [Bahturin, 1974]. The cases of restricted Lie algebras and superalgebras has been settled in [Petrogradsky, 1988a], [Petrogradsky, 1991a] and [Petrogradsky, 1991b], respectively. By somewhat different methods the restricted Lie algebra case is also treated in [Passman, 1991]; in particular the idea exploited in 3.3 in due to him. For later developments of those methods for Lie superalgebras see [Bergen, Passman, 1991]. We would like to compare Theorem 1.1 with:

Theorem. ([Passman, 1972]). *A group ring $K[H]$ over a field K, char $K = p > 0$, satisfies a non-trivial identity if, and only if, there exist normal subgroups $B \subset A \subset H$ such that $[H:A] < \infty$, A/B is abelian and B is a finite p-group.* □

Thus, a restricted envelope of a colour Lie p-superalgebra resembles a group ring in positive characteristic. We shall return to this analogy in Section 3 of the next chapter.

Recall that the *PI-degree* for a *PI*-algebra R is the least integer m such that R satisfies some power of a standard identity $(S_{2m})^l \equiv 0$, $l \in \mathbb{N}$.

Theorem ([Lichtman, 1989]). *Let R be a PI-subring of $U(L)$, L being a Lie algebra. Then R is commutative if char $K = 0$ and the PI-degree of R is a power of p if char $K = p > 0$.* □

Question. Is an analogous statement true for the restricted enveloping algebra of a Lie p-algebra?

Lemma 1.10 for ordinary Lie p-algebras was proved in [Kukin, 1970].

Chapter 5

Irreducible representations of Lie superalgebras

§1. The Jacobson radical of universal enveloping algebras

1.1. The Jacobson radical for the semidirect product. A general description of the Jacobson radical for universal enveloping algebras of Lie superalgebras is a rather difficult problem and has not been settled yet. We will later need the following fact.

Theorem. *Let $L = \bigoplus_{g \in G} L_g$ be a colour Lie superalgebra over an arbitrary field. Suppose that $M \subset L_-$ is a homogeneous L_+-submodule such that $[M, L_-] = 0$. Then $U(L) \cdot M \subset J(U(L))$. Moreover, $U(L)M$ is a locally nilpotent algebra.*

Proof. Evidently, M is an ideal in L. We will prove that $R = U(L)M \triangleleft U(L)$ is a locally nilpotent algebra. The remaining statement will follow from the fact that any nil-ideal is contained in the Jacobson radical [Jacobson, 1956].

Any finite set of elements in R lies in a finite-dimensional subspace spanned by monomials of the type

$$u = x_{i_1} \ldots x_{i_a} y_{j_1} \ldots y_{j_b} z_{k_1} \ldots z_{k_c}, \tag{1}$$

$$i_1 \leq \cdots \leq i_a, \quad j_1 < \cdots < j_b, \quad k_1 < \cdots < k_c,$$

where $b \geq 1$; $a, b \geq 0$, and $\{x_i\}, \{y_j\}, \{z_k\}$ are homogeneous bases for L_+, M, L_-/M respectively.

It is enough to show that for any $u_1, \ldots, u_n$ of type (1) there exists an integer N such that any product of the form

$$u_{l_1} \cdot u_{l_2} \cdot \ldots \cdot u_{l_N}; \quad l_1, \ldots, l_N \in \{1, 2, \ldots, N\}, \tag{2}$$

equals zero. We will be looking for N in the form $N = (s + 1)n$. Then one of the monomials $u_1, \ldots, u_n$, let it be u_1, enters the product (2) at least $s + 1$ times. Suppose that y_1 enters the monomial u_1. Let us denote by r the number of the pairwise distinct x's, $x_{i_1} \ldots x_{i_r}$, which enter the monomials $u_1, \ldots, u_n$. We introduce the following notation: for $a = x_{i_1} \cdot \ldots \cdot x_{i_t}$ and homogeneous $y \in M$ we set

$$a \circ y = [x_{i_1}, \ldots, [x_{i_t}, y] \ldots] \in M. \qquad (3)$$

For any homogeneous $x \in L_+$, $y \in M$, $z \in L_-/M$ one has the commutation relations

$$(a \circ y) \cdot x = \lambda x \cdot (a \circ y) - \lambda (x \cdot a) \circ y,$$

$$(a \circ y) \cdot z = \mu z \cdot (a \circ y), \qquad (4)$$

$$x \cdot z = vz \cdot x + [x, z], \quad [x, z] = \sum_j \beta_j y_j + \sum_k \gamma_k z_k,$$

for appropriate scalars λ, μ, v. A sequence of these transformations enables us to reduce the product (2) to a linear combination of products

$$x_{p_1} \ldots x_{p_u}(a_1 \circ y_{q_1}) \ldots (a_v \circ y_{q_v}) z_{q_1} \ldots z_{q_w}, \qquad (5)$$

where $a_1, \ldots, a_v$ are monomials in $\{x_{i_1}, \ldots, x_{i_r}\}$ with x's and z's in (5) not necessarily ordered. An important remark is that the product (5) vanishes if $a_1 \circ y_{q_1}, \ldots, a_v \circ y_{q_v} \in M$ are linearly dependent. This follows from the fact that $U(M)$ is isomorphic to a colour Grassmann algebra.

Now we return to u_1 which enters (2) at least $s + 1$ times. Then y_1 also enters (2) at least $s + 1$ times. Since transformations in (4) do not kill elements of the type $a \circ y$, one easily observes that the product (5) contains homogeneous factors

$$a_1 \circ y_1, \quad a_2 \circ y_1, \ldots, a_{s+1} \circ y_1 \in M. \qquad (6)$$

We will be looking for s in the form $s = d_r(1) + \cdots + d_r(m)$, where m is some integer, and $d_r(m)$ is the dimension of the space of homogeneous elements of degree m in a free associative algebra in r variables. Let C be the maximum of the degrees of $u_1, \ldots, u_n$ with respect to $\{x_i\}$. Then the degree of (2) with respect to $\{x_i\}$ is bounded by

$$Nc = (s + 1)nc = (1 + d_r(1) + \cdots + d_r(m))nc. \qquad (7)$$

Let us estimate the degree of $a_1 \cdot a_2 \cdot \ldots \cdot a_{s+1}$ with respect to $\{x_i\}$ provided that the product of elements in (6) is non-zero. The latter condition implies that (6) is a linearly independent set, therefore, $a_1, \ldots, a_{s+1}$ are also linearly independent. Suppose that $x_{i_1}, \ldots, x_{i_r} \in U(L_+)$ generate a free associative algebra B in r variables. Then the minimal value for the degree in question is

§1. The Jacobson radical of universal enveloping algebras   149

achieved if $\{a_1, \ldots, a_{s+1}\}$ is the set of all monomials in B of degree not greater then m. In this case the desired number is

$$0 + 1 \cdot d_r(1) + 2 \cdot d_r(2) + \cdots + m \cdot d_r(m). \tag{8}$$

If $x_{i_1}, \ldots, x_{i_r}$ do not generate a free associative algebra in r variables then it follows from the condition of linear independence of $a_1, \ldots, a_{s+1}$ that the number under consideration is greater then (8). Now recall that $d_r(m) = r^m$. It is not difficult to verify that with m tending to infinity (7) and (8), up to scalars, are equivalent to r^{m+1} and $r^m(m(r-1) - 1)$, respectively. For m sufficiently large (8) exceeds (7), a contradiction. Hence (5) and (2) are equal to zero and the result follows.   □

Corollary. *Let $L = \bigoplus_{g \in G} L_g$ be a colour Lie superalgebra over an arbitrary field. Suppose that $[L_-, L_-] = 0$. Then $J(U(L)) = U(L)L_-$. Moreover, $J(U(L))$ is a locally nilpotent algebra.*

Proof. By setting $M = L_-$ and applying the theorem one has $J(U(L)) \supset U(L)L_-$. The converse inclusion follows from the fact that $U(L)/U(L)L_- \cong U(L_+)$ is a prime algebra (2.8 of Chapter 3).   □

1.2. Example. Let $L = \Gamma_n(1, 1)$ be the Heisenberg Lie superalgebra with

$$L_0 = \langle z \rangle, \quad L_1 = \langle x_1, \ldots, x_n, y_1, \ldots, y_n \rangle, \quad [x_i, y_j] = \delta_{i,j} z,$$

where z is central and the ground field K is arbitrary. Then $J(U(L)) = 0$.

Proof. Consider a localization of $U(L)$ by $Z = \{z^i | i \in \mathbb{N}\}$ and set $\overline{K} = K(z)$, which is a transcendental extension. Then it is easy to verify that $U(L)_Z$ is isomorphic to the Clifford algebra B of the vector space $V = \langle a_1, \ldots, a_n, b_1, \ldots, b_n \rangle_{\overline{K}}$ with a bilinear form

$$\Phi: V \times V \to \overline{K}, \quad \Phi(a_i, b_j) = \delta_{i,j} z, \quad \Phi(a_i, a_j) = \Phi(b_i, b_j) = 0.$$

Since Φ is non-degenerate B is a simple algebra. This implies $J(U(L)) = 0$ because Z is a central set.   □

1.3. The Generic Flatness Lemma. Let n be an integer. For any tuple $v = (v_1, \ldots, v_n) \in \mathbb{N}^n$ we set $|v| = v_1 + \cdots + v_n$. There exists a unique total order $\leq$ on $\mathbb{N}^n$ such that: 1. $v \leq \mu$ provided that $|v| < |\mu|$; 2. among all elements with $|v|$ fixed $\leq$ coincides with the lexicographical order. It is easy to see that for any $v, v', v'' \in \mathbb{N}^n$ one has $v \leq v'$ if, and only if, $v + v'' \leq v' + v''$.

We will also use another order on $\mathbb{N}^n$.

Proposition. *Introduce on $\mathbb{N}^n$ a partial order which is equal to the product of natural orders on the factors, i.e. $(a_1, \ldots, a_n) \prec (b_1, \ldots, b_n)$ if, and only if, $a_i \leq b_i$ for all $i = 1, 2, \ldots, n$. Suppose that $S \subset \mathbb{N}^n$ and let S_0 be the set of minimal elements of S. Then S_0 is a finite set and every element in S is larger than some element in S_0.*

Proof. Suppose that the contrary holds. Then S_0 may be written as an infinite sequence of pairwise distinct elements $(s_1, s_2, \ldots)$. Obviously, any infinite sequence of integers contains an ascending subsequence. Applying this fact to the first component of elements $s_1, s_2, \ldots$, to the second component, etc., we find an infinite ascending subsequence in $(s_1, s_2, \ldots)$, a contradiction. Hence S_0 is finite. The second claim is immediate. □

Lemma. *Let A be a commutative principal ideal domain, B a finitely generated commutative A-algebra with unity and M some B-module of finite type. Then there exists an element $0 \neq f \in A$ such that $M \otimes_A A_f$ is a free A_f-module.*

Proof. Using induction on the number of generators for M we may assume that M is a cyclic module, hence it is isomorphic to B modulo some ideal. Since any such quotient module is, in fact, a finitely generated commutative algebra, it is sufficient to prove that there exists $0 \neq f \in A$ such that $B_f = B \otimes_A A_f$ is a free A_f-module.

Let $x_1, \ldots, x_n$ be generators for the A-algebra B. For $v = (v_1, \ldots, v_n) \in \mathbb{N}^n$ it is convenient to set $x^v = x_1^{v_1} \ldots x_n^{v_n}$. Using the order introduced in the beginning of this subsection we set

$$B_v = \sum_{v' \leq v} x^{v'} A, \qquad B_v^- = \sum_{v' < v} x^{v'} A.$$

Let x_*^v stand for the image of x^v in B_v/B_v^- and I_v for the annihilator of x_*^v in A. For $i = 1, \ldots, n$ we denote by ε_i an n-tuple with all components zero except for the i-th which is 1. If $a \in I_v$ then $x^v a \in B_v^-$. Therefore, $x^{v+\varepsilon_i} a = x_i x^v a \in x_i B_v^- \subset B_{v+\varepsilon_i}^-$ and hence $I_v \subset I_{v+\varepsilon_i}$. Let Λ be the set of all $v \in \mathbb{N}^n$ such that $I_v \neq 0$. By the proposition there exist $v_1, \ldots, v_r \in \Lambda$ such that any non-zero ideal I_v contains some ideal I_{v_i}.

Pick a non-zero element $f \in I_{v_1} \cap \cdots \cap I_{v_r}$. If $v \in \Lambda$ then $f \in I_v$, hence $(B_v/B_v^-) \otimes_A A_f = 0$. If $v \notin \Lambda$ then B_v/B_v^- is isomorphic to A and therefore $(B_v/B_v^-) \otimes_A A_f$ is isomorphic to A_f. Thus we have shown that B_f is a multiple extension of free A_f-modules, hence it is a free A_f-module itself. □

1.4. Centralizers of irreducible modules.

Proposition. *Let C be a K-algebra endowed with an ascending filtration C^i, $i \in \mathbb{N}$. Suppose that its associated graded algebra $\operatorname{gr} C$ is isomorphic to the ring of colour superpolynomials in a finite number of variables with a finite group*

§ 1. The Jacobson radical of universal enveloping algebras

G. *Let M be a simple C-module and $D = \mathrm{End}_C M$ be an endomorphism ring (it is a skew field by Schur's Lemma). Then all elements $x \in D$ are algebraic over the ground field K.*

Proof. By way of contradiction suppose that $x \in D$ is transcendental over K. Then the algebra $A = K[x] \subset D$, generated by K and x, is isomorphic to the polynomial ring in one variable x over K.

Denote $V = A \otimes_K C$. Then M can be uniquely endowed with the structure of a V-module such that $(p \otimes c)m = pcm = cpm$ for all $m \in M$, $p \in A$, $c \in C$. Pick some $0 \neq m_0 \in M$. Then $M = Vm_0$. For $r \in \mathbb{N}$ we set $V_r = A \otimes C_r$ and $M_r = V_r m_0$. Then M is a filtered module over a filtered ring V. It is obvious that the graded module $\mathrm{gr}\, M$ over $\mathrm{gr}\, V$ is cyclic. By hypothesis,

$$\mathrm{gr}(V) \cong A \otimes K^\varepsilon[X_1, \ldots, X_n] \otimes \Lambda^\varepsilon[Y_1, \ldots, Y_l],$$

where the second and the third factor are colour polynomial and colour Grassmann algebras in finitely many variables, respectively. We have the relation $X_i X_j = \varepsilon(\alpha, \beta) X_j X_i$ provided that X_i, X_j have weights $\alpha, \beta \in G$ respectively. Moreover $(\varepsilon(\alpha, \beta))^N = 1$, $N = |G|$, for all $\alpha, \beta \in G$. Therefore, the subalgebra $B \subset \mathrm{gr}(V)$, generated by A and $X_1^N, \ldots, X_n^N$, is isomorphic to an ordinary polynomial ring. The B-module $\mathrm{gr}\, M$ is finitely generated because it possesses the following finite set of generators:

$$\{X_1^{a_1} \ldots X_n^{a_n} Y_1^{b_1} \ldots Y_l^{b_l} m_0 | 0 \leq a_i < N, 0 \leq b_l \leq 1\}.$$

By Lemma 1.3 there exists $0 \neq f \in A$ such that $\mathrm{gr}(M) \otimes_A A_f$ is a free A_f-module. Since A_f is the principle ideal domain all $(M_r/M_{r-1}) \otimes_A A_f$ are free A_f-modules. Then $M \otimes_A A_f$ is a multiple extension of free A_f-modules, so it is a free A_f-module itself.

Pick any $0 \neq g \in A$ which does not divide a power of f. Then the mapping $A_f \to A_f$ defined as mutliplication by g, is not surjective. Therefore another homotety v with coefficient g of the free A_f-module $M \otimes_A A_f$ is also not surjective. On the other hand $v(m \otimes a) = m \otimes ga = gm \otimes a$ for any $m \in M$, $a \in A_f$. Since D is a field the mapping $m \mapsto gm$ of the module M into itself is bijective. This contradiction yields the result. □

Corollary. *Let G be a finite group and let $L = \bigoplus_{g \in G} L_g$ be a finite-dimensional colour Lie superalgebra over an arbitrary field K. Then for any simple L-module M the elements of the endomorphism skew field $D = \mathrm{End}_L M$ are algebraic over K.*

Proof. The assertion follows immediately using the standard filtration for $C = U(L)$ (see 2.8 of Chapter 3). □

1.5. Let $A = \bigoplus_{g \in G} A_g$ be a graded associative algebra. Recall that a *graded irreducible A-module* is a graded A-module having no proper non-zero graded submodules.

Lemma. *Let G be a finite group and let $A = \bigoplus_{g \in G} A_g$ be a graded associative algebra with unity. Then the dimensions of all irreducible A-modules are bounded uniformly if, and only if, the dimensions of all graded irreducible A-modules are bounded uniformly.*

Proof. Suppose that V is an irreducible A-module. Consider a collection of vector spaces W_g, $g \in G$ with attached K-isomorphisms $\tau_g \colon W_g \to V$, $g \in G$. Set $W = \bigoplus_{g \in G} W_g$. We define the structure of a graded A-module on W as follows:

$$a \circ w = \tau_{\alpha+g}^{-1}(a \circ \tau_g(w)); \qquad a \in A_\alpha, \qquad w \in W_g; \qquad \alpha, g \in G.$$

The mappings τ_g, $g \in G$, can be extended to an epimorphism $\tau \colon W \to V$, which is a homomorphism of (non-graded) A-modules. Pick any homogeneous element $0 \neq w \in W$; let S be a graded A-module generated by w, and let T be a maximal graded A-module not containing w. Then, evidently, $U = (S + T)/T$ is a graded irreducible A-module. Since V is irreducible and $\tau(w) \neq 0$ one has $\tau(T) = 0$ and $\tau(S + T) = V$. So, we obtain an epimorphism $\bar{\tau} \colon U \to V$ of (non-graded) A-modules. This argument proves the statement in one direction.

For the converse we suppose that $W = \bigoplus_{g \in G} W_g$ is a graded irreducible A-module. Let $\pi_g \colon W \to W_g$, $g \in G$, be the natural projections. Pick any homogeneous element $0 \neq w \in W_\alpha$. Then $\tilde{W} = \bigoplus_{g \in G} A_g w$ is a non-zero graded submodule, hence $A_g w = W_{\alpha+g}$. Thus the W_g, $g \in G$, are irreducible A_0-modules and the A_0-module W is of finite type. This implies that there exists a minimal (non-graded) A-submodule $V \subset W$. Suppose that $0 \neq v = \sum_{\beta \in G} v_\beta \in V$, $v_\beta \in V_\beta$, with some v_α, $\alpha \in G$, being non-zero. The argument above implies that $\pi_{\alpha+g}(A_g v) = W_{\alpha+g}$. Therefore $\pi_g(V) = W_g$ for all $g \in G$. Now we conclude that $\dim W = \sum_{g \in G} \dim W_g \leq |G| \dim V$. This proves the converse implication. $\square$

§2. Dimensions of irreducible representations

The methods of Chapter 4 prove to be useful for the study of irreducible representations. The main goal of this section is to prove the following two theorems.

§2. Dimensions of irreducible representations

2.1. Theorem. *Let G be a finite group and let $L = \bigoplus_{g \in G} L_g$ be a colour Lie superalgebra over an algebraically closed field K of positive characteristic. Then the dimensions of all irreducible L-representations are bounded by some finite number if, and only if, there exists a homogeneous ideal $R \subset L$ such that:*
1) $\dim L/R < \infty$;
2) $R^2 \subset R_-$;
3) *All inner derivations* $\text{ad}\, x|_{L_+}$, $x \in L_g$, $g \in G_+$, *are algebraic of bounded degree*;
4) $\dim_K L_+ < |K|$.

Note that 2) implies that $R^2 = [R_+, R_-]$. It is easy to see that the ideals R and R^2 in the theorem are analogous to the ideals A and B in Corollary 1.9 in Chapter 4 on the existence of an identity in universal enveloping algebras for the modular case. An essential difference to Corollary 1.9 is that we do not require, first, that $[R_+, R_-] < \infty$, and, second, that $\text{ad}\, x|_{L_-}$, $x \in L_g$, $g \in G_+$ is algebraic. In other words, restrictions on the structure of the L_+-module L_- are lifted. The sufficiency conditions in the theorem will be proved with R being a subalgebra rather than an ideal.

Example 1.6 of Chapter 4 explains also why we restrict ourselves to the case of G being finite.

Using Lemma 1.5, in this theorem by "irreducible representations" we mean either graded irreducible representations or ungraded irreducible representations. These remarks hold for the next result as well.

2.2. Theorem. *Let G be a finite group and let $L = \bigoplus_{g \in G} L_g$ be a colour Lie superalgebra over an algebraically closed field K of characteristic zero. Then the dimensions of all irreducible L-representations are bounded by some finite number if, and only if, there exists a homogeneous L_+-submodule $M \subset L_-$ such that:*
1) $\dim L_-/M < \infty$;
2) L_+ *is abelian and* $[M, M] = 0$;
3) $\dim_K L_+ < |K|$.

The main and essential difference to the conditions of Theorem 1.2 of Chapter 4 on the existence of a non-trivial identity in the universal enveloping algebra is that we do not need the condition $\dim[L_+, M] < \infty$.

2.3. Lemma. *Let L be a colour Lie superalgebra over an arbitrary field. Suppose that $\dim L_+ < \infty$ and that there exists a homogeneous subspace $M \subset L_-$ of finite codimension with $[M, M] = 0$. Then their exists also an L_+-submodule $\tilde{M} \subset L_-$ of finite codimension with $[\tilde{M}, L_-] = 0$.*

Proof. Since $[M, M] = 0$ we may consider $M \subset L$ as a subalgebra. Let N be the kernel of the natural action of M on the finite-dimensional space L/M. Then, evidently, $\dim L/N < \infty$ and $[N, L_-] \subset L_+ \cap M = 0$. We claim that the L_+-submodule $\tilde{M} \subset L_-$ generated by N is as required. Indeed, for any homogeneous $x \in L_+, a \in N, y \in L_-$, one has

$$[[x, a], y] = \lambda[[x, y], a] + [x, [a, y]] \subset [L_-, N] + [x, [N, L_-]] = 0.$$

Hence $[\tilde{M}, L_-] = 0$. □

2.4. *Proof of the sufficiency in Theorem 2.1.* In this proof we will not assume that $R \subset L$ is an ideal, as in the wording of the theorem, but merely a subalgebra. Choose a homogeneous basis $L_+ = \langle a_1, \ldots, a_m \rangle \oplus R_+$. By hypothesis each $\text{ad } a_j|_{L_+}, j = 1, \ldots, m$, is algebraic. By Proposition 1.9 of Chapter 4 we may assume without loss of generality that $(|G|, p) = 1$. Choose k such that $p^k \equiv 1 \pmod{|G|}$. Lemma 1.9 of Chapter 4 shows that these operators are annihilated by some p^k-polynomials $f_1(X), \ldots, f_m(X)$. Notice that each $z_j = f_j(a_j), j = 1, \ldots, m$, is homogeneous. We have $0 = f_j(\text{ad } a_j)|_{L_+} = \text{ad}(f_j(a_j))|_{L_+}$, so $z_j = f_j(a_j), j = 1, \ldots, m$, is supercentral in $U(L_+)$.

By condition 2) R_+ is abelian. Therefore the associative subalgebra $\tilde{D} \subset U(L_+)$ generated by $U(R_+)$ together with $z_1, \ldots, z_m$ is isomorphic to a colour polynomial algebra. Then Corollary 1.6 yields that $\tilde{D}$ satisfies some non-trivial identity.

Let D be the algebra generated by $\tilde{D}$ and R_-. It is not difficult to see that D is isomorphic to the universal enveloping algebra of some colour Lie superalgebra. Moreover $D/(D \cdot R_-) \cong \tilde{D}$ and, by Corollary 1.1, $J(D) = D \cdot R_-$ because $[R_-, R_-] = 0$. Observe that $_D U(L)$ is a free module of finite rank $n = \deg f_1 \cdot \ldots \cdot \deg f_m \cdot 2^{\dim L_-/R_-}$. A standard argument shows that there exists an embedding $U(L) \subset M_n(D)$. Thus we have an ideal $I = M_n(J(D)) \cap U(L)$ in $U(L)$. This ideal is nil because $J(D)$ is a locally nilpotent algebra by Corollary 1.1. Consequently I lies in $J(U(L))$. Hence $A = U(L)/J(U(L))$ is a homomorphic image of $U(L)/(M_n(J(D)) \cap U(L)) \subset M_n(D)/M_n(J(D)) \cong M_n(\tilde{D})$ since $D/J(D) \cong D/D \cdot R_- \cong \tilde{D}$.

By Theorem 1.5 of Chapter 4 we conclude that A is a PI-algebra. Each simple $U(L)$-module is also a simple A-module. Let $\rho: A \to \text{End } V$ be an irreducible representation and $k = \text{End}_A V$ its endomorphism ring. By Lemma 1.3 of Chapter 4 $\dim_k V \leq N$ for some integer N which is the same for all ρ.

Now it remains to prove that $k = K$. By way of contradiction suppose that $\lambda \in k$ is a transcendental element over K. It is easy to verify that $K(\lambda)$ contains a linearly independent set $\{(\lambda - \alpha)^{-1} | \alpha \in K\}$ of cardinality $|K|$. We have $V \cong A/I$ for some maximal left ideal I. Suppose that B is a maximal subalgebra of A containing I as two-sided ideal. It is well-known that $B/I \cong k$

[Jacobson, 1956]. One has the following bounds for dimensions over K

$$|K| \leq \dim K(\lambda) \leq \dim k \leq \dim A \leq \dim M_n(\tilde{D}) = \dim \tilde{D}, \tag{1}$$

the latter equality being true since $\tilde{D}$ is infinite-dimensional. If $\dim L_+ = \infty$ then $\dim \tilde{D} = \dim L_+$, and this fact together with (1) contradicts condition 4) of the theorem.

Now it remains to settle the case of L_+ being finite-dimensional. By applying Lemma 2.3 to $M = R_-$ one arrives at an L_+-submodule $\tilde{M} \subset L_-$ of finite codimension with $[\tilde{M}, L_-] = 0$. By Theorem 1.1 $U(L) \cdot \tilde{M} \subset J(U(L))$ and the proof reduces to the study of the finite-dimensional algebra $L/\tilde{M}$.

By Corollary 1.4, in the case of finite-dimensional colour Lie superalgebras, the endomorphism field of any simple module is algebraic over the ground field. Recall that, by hypothesis, K is algebraically closed. Thus $k = K$, and the proof is complete. □

In what follows we will use a lemma, in which K is not necessarily an algebraically closed field.

Lemma. *Let L be a finite-dimensional colour Lie superalgebra over a field of positive characteristic. Then all simple L-modules are finite-dimensional.*

Proof. Let $L_+ = \langle a_1, \ldots, a_m \rangle$ be a homogeneous basis. As above, we construct in $U(L)$ certain supercentral elements $z_j = f_j(a_j), j = 1, \ldots, m$. Then, for $N = |G|$, the elements $v_j = z_j^N, j = 1, \ldots, m$, are central in $U(L)$. Suppose that $\rho: L \to \text{End } V$ is an irreducible representation and let $k = \text{End}_L V$. Then $\rho(v_j) = \lambda_j \cdot 1, \lambda_j \in k$. By Corollary 1.4 each λ_j is a root of some polynomial $g_j(X)$ over K. Thus ρ is in fact a representation of the finite-dimensional algebra $U(L)/(g_j(v_j)|j = 1, \ldots, m)$, and the proof is complete. □

2.5. *Proof of the sufficiency in Theorem 2.2.* We consider subalgebras $\tilde{D} = U(L_+)$ and $D = U(L_+ \oplus M)$, then $J(D) = D \cdot M$. Now $_D U(L)$ is a free module of finite rank n; thus there exists an embedding $U(L) \subset M_n(D)$. To complete the proof we apply an argument which is similar to that used in the preceding subsection. □

Lemma. *Let L be an abelian finite-dimensional colour Lie superalgebra over a field K of characteristic zero. Then all simple L-modules are finite-dimensional.*

Proof. Set $N = |G|$; let $L_+ = \langle a_1, \ldots, a_m \rangle$ be a homogeneous basis. Then we apply the argument as in the lemma of the subsection above to $v_j = a_j^N, j = 1, \ldots, m$. □

2.6. Lemma. *Let A be a subalgebra with unity in an associative algebra B with the same unity over a field K. Suppose that B_A is free. Moreover, assume we have a decomposition $B_A = \bigoplus_{j \in J} b_j A \oplus 1 \cdot A$.*
1) If the degrees of all simple B-modules are bounded by some finite number, then the same number bounds the dimensions of all simple A-modules.
2) $J(B) \cap A \subset J(A)$.

Proof. 1) Let $_A V$ be a simple module. We claim that there exists a simple module $_B T$ with $\dim T \geq \dim V$. We construct an induced B-module $W = B \otimes_A V$, where V is identified with $1 \otimes V \subset W$. Pick $0 \neq v \in V$ and suppose that W is a B-submodule generated by v. Since $_A V$ is simple and $A \subset B$, this cyclic module coincides with W. Suppose that M is a maximal B-submodule in W non-containing v. The irreducibility of $_A V$ implies that $M \cap V = 0$. Evidently, $T = W/M$ is a simple B-module. On the other hand, for A-modules, one has

$$V \cong V/(V \cap M) \cong (V + M)/M \subset W/M = T,$$

completing the proof.

2) Suppose that $r \in J(B) \cap A$. Then for any $s \in A$ there exists an inverse element t for $x = 1 - rs$. By hypothesis $t = \sum_j b_j a_j + a$, $a_j, a \in A$. Then $1 = tx = \sum_j b_j(a_j x) + ax = ax$ because $x \in A$ and since B_A is a free module. So, $t = a \in A$ and $r \in J(A)$ by definition of the Jacobson radical. □

If $H \subset L$ is a homogeneous subalgebra in a colour Lie superalgebra L, then by 2.2 of Chapter 3 $U(L)_{U(H)}$ is a free module and one can apply the lemma. The same is true for restricted enveloping algebras.

2.7. Lemma. *Let K be an algebraically closed field, $A = K[X]$ an ordinary polynomial ring in some set X of variables. Then A has an infinite-dimensional irreducible representation if, and only if, $|X| \geq |K|$. Moreover, if $|X| < |K|$ then all simple A-modules are finite-dimensional.*

Proof. Note that if M is a simple A-module, then $M \cong A/I$ for some maximal ideal I in A. Consequently, M is a field and finitely generated as K-algebra provided that X is finite. In this case by Corollary 1.4 M is algebraic over K, so $M = K$. Thus, without loss of generality, we may assume that X is infinite. Observe that

$$\dim M = \dim(A/I) \leq \dim A = |X|.$$

Suppose $u \in A/I$ and set $B = K(u)$. We claim that if u is transcendental over K then $\dim B = |K|$. Indeed, $K(u)$ contains $\{(u - \alpha)^{-1} | \alpha \in K\}$ which is linearly independent and has cardinality $|K|$. So $\dim B \geq |K|$. The reverse inequality follows from the fact that B, as an algebra, is generated by the set $R = \{u, (u - \alpha)^{-1} | \alpha \in K\}$ whose cardinality is equal to $|K|$. Thus, for any irreducible representation $M = A/I$, if $|X| < |K|$ then every $u \in A/I$ is algebraic over K, i.e. $\dim_K(A/I) = 1$.

Now if $|X| \geq |K|$ then any mapping of X onto R can be extended to an epimorphism of $A = K[X]$ onto $B = K(u)$. This implies that A has an infinite-dimensional irreducible representation. □

If L is an abelian Lie algebra, then $U(L)$ is isomorphic to a polynomial ring and one can apply the lemma in Section 2.5.

Corollary. *Let G be a finite group and let $L = \bigoplus_{g \in G} L_g$ be an abelian colour Lie algebra ($G = G_+$) over an algebraically closed field K. Suppose that the dimensions of all simple L-modules are bounded by some finite number. Then $\dim L_K < |K|$.*

Proof. Let $L = \langle X_\alpha | \alpha \in \Xi \rangle$ be a homogeneous basis. Remark that $X_\alpha^{|G|}$, $\alpha \in \Xi$ are central in $U(L)$. They generate an algebra A isomorphic to a polynomial ring and the module $U(L)_A$ is free. It remains to apply Lemma 2.6. □

2.8. Now we introduce some more delta-sets which are useful for handling problems is this section.

Definition. Let $L = \bigoplus_{g \in G} L_g$ be a colour Lie superalgebra, $\alpha \in G_-$, $m \in \mathbb{N}$. We set
1) $D_\alpha^m(L) = \bigcap_{\beta \in G_-} \delta_{\alpha,\beta}^m(L)$,
2) $D_\alpha(L) = \bigcup_{m \in \mathbb{N}} D_\alpha^m(L)$,
3) $D(L) = \bigoplus_{\alpha \in G_-} D_\alpha(L)$.

If it is clear what superalgebra is being dealt with, the argument L will be omitted. Obviously the set $D_\alpha^m \subset L_\alpha$, $\alpha \in G_-$ is analogous to $\Delta_\alpha^m \subset L_\alpha$, $\alpha \in G$ of Subsection 2.2 of Chapter 4.

Lemma. *Suppose that $\alpha \in G_-$.*
1) *If $x_1, \ldots, x_n \in D_\alpha^m$ then $\lambda_1 x_1 + \cdots + \lambda_n x_n \in D_\alpha^{nm}$ for any $\lambda_1, \ldots, \lambda_n \in K$.*
2) *If $x \in D_\alpha^m$ and $y \in L_\beta$, $\beta \in G_+$ then $[y, x] \in D_{\alpha+\beta}^{2m}$.* □

(The proof follows from the same inclusions as in Subsection 2.2 of Chapter 4.)

In the general case, $D_\alpha^m \subset L_\alpha$, $\alpha \in G_-$, are no subspaces, but according to the lemma the $D_\alpha(L)$ are subspaces. Hence, equation 3) in our definition is unambiguous. This lemma implies the following.

Corollary. *The subspace $D = \bigoplus_{\alpha \in G_-} D_\alpha \subset L_-$ is a homogeneous L_+-submodule.* □

2.9. Proposition. *Let G be a finite group and let $L = \bigoplus_{g \in G} L_g$ be a colour Lie superalgebra over a field K. Suppose the dimensions of all irreducible L-modules are bounded by some finite number. Then there exists a homogeneous L_+-submodule $M \subset L_-$ such that*
1) $\dim L_-/M < \infty$,
2) $[M, M] = 0$.

Proof. There exists an embedding of $R = U(L)/J(U(L))$ into the Cartesian product of primitive algebras. Since the dimensions of all simple modules are bounded, each of these primitive algebras can be embedded in a matrix algebra over K of fixed degree. By Kaplansky's Theorem R satisfies some non-trivial identity. Fix some $\alpha, \beta \in G_-$; $\theta = \varepsilon(\alpha, \beta) \in K$. Recall that in this case $\alpha + \beta \in G_+$. By Proposition 1.11 of Chapter 4 there exists an n independent of θ such that $U(L)$, modulo $J(U(L))$, satisfies some non-trivial θ-identity. Let us substitute in it arbitrary elements $x_1, \ldots, x_n \in L_\alpha$, $y_1, \ldots, y_n \in L_\beta$. Observe that $[x_i, y_j] \in L_{\alpha+\beta} \subset L_+$. Therefore one has

$$\sum_{\pi \in \mathrm{Sym}(n)} \lambda_\pi [x_1, y_{\pi(1)}]_\theta \cdot \ldots \cdot [x_n, y_{\pi(n)}]_\theta = 0, \qquad \lambda_\pi \in K, \tag{2}$$

since the left hand side belongs to $U(L_+) \cap J(U(L)) \subset J(U(L_+))$ by Lemma 2.6 while $J(U(L_+)) = 0$ (2.8 of Chapter 3). Now we are in the situation of Theorem 2.3 of Chapter 4. By this theorem any n elements in L_α are linearly dependent modulo $\delta_{\alpha,\beta}^{n^2}$. Since the pair $\alpha, \beta \in G_-$ was taken arbitrarily this statement holds for any such pair.

This, in particular, implies that in order to get a linear span $\langle \delta_{\alpha,\beta}^{n^2} \rangle$, $\alpha, \beta \in G_-$ we can only take sums of at most n elements; by Lemma 2.2 of Chapter 4 we have $\langle \delta_{\alpha,\beta}^{n^2} \rangle \subset \delta_{\alpha,\beta}^{n^3}$. Moreover, it is evident that $\dim L_\alpha / \langle \delta_{\alpha,\beta}^{n^2} \rangle < n$. For any $\alpha \in G_-$ we set $W_\alpha = \bigcap_{\beta \in G_-} \langle \delta_{\alpha,\beta}^{n^2} \rangle \subset L_\alpha$. Then $\dim L_\alpha/W_\alpha < |G|n$ and by Lemma 2.8 $W_\alpha \subset D_\alpha^{n^3}$. Denote $W = \bigoplus_{\alpha \in G_-} W_\alpha$. We want to show that $W \subset L_-$ may be considered as L_+-submodule. If it is not the case we take another subspace $\tilde{W} = \bigoplus_{\alpha \in G_-} \tilde{W}_\alpha$ as follows:

$$\tilde{W} = \langle [x, y] | x \in L_\alpha, \alpha \in G_+, y \in W_\beta, \beta \in G_- \rangle.$$

Again we can restrict ourselves to the sums of at most n elements, by Lemma 2.8, and one has $\tilde{W}_\alpha \subset D_\alpha^m$, $\alpha \in G_-$, for some m. Note that $\dim L_-/\tilde{W} < \dim L_-/W$ and after finitely many steps we get the desired L_+-submodule W.

Now the bilinear mappings $W_\alpha \times W_\beta \to L_{\alpha+\beta}$, $\alpha, \beta \in G_-$ satisfy the conditions of Theorem 2.1 (Chapter 4) and, as a result, $\dim[W, W] < \infty$. Now we want to get rid of the finite-dimensional space $V = [W, W] \subset L_+$. Let

$\{v_1, \ldots, v_t\}$ be a homogeneous basis of V. Suppose that s, $1 \leq s \leq t$, is fixed and $\rho_s \colon V \to K$ is the operator which maps $v \in V$ to the coefficient of v_s. For any $\alpha, \beta \in G_-$ we introduce bilinear mappings

$$\psi^s_{\alpha,\beta} \colon W_\alpha \times W_\beta \to K, \qquad \psi^s_{\alpha,\beta}(x,y) = \rho_s([x,y]), \qquad x \in W_\alpha, \qquad y \in W_\beta.$$

We claim that $\dim W_\alpha/\operatorname{Ker}_l \psi^s_{\alpha,\beta} < n$ and that the same bound holds for the right kernel.

For, otherwise there exist $a_1, \ldots, a_n \in W_\alpha$, $b_1, \ldots, b_n \in W_\beta$ such that $\psi^s_{\alpha,\beta}(a_i, b_j) = \delta_{i,j}$. After substituting into (2), with $\theta = \varepsilon(\alpha, \beta)$, we get

$$\sum_{\pi \in \operatorname{Sym}(n)} \lambda_\pi [a_1, b_{\pi(1)}]_\theta \cdot \ldots \cdot [a_n, b_{\pi(n)}]_\theta = 0. \tag{3}$$

The identity permutation, as one of the terms in (3), yields v_s^n, which cannot be cancelled because each monomial of degree n in (3) contains a factor v_j, $j \neq s$. This is a contradiction, and the desired bounds for the codimensions of the kernels have been established. Now the subspace $M = \bigoplus_{\alpha \in G_-} M_\alpha$, where $M_\alpha = \bigcap_{1 \leq s \leq t, \beta \in G_-} \operatorname{Ker}_l \psi^s_{\alpha,\beta}$, is as required. Indeed, by construction $[M, M] = 0$ and $\dim L_-/M < \infty$. Notice that M is a kernel of an L_+-invariant bilinear map $\Psi \colon W \times W \to V$; hence $M \subset L_-$ is an L_+-submodule. Now the proof is complete. $\square$

2.10. Lemma. *Let L be a colour Lie superalgebra. Suppose that $R \subset L$ is a homogeneous subalgebra of finite codimension with $[L_+, R] \subset R$. Then there exists a homogeneous ideal $\tilde{R} \subset L$ of finite codimension lying in R.*

Proof. We argue as in the proof of Lemma 1.10, Chapter 4. We define a descending sequence of homogeneous subalgebras R^i, $i \in \mathbb{N}$ as follows. Start with $R^0 = R$. Suppose that all R^j, $0 \leq j \leq i-1$ have been defined. Then R^i is the kernel of the adjoint action of R^{i-1} on L/R^{i-1}. Obviously, the codimension is finite. One has

$$[[L_+, R^i], L] \subset [L, R^i] + [L_+, [R^i, L]] \subset R^{i-1} + [L_+, R^{i-1}] \subset R^{i-1},$$

hence $[L_+, R^i] \subset R^i$, provided that R^{i-1} is L_+-invariant. By induction it follows that each R^i, $i \in \mathbb{N}$ is an L_+-invariant subalgebra.

Since $R^i = \{x \in L \mid x \in R^{i-1}, [L_-, x] \subset R^{i-1}\}$ a further induction shows that

$$R^i = \{x \in L \mid x \in R, [L_-, x] \subset R, \ldots, [L_-, \ldots, L_-, x] \subset R\},$$

where the latter commutator is i-fold. In other words

$$R^i = \{x \in L \mid x \in R, [y_1, x] \in R, \ldots, [y_1, \ldots, y_i, x] \in R, \forall y_i \in L\}.$$

Let $v_1, \ldots, v_n \in L_-$ be a homogeneous basis for L_- modulo R_-. As in the proof of the Poincaré-Birkhoff-Witt Theorem it is easy to show that R^i consists of x such that

$$x \in R, \quad [v_{\alpha_1}, x] \in R, \quad [v_{\alpha_1}, v_{\alpha_2}, x] \in R, \ldots,$$

$$[v_{\alpha_1}, \ldots, v_{\alpha_i}, x] \in R, \quad 1 \leq \alpha_1 < \alpha_2 < \cdots < \alpha_i \leq n.$$

After at most n steps our chain must stabilize. Thus $\tilde{R} = R^n$ is an ideal, as required. □

2.11. *Proof of necessity in Theorem* 2.1. First we consider the case of colour Lie algebras, i.e. $L = L_+$. By 2.8 of Chapter 3 $J(U(L)) = 0$. As in Subsection 2.9 $U(L)$ is embedded in the Cartesian product of matrix rings of bounded degree over K. Therefore, $U(L)$ satisfies some non-trivial identity. By Corollary 1.9 of Chapter 4 there exists an abelian ideal $H \subset L$ of finite codimension and all inner derivations $\mathrm{ad}\,x|_L$, $x \in L_g$, $g \in G_+$, are algebraic. By applying Lemma 2.6 and Corollary 2.7 one derives $\dim L = \dim H < |K|$.

Now we treat the general case. Combining 2.6 with the argument above for $L_+ \subset L$ one has an abelian ideal H in L_+ with $\dim L_+/H < \infty$; moreover, Claims 3), 4) of the theorem are thereby proved. On the other hand, Proposition 2.9 yields an L_+-submodule $M \subset L_-$. Then $R = H \oplus M$ is a subalgebra satisfying Claims 1), 2) of the theorem, although R need not be necessarily an ideal in L. In this case we remark that, by construction, $[L_+, R] \subset R$ and using Lemma 2.10 R can be replaced by an ideal $\tilde{R} \subset R$ of finite codimension in L. Now the theorem is completely proved. □

2.12. *Proof of the necessity in Theorem* 2.2. Applying Theorem 4.3 of Chapter 4 to $L_+ \subset L$ we conclude that L_+ is abelian since $U(L_+)$ satisfies some non-trivial identity as in the subsection above. It follows from 2.6 and 2.7 that $\dim L_+ < |K|$. An L_+-submodule $M \subset L_-$ was found in 2.9 and the theorem is proved. □

§ 3. More on restricted enveloping algebras

In this section we study restricted enveloping algebras $u(L)$ with L being a colour Lie p-superalgebra. If L is finite-dimensional then $u(L)$ provides an interesting example of a finite-dimensional Frobenius algebra (Lemma 3.4). Our main goal is to specify those L where $u(L)$ is a regular ring (Theorem 3.9). We also prove that $u(L)$ is a self-injective ring if, and only if, L is finite-dimensional (Theorem 3.10).

3.1. Regular rings. For a ring R with unity the following properties are equivalent:
1. For any $a \in R$ there exists $x \in R$ with $axa = a$.
2. Each left principal ideal I of R is generated by an idempotent (i.e. I is a direct summand of R as a left R-module).
3. Each finitely generated left ideal is generated by an idempotent.

Such rings are called (*von Neumann*) *regular*. Obviously, one can complete the list above by properties 2'., 3'. which are the same as 2., 3. with "left" replaced by "right".

Lemma. *Let R be a regular ring and $r_1, \ldots, r_n \in R$. Suppose that $Rr_1 + \cdots + Rr_n \neq R$. Then there exists $b \in R$, $b \neq 0$ such that $r_1 b = \cdots = r_n b = 0$.*

Proof. We set $J = Rr_1 + \cdots + Rr_n$. Then $J \neq R$ and $J = Re$ for some idempotent $e \neq 1$. Notice that J is annihilated on the right by $b = 1 - e \neq 0$. □

3.2. Injective modules. A module Q is called *injective* if any diagram

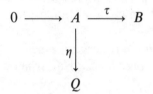

with exact upper sequence can be extended to a commutative diagram by some homomorphism $\sigma \colon B \to Q$. This definition is equivalent to the following: injective modules are those ones which, being embedded into some other module, can be directly complemented in this latter.

A criterion, due to R. Baer, claims that Q is injective provided that the above diagram is extendable if $A = I$, $B = R$, where I is a left ideal in R.

Lemma. *Let $_R V$ be an injective module and $r_1 V + r_2 V + \cdots + r_n V \neq V$ for some $r_1, \ldots, r_n \in R$. Then $r_1, \ldots, r_n$ have a common non-zero left annihilator.*

Proof. Consider the diagram

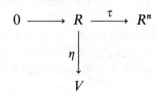

where $\tau: R \to R^n$ is defined as follows: $\tau(r) = (rr_1, \ldots, rr_n)$, $r \in R$, and $\eta(r) = rv$ for some $0 \neq v \notin r_1 V + r_2 V + \cdots + r_n V$. By way of contradiction suppose that $r_1, \ldots, r_n$ do not have any common left annihilator. Then τ is injective and the diagram can be extended to a commutative one by some homomorphism $\sigma: R^n \to V$. We denote $e_i = (0, \ldots, 0, 1, 0, \ldots, 0) \in R^n$ with 1 on the i-th place. Then

$$v = \eta(1) = \sigma\left(\sum_{i=1}^n r_i e_i\right) = \sum_{i=1}^n r_i \sigma(e_i) \in \sum_{i=1}^n r_i V,$$

a contradiction. □

Definition. Let A be an associative algebra. If $_A A$ is injective then A is called a (*left-*) *self-injective* algebra.

3.3. Locally Wedderburn algebras.

Definition. Let A be an associative algebra with unity over K. Then A is called *locally Wedderburn* if every finite set of elements in A is contained in a semisimple subalgebra which is finite-dimensional over K.

Lemma. *Let A be a locally Wedderburn algebra. Then every simple module which is finite-dimensional over its centralizer must be injective.*

Proof. Let V be a simple A-module which is finite-dimensional over its centralizer C. If V is not injective, by Baer's criterion, there exists a left ideal $I \subset A$ and a homomorphism of left A-modules $f: I \to A$ which cannot be lifted to A. Setting $D(B) = \{v \in V | iv = f(i), \forall i \in I \cap B\}$ for any subalgebra $B \subset A$, our assumption is $\bigcap_{B \in X} D(B) = \emptyset$ where the intersection is taken over all finite-dimensional semisimple subalgebras B. Since all modules over such algebras are injective, $D(B) \neq \emptyset$ for all $B \in X$.

Denote $\mathrm{Ann}(T) = \{v \in V | tv = 0, \forall t \in T\}$ for any subset $T \subset A$. Choose $B_0 \in X$ so that $d = \dim_C \mathrm{Ann}(I \cap B_0)$ is minimal. Since the intersection is empty there exists $B_1 \in X$ with $D(B_0) \not\subset D(B_1)$. If $B_2 \in X$ contains a basis for B_0 and B_1 then $B_0 \subset B_2$ implies $\emptyset \neq D(B_2) \subset D(B_0)$ where the inclusion is strict. Thus, if $w \in D(B_2)$ then the inclusion $D(B_2) = w + \mathrm{Ann}(I \cap B_2) \subset w + \mathrm{Ann}(I \cap B_0) = D(B_0)$ is strict, contradicting the minimality of d. □

3.4. Finite-dimensional Frobenius algebras.
For a finite-dimensional associative algebra with unity A the following conditions are equivalent.
1. There is a linear function $\Phi: A \to K$ such that $\mathrm{Ker}\,\Phi$ does not contain any non-zero left ideal.

§ 3. More on restricted enveloping algebras

1'. There is a linear function $\Phi: A \to K$ such that Ker Φ does not contain any non-zero right ideal.
2. There is a linear function $\Phi: A \to K$ such that the bilinear form $\Psi: A \times A \to K$, $\Psi(x, y) = \Phi(x \cdot y)$, $x, y \in A$ is non-degenerate.
3. $_A A \cong (A_A)^*$, where $(A_A)^*$ is the set of linear functions $f: A \to K$ satisfying $(a \cdot f)(x) = f(xa)$, $x, a \in A$.

Definition. A finite-dimensional algebra satisfying any of these conditions is called a *Frobenius algebra*.

Lemma. *Let L be a finite-dimensional colour Lie p-superalgebra. Then $u(L)$ is a finite-dimensional Frobenius algebra.*

Proof. Let $L_+ = \langle a_1, \ldots, a_n \rangle$, $L_- = \langle a_{n+1}, \ldots, a_m \rangle$, $m \geq n$, be homogeneous bases. It is conventional to denote $a^I = a_1^{i_1} \ldots a_n^{i_n} \ldots a_m^{i_m}$, where

$$I = (i_1, \ldots, i_n, \ldots, i_m); \quad 0 \leq i_j < p, \quad 1 \leq j \leq n; \quad i_j \in \{0, 1\}, \quad n < j \leq m.$$

Suppose that X is the set of all these tuples and note that we can add the tuples. Set $I_0 = (p - 1, \ldots, p - 1, 1, \ldots, 1) \in X$, where 1's are on the places $n + 1$-th onwards. An algebra $u(L)$ has a natural filtration and $a^{I_0} \in u^N$, $a^{I_0} \notin u^{N-1}$, where $N = (p-1)n + (m-n)$. One has a linear function $\Phi: u(L) \to K$ by setting Ker $\Phi = \langle a^I | I \in X, I \neq I_0 \rangle$, $\Phi(a^{I_0}) = 1$.

Suppose we are given $I, J \in X$. If $\sum_{s=1}^m i_s + j_s < N$, then it is not difficult to verify that $a_I \cdot a_J \in$ Ker Φ. If $\sum_{s=1}^m (i_s + j_s) = N$ then only in the case $I + J = I_0$ one has $a^I \cdot a^J = \mu a^{I_0} (\mathrm{mod}\, \mathrm{Ker}\,\Phi)$, $0 \neq \mu \in K$, whereas in all other cases $a^I \cdot a^J \in$ Ker Φ.

Now suppose that $0 \neq h \in$ Ker Φ and that h belongs to some non-zero left ideal. One has $h = \sum_{I \in X} \lambda_I a^I$, $\lambda_I \in K$. Let a^{I_1} be one of the leading terms of h. The above remark shows that $\Phi(a^{I_0 - I_1} h) = \mu \neq 0$. On the other hand $a^{I_0 - I_1} h \in$ Ker Φ by the choice of h. This contradiction proves the lemma. □

3.5. Lemma. *Let L be a colour Lie p-superalgebra. Suppose that $H \subset L$ is a homogeneous restricted subalgebra and V is an injective $u(L)$-module. Then V is also an injective $u(H)$-module.*

Proof. With any diagram of $u(H)$-modules

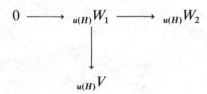

we associate a diagram of $u(L)$-modules as follows.

$$\begin{array}{ccccc} 0 & \longrightarrow & u(L) \otimes_{u(H)} W_1 & \longrightarrow & u(L) \otimes_{u(H)} W_2 \\ & & \downarrow & & \\ & & {}_{u(L)}V & & \end{array}$$

Since $u(L)_{u(H)}$ is free (2.5, Chapter 3), the horizontal line in the second diagram is exact. Now by extending the second diagram we extend the first one. □

3.6. Left and right integrals. Let A be an associative algebra and $T \subset A$ some subset. Then it is conventional to denote left and right annihilators as follows:

$$l_A(T) = \{x \in A \mid xT = 0\}; \qquad r_A(T) = \{x \in A \mid Tx = 0\}.$$

Suppose that L is a colour Lie p-superalgebra. Then $u = u(L)$ has a two-sided ideal $u^+ = u^+(L)$ of codimension one. For the annihilators of this ideal it is convenient to introduce a special notation:

$$l(L) = l_{u(L)}(u^+(L)), \qquad r(L) = r_{u(L)}(u^+(L)).$$

It is obvious that $l(L)$ and $r(L)$ are two-sided homogeneous ideals of $u(L)$.

Lemma. *Suppose that* $\dim L < \infty$. *Then* $l(L)$ *and* $r(L)$ *are two-sided one-dimensional homogeneous ideals of* $u(L)$.

Proof. We want to prove that $r(L)$ is one-dimensional. By 3.4 there is a linear function $\Phi: u \to K$ with a non-degenerate bilinear form $\Psi: u \times u \to K$, $\Psi(x, y) = \Phi(x \cdot y)$. We consider the right annihilator for Ψ given by $r_\Psi(u^+) = \{x \in u \mid \Psi(u^+, x) = 0\}$. Since Ψ is non-degenerate one has $\dim r_\Psi(u^+) = 1$. Observe that now $u^+ \cdot r_\Psi(u^+)$ is a left ideal in $\operatorname{Ker} \Phi$, and by definition of a finite-dimensional Frobenius algebra this ideal must be zero. Thus we have proved that $r(L) \supset r_\Psi(u^+)$. The converse inclusion being obvious, we derive that $r(L) = r_\Psi(u^+)$ is a one-dimensional ideal. The case of $l(L)$ can be treated similarly. □

Owing to the theory of Hopf algebras $l(L)$ and $r(L)$ are called *left* and *right integrals*. Note that in general $r(L) \neq l(L)$.

Proposition. *Let* $H \subset L$ *be a homogeneous restricted subalgebra in a colour Lie p-superalgebra L. Then the following holds.*

1) If $\dim H = \infty$, then $l_{u(L)}(u^+(H)) = r_{u(L)}(u^+(H)) = 0$.
2) If $\dim H < \infty$, then

$$l_{u(L)}(u^+(H)) = u(L)l(H),$$

$$r_{u(L)}(u^+(H)) = r(H)u(L),$$

$$r_{u(L)}(l(H)) = u^+(H)u(L),$$

$$l_{u(L)}(r(H)) = u(L)u^+(H).$$

Proof. Suppose that $x \in l_{u(L)}(u^+(H))$. Then, by 2.5 of Chapter 3, $x = \sum_i a_i h_i$, where $h_i \in u(H)$ and a_i are standard monomials of a homogeneous basis for L modulo H. It follows that $h_i \in l(H)$, and this implies that $x \in u(L) \cdot l(H)$. If H is finite-dimensional then this proves the first equality of Claim 2 because the inverse inclusion is obvious.

Now suppose that $\dim H = \infty$. Only finitely many basis elements for H enter finitely many elements h_i taken from the above decomposition. By the hypothesis there is a basis element y for H which is different from these basis elements. One has $0 = xy = \sum_i a_i h_i y$. The basis for L may be considered appropriately ordered, and then the latter equality implies $x = 0$. Thus $l_{u(L)}(u^+(H)) = 0$.

The other equalities can be obtained by similar arguments. □

3.7. Proposition. *Let L be a finite-dimensional colour Lie p-superalgebra over an arbitrary field K. Suppose that the trivial one-dimensional $u(L)$-module K is injective. Then*
1) *L is abelian and $L = L_+$,*
2) *the p-map is non-degenerate in the sense that $L = \langle x^{[p]} | x \in L_g, g \in G_+ \rangle$.*

Proof. Suppose that $L_+ = \langle x_1, \ldots, x_n \rangle$ is a homogeneous basis which is the union of some bases for $L_g, g \in G_+$ taken in some order. Let $Z = \langle z_1, \ldots, z_n \rangle$ be a vector space. We obtain a p-semilinear map on the components $l: L_g \to Z, g \in G_+$ if we set

$$l(\lambda_s x_s + \cdots + \lambda_t x_t) = \lambda_1^p z_s + \cdots + \lambda_t^p z_t, \quad L_g = \langle x_s, \ldots, x_t \rangle, \quad 1 \leq s \leq t \leq n.$$

Employing this notation we notice that $z_s, \ldots, z_t$ have the weight $pg \in G_+$, so $Z = Z_+$. There arises a central extension $\tilde{L} = L \oplus Z$ if one sets

$$[(x, z), (x', z')] = ([x, x'], 0); \quad x, x' \in L; \quad z, z' \in Z;$$

$$(x, z)^{[p]'} = (x^{[p]}, -l(x)); \quad x \in L_g, \quad z \in Z_g, \quad g \in G_+.$$

It is easy to verify that the axioms of colour Lie p-superalgebras are satisfied.

We observe that $A = u^+(\tilde{L})/(u^+(\tilde{L})Z)$ is an associative algebra (without unity) since $z_1, \ldots, z_n$ are supercentral in $u(\tilde{L})$. Left multiplication in A provides the structure of a restricted $\tilde{L}$-module in A. Since Z acts on A trivially, in fact, A is an $L \cong \tilde{L}/Z$-module, i.e. A is a $u(L)$-module. Since $[p]'$ is trivial on Z, we claim that $Z \cap u^+(L)Z = 0$; for this we only make use of a standard basis of $u^+(\tilde{L})$ and the supercentrality of Z. Hence, we assume that $Z \subset A$. Now Z is a submodule, moreover, the action of L on Z is trivial. Then Z is an injective $u(L)$-module being the direct sum of n one-dimensional trivial modules which are injective by hypothesis. Therefore there is a $u(L)$-submodule $W \subset A$ with $A = W \oplus Z$. Since A is generated by $\tilde{L}$ and Z acts trivially on A, W is a left ideal in A. So, $A^2 \subset AW + AZ = AW \subset W$.

Now we consider a linear map $f: L \to Z$, which is a composition of canonical homomorphisms of vector spaces:

$$L \xrightarrow{i} \tilde{L} \xrightarrow{\alpha} A = W \oplus Z \xrightarrow{\pi} Z.$$

Observe that α is the restriction of a natural epimorphism of associative algebras $\bar{\alpha}: u^+(\tilde{L}) \to A$. One has $\bar{\alpha}(\tilde{L}^2) \subset A^2 \subset W$, hence $f(L^2) = 0$. Suppose that $x \in L_g, g \in G_+$, then $\alpha(x^{[p]'}) = (\bar{\alpha}(x))^p \subset A^2 \subset W$, so

$$f(x^{[p]}) = \pi \alpha i(x^{[p]}) = \pi \alpha(x^{[p]'} + l(x)) = \pi \alpha l(x) = l(x).$$

This proves that $x_1^{[p]}, \ldots, x_n^{[p]} \in L_+$ are linearly independent because $f(x_i^{[p]}) = l(x_i) = z_i$, $i = 1, \ldots, n$, are linearly independent. Hence $\langle x^{[p]} | x \in L_g, g \in G_+ \rangle = L_+$ and the restriction $f: L_+ \to Z$ is an isomorphism of vector spaces. Since $f(L^2) = 0$ it follows that $[L_+, L_+] = 0$ and $[L_-, L_-] = 0$.

Now it only remains to prove that $L_- = 0$. By way of contradiction suppose that this is not the case. Since L_- is a subalgebra by 3.5 the trivial $u(L_-)$-module K is injective. Remark that $u(L_-)$ is a colour Grassmann algebra. Suppose that $L_- = \langle y_1, \ldots, y_m \rangle$ is a homogeneous basis; then $\langle y_1 \cdot y_2 \cdot \ldots \cdot y_m \rangle \subset u(L_-)$ is a trivial one-dimensional module. We claim that it cannot be complemented by a direct summand. Indeed, let $M \subset u(L)$ be a submodule and $0 \neq a \in M$. If y_i does not enter any monomial of a in a standard decomposition then $0 \neq y_i a = a' \in M$. By iterating this procedure on has $\langle y_1 \cdot y_2 \cdot \ldots \cdot y_m \rangle \subset M$, a contradiction. $\square$

3.8. Let A be a finite-dimensional Frobenius algebra with a linear function $\Phi: A \to K$ and corresponding bilinear form $\Psi: A \times A \to K$, $\Psi(x, y) = \Phi(x \cdot y)$, $x, y \in A$. Two bases $\{a_i | i = i, \ldots, n\}$ and $\{b_j | j = 1, \ldots, n\}$ of A are called *dual* if $\Psi(a_i, b_j) = \delta_{ij}$, $1 \leq i, j \leq n$.

Lemma. *Suppose that M is an A-module and $\pi \in \operatorname{Hom}_K(M, M)$. Then using the above notation one has*

$$\sum_{i=1}^{n} b_i \pi a_i \in \operatorname{Hom}_A(M, M).$$

Proof. Suppose that

$$a_i a = \sum_{j=1}^{n} \lambda_{ij}(a) a_j, \qquad a \in A, \qquad \lambda_{ij} \in K, \qquad 1 \leq i \leq n.$$

We claim that

$$a b_i = \sum_{j=1}^{n} \lambda_{ji}(a) b_j, \qquad a \in A, \qquad 1 \leq i \leq n.$$

Indeed, there is a decomposition

$$a b_i = \sum_{j=1}^{n} \xi_{ji}(a) b_j, \qquad a \in A, \qquad \xi_{ji}(a) \in K, \qquad 1 \leq i \leq n.$$

Since $\Psi(a_i a, b_j) = \Psi(a_i, a b_j)$ and $\Psi(a_i, b_j) = \delta_{ij}$, one has $\lambda_{ij}(a) = \xi_{ij}(a)$, which proves the desired decomposition. For $\pi \in \operatorname{Hom}_K(M, M)$, using these two decompositions, one has

$$a\left(\sum_{i=1}^{n} b_i \pi a_i\right) = \sum_{i=1}^{n} \sum_{j=1}^{n} \lambda_{ji}(a) b_j \pi a_i, \qquad a \in A,$$

$$\left(\sum_{i=1}^{n} b_i \pi a_i\right) a = \sum_{i=1}^{n} \sum_{j=1}^{n} \lambda_{ij}(a) b_i \pi a_j, \qquad a \in A.$$

Since the right hand sides of these equalities coincide, the result follows. □

Example. Let $L = \langle e | e^{[p]} = e \rangle$ be a one-dimensional Lie p-algebra and let $\Phi: u(L) \to K$ be as in 3.4: $\Phi(e^{p-1}) = 1$, $\operatorname{Ker} \Phi = \langle 1, e, \ldots, e^{p-2} \rangle$. It is not difficult to verify that a dual basis for $u(L) = \langle 1, e, e^2, \ldots, e^{p-1} \rangle$ is $u(L) = \langle e^{p-1} - 1, e^{p-2}, e^{p-3}, \ldots, e, 1 \rangle$.

We remark that in this case with the above notation $\sum_{i=1}^{p} b_i a_i = -1 + p \cdot e^{p-1} = -1$.

Theorem. *Let L be a finite-dimensional colour Lie p-algebra (i.e. $L = L_+$) over an algebraically closed field K. Suppose that L is abelian and $\langle x^{[p]} | x \in L_g, g \in G_+ \rangle = L$. Then there exist homogeneous elements $a_1, \ldots, a_l$ with $a_j^{[p^{r_j}]} = a_j$*

for some integers $n_j, j = 1, \ldots, l$, and the basis for $u(L)$ is as follows:

$$u(L) = \langle a^\alpha = a_1^{\alpha_1} \cdot \ldots \cdot a_l^{\alpha_l} | \alpha \in \Xi \rangle; \qquad \Xi = \{\alpha = (\alpha_1, \ldots, \alpha_l) | 0 \leq \alpha_j < p^{n_j}\}.$$

Proof. By Proposition 1.9 of Chapter 4, without loss of generality $(|G|, p) = 1$. With respect to the map $\tau: G_+ \to G_+$, $\tau(g) = pg$, the group G_+ is a disjoint union of orbits $G_+ = \bigcup_{i=1}^m X_i$. Pick a representative in each orbit $g_i \in X_i$, $i = 1, \ldots, m$. One has a direct decomposition of restricted subalgebras

$$L = \bigoplus_{i=1}^m H^i, \qquad H^i = \bigoplus_{g \in X_i} H_g.$$

Fix some $X = X_i$, $H = H^i$, $g = g_i$. Since $\varepsilon(\alpha, \beta) = 1$ for all $\alpha, \beta \in X$, it follows that H is an ordinary Lie p-algebra. By hypothesis, $\text{Rad}(H) = 0$ and by Theorem 3.2 of Chapter 4

$$H = \langle e_1, \ldots, e_s | e_j^{[p]} = e_j \rangle, \tag{1}$$

where $e_1, \ldots, e_s$, in general, are non-homogeneous. Suppose that $X = \{g_1, \ldots, g_n\}$ with $pg_1 = g_2$, $pg_2 = g_3$, $\ldots$, $pg_n = g_1$. Let us decompose an e in (1):

$$e = v_1 + \cdots + v_n, \qquad v_j \in H_{g_j}, \qquad j = 1, \ldots, n. \tag{2}$$

Since $e^{[p]} = e$ one has $v_1^{[p]} = v_2, v_2^{[p]} = v_3, \ldots, v_n^{[p]} = v_1$. Without loss of generality we assume that $g = g_1$, where g was fixed above in the orbit X. The elements analogous to v_1 in (2) supply us with a basis $\{a_1, \ldots, a_t\}$ of H_g such that

$$\{a_1^{\alpha_1} \cdot \ldots \cdot a_t^{\alpha_t} | 0 \leq \alpha_j < p^n\}$$

is a basis of $u(H)$. Note that $a_j^{[p^n]} = a_j, j = 1, \ldots, t$.

By collecting together all bases for L_{g_i}, $i = 1, \ldots, n$, corresponding to a fixed g_i in X_i one has

$$\bigoplus_{i=1}^m L_{g_i} = \langle a_1, \ldots, a_l \rangle.$$

Let n_j denote the order of the orbit which contains a_j. Then we have a basis for the whole restricted enveloping algebra of L:

$$u(L) = \langle a^\alpha = a_1^{\alpha_1} \cdot \ldots \cdot a_l^{\alpha_l} | \alpha \in \Xi \rangle; \qquad \Xi = \{\alpha = (\alpha_1, \ldots, \alpha_l) | 0 \leq \alpha_j < p^{n_j}\},$$

and by construction $a_j^{[p^{n_j}]} = a_j, j = 1, \ldots, l$. □

§ 3. More on restricted enveloping algebras

We remark that this theorem is a colour generalization of Theorem 3.2 of Chapter 4.

Proposition. *Let L be a finite-dimensional colour Lie p-algebra (i.e. $L = L_+$). Suppose that L is abelian and $L = \langle x^{[p]} | x \in L_g, g \in G_+ \rangle$. Then $u(L)$ is semisimple.*

Proof. Suppose that $V \subset u(L)$ is a left ideal. Let W be a vector subspace such that $u(L) = V \oplus W$, and let $\pi: u(L) \to V$ be the projection on V with the kernel W. Suppose that $\{a_i | i = 1, \ldots, n\}$ and $\{b_i | i = 1, \ldots, n\}$ are dual bases. Then by the lemma $\bar{\pi} = \sum_{i=1}^{n} b_i \pi a_i$ is a homomorphism of $u(L)$-modules with $\bar{\pi} \cdot u(L) \subset V$. The restriction of $\bar{\pi}$ to V coincides with the multiplication by $\gamma = \sum_{i=1}^{n} b_i a_i \in u(L)$. Next we prove that, for an appropriate basis, γ is a non-zero scalar, thus completing the proof.

Without loss of generality we assume K is algebraically closed. Indeed, let $K \subset K'$ be an algebraic closure. Then any non-zero nilpotent ideal $I \subset u(L)$ induces a non-zero nilpotent ideal $I \otimes_K K' \subset u(L \otimes_K K')$.

In the notations of the theorem above we define a linear function $\Phi: u(L) \to K$ which is equal to 1 on the monomial

$$a^\theta = a_1^{\theta_1} \cdot \ldots \cdot a_l^{\theta_l}, \quad \theta = (\theta_1, \ldots, \theta_l) = (p^{n_1} - 1, \ldots, p^{n_l} - 1) \in \Xi,$$

and equal to zero on all other monomials. The same argument as in 3.4 implies that $u(L)$ is a finite-dimensional Frobenius algebra with respect to Φ. There is a corresponding bilinear function $\Psi(x, y) = \Phi(x \cdot y); x, y \in u(L)$.

Thus we have obtained a basis $\{a^\alpha = a_1^{\alpha_1} \cdot \ldots \cdot a_l^{\alpha_l} | \alpha \in \Xi\}$. We set $b_t^{\bar{\alpha}_t} = a_t^{\theta_t - \alpha_t}$ if $0 < \alpha_t \leq \theta_t$ and $b_t^{\bar{\alpha}_t} = a_t^{\theta_t} - 1$ if $\alpha_t = 0$. Observe that $a_t^{\theta_t + 1} = a_t^{[p^{n_t}]} = a_t$ implies that the weight of $a_t^{\theta_t}$ equals $0 \in G$. Therefore, $a_t^{\theta_t} - 1$ is homogeneous and central in $u(L)$ for each t. Since L is abelian for any homogeneous x, $y \in u(L)$ one has $xy = \lambda yx$ for some $\lambda \in K$. It is not difficult to verify that for $\{a^\alpha | \alpha \in \Xi\}$ with respect to Ψ the dual basis is $\{b^{\bar{\alpha}} = \mu_\alpha b_1^{\bar{\alpha}_1} \cdot \ldots \cdot b_l^{\bar{\alpha}_l} | \alpha \in \Xi\}$. It is essential to have some μ_α since commuting of $a_1, \ldots, a_l$ leads to some scalars. Hence, we choose $\mu_\alpha \in K$ in such a way that $a^\alpha \cdot b^{\bar{\alpha}} = 1 \cdot a^\theta + v_\alpha$, v_α being of lower degree then a^θ. On the other hand, by commuting, one has $a^\alpha \cdot b^{\bar{\alpha}} = \lambda_\alpha(a_1^{\alpha_1} \cdot b_1^{\bar{\alpha}_1}) \cdot \ldots \cdot (a_l^{\alpha_l} \cdot b_l^{\bar{\alpha}_l})$, $\lambda_\alpha \in K$, hence $\lambda_\alpha = 1$. If a^α is of weight $g \in G$ then, by construction, $b^{\bar{\alpha}}$ has weight $-g \in G$, so $a^\alpha \cdot b^{\bar{\alpha}} = \varepsilon(g, -g) b^{\bar{\alpha}} \cdot a^\alpha = b^{\bar{\alpha}} \cdot a^\alpha$. Now we are able to compute the desired sum

$$\gamma = \sum_{\alpha \in \Xi} b^{\bar{\alpha}} \cdot a^\alpha = \left(\sum_{0 \leq \alpha_1 \leq \theta_1} a_1^{\alpha_1} \cdot b_1^{\bar{\alpha}_1} \right) \cdot \ldots \cdot \left(\sum_{0 \leq \alpha_l \leq \theta_l} a_l^{\alpha_l} \cdot b_l^{\bar{\alpha}_l} \right)$$

$$= (p^{n_1} \cdot a_1^{\theta_1} - 1) \cdot \ldots \cdot (p^{n_l} \cdot a_l^{\theta_l} - 1) = \pm 1,$$

thus completing the proof. □

3.9. Theorem. *Let G be a finite group and $L = \bigoplus_{g \in G} L_g$ a colour Lie p-superalgebra. The following conditions are equivalent.*
1) *The ring $u(L)$ is regular.*
2) *The trivial one-dimensional $u(L)$-module K is injective.*
3) *All simple $u(L)$-modules are injective.*
4) *$L = L_+$ is abelian, locally finite-dimensional, and possesses a non-degenerate p-map in the following sense: $H = \langle x^{[p]} | x \in H_g, g \in G_+ \rangle$ for any homogeneous restricted subalgebra $H \subset L$.*

Proof. (1) $\Rightarrow$ (4). Suppose that $u(L)$ is regular. Pick any homogeneous $x_1, \ldots, x_n \in L$. They generate some restricted subalgebra H. By Lemma 3.1 they have a non-zero common right annihilator. Note that $r_{u(L)}(\{x_1, \ldots, x_n\}) = r_{u(L)}(u^+(H))$ because $u^+(H)$ is generated by $x_1, \ldots, x_n$. Now by Proposition 3.6 one has $\dim H < \infty$. Thus, L is locally finite-dimensional.

Suppose that $H \subset L$ is a finite-dimensional restricted homogeneous subalgebra. By hypothesis, there exists $x \in u(L)$ with $rxr = r$, where $0 \neq r \in r(H)$ (see Lemma 3.6). So $(1 - rx) \in l_{u(L)}(r(H)) = u(L)u^+(H)$ by Proposition 3.6. Using the augmentation map $\rho: u(L) \to K$, one has $1 = \rho(r)\rho(x)$, which implies $r(H) \not\subset u^+(H)$. By a dimension argument $u(H) = u^+(H) \oplus r(H)$. Now $_{u(H)}r(H)$ is isomorphic to the trivial one-dimensional $u(H)$-module K. Since by the theory of finite-dimensional Frobenius algebras $_{u(H)}u(H)$ is injective it follows that $_{u(H)}K$ is also injective. Now, by 3.7, $H = H_+$ is abelian, and $H = \langle x^{[p]} | x \in H_g, g \in G_+ \rangle$. The implication follows because we can choose H to be a subalgebra generated by arbitrary homogeneous $x_1, \ldots, x_n \in L$.

(2) $\Rightarrow$ (4). Suppose that the trivial $u(L)$-module K is injective. Pick any homogeneous $x_1, \ldots, x_n \in L$. Since they act trivially on K, by Lemma 3.2, they have a common non-zero left annihilator. As above, $x_1, \ldots, x_n$ generate a finite-dimensional subalgebra H. It follows that L is locally finite-dimensional.

Let $H \subset L$ be any finite-dimensional restricted homogeneous subalgebra. By 3.5 $_{u(H)}K$ is injective and an application of 3.7 yields the desired properties of L.

(4) $\Rightarrow$ (3). Let $a_1, \ldots, a_n \in u(L)$ be arbitrary. They depend on finitely many basis elements, which, by hypothesis, generate a finite-dimensional restricted subalgebra H. Now $a_1, \ldots, a_n \in u(H)$ and $B = u(H)$ is semisimple by 3.8. This, by definition, implies that $u(L)$ is a locally Wedderburn algebra.

By Theorem 1.1 (or Corollary 1.6) of Chapter 4 $u(L)$ is a PI-algebra. Then any simple $u(L)$-module V is finite-dimensional over its centralizer by Lemma 1.3 of Chapter 4. Now the statement follows from Lemma 3.3.

(4) $\Rightarrow$ (1). As above, $u(L)$ is a locally Wedderburn algebra. If we want to solve the equation $axa = a$ for $a \in u(L)$ then we find a finite-dimensional semisimple subalgebra $B \subset u(L)$, containing a. Now this equation can be solved in B because all finite-dimensional semisimple algebras are regular. □

3.10. Theorem. *Let L be a colour Lie p-superalgebra. Then $u(L)$ is self-injective if, and only if, $\dim L < \infty$.*

Proof. If $\dim L < \infty$ then $u(L)$ is a finite-dimensional Frobenius algebra (3.4) and the statement follows from the theory of such algebras.

Suppose that $V =_{u(L)} u(L)$ is injective. Pick arbitrary homogeneous $r_1, \ldots, r_n \in L$. Evidently, $r_1 V + \cdots + r_n V \subset u^+(L) \neq V$, and, by Lemma 3.2, $r_1, \ldots, r_n$ have a common non-zero left annihilator. Suppose that $r_1, \ldots, r_n$ generate some restricted subalgebra. We remark that $l_{u(L)}(u^+(H)) = l_{u(L)}(\{r_1, \ldots, r_n\})$. Proposition 3.6 now yields $\dim H < \infty$. Thus, L is locally finite-dimensional.

By way of contradiction suppose that $\dim L = \infty$. Then there is an infinite strictly ascending sequence of finite-dimensional restricted subalgebras $H^0 \subset H^1 \subset \cdots \subset H^n \subset \cdots$, and $H = \sum_{i=0}^{\infty} H^i$. Let $f(H)$, in this subsection, stand for some non-zero element in the one-dimensional ideal $r(H)$, where H is a finite-dimensional restricted subalgebra. For any $n \geq 1$ we define an element $d_n \in u(H^{n-1})$ by $d_n = f(H^0) + f(H^1) + \cdots + f(H^{n-1})$. For any $n \geq 1$ pick some $h_n \in H^n \backslash H^{n-1}$. Note that for $i \geq j$ one has $h_j f(H^i) = 0$. Then for $n \geq j$ we have $h_j d_n = h_j d_j$. Now we define a mapping $\tau : I = \sum_{i=1}^{\infty} u(L) h_i \to u(L)$ as follows: $\tau(\sum_{i=1}^{\infty} \alpha_i h_i) = \sum_{i=1}^{\infty} \alpha_i h_i d_i$, where $\alpha_i \in u(L)$. If $\sum_{i=1}^{n} \alpha_i h_i = 0$ then $0 = \sum_{i=1}^{n} \alpha_i h_i d_n = \sum_{i=1}^{n} \alpha_i h_i d_i$, thus τ is well-defined. By hypothesis, τ can be extended to a homomorphism of modules $\bar\tau : u(L) \to u(L)$ and we may set $d = \bar\tau(1)$. Then for $i \geq 1$

$$h_i d = h_i \bar\tau(1) = \tau(h_i) = h_i d_i.$$

There is a decomposition $d = a + b$, where $a \in u(H)$ is the sum of basis monomials of $u(H)$, and b is the sum of monomials, some factors of which are basis elements of L modulo H. Evidently, $h_i a = h_i d_i$. For some $n \in \mathbb{N}$ one has $a \in u(H^n)$. Then for $m > n$ we obtain $h_m(a - d_m) = 0$, where $h_m \in H^m \backslash H^{m-1}$ and $a - d_m \in u(H^{m-1})$. This is possible only in the case where $a = d_m$ for $m > n$. Therefore $0 = d_{n+2} - d_{n+1} = f(H^{n+1}) \neq 0$, a contradiction. As a result, L must be finite-dimensional. □

§4. Examples

Here we give two interesting examples of algebras which are somewhat close to enveloping algebras of colour Lie superalgebras.

4.1. Let $A = K[x, y, z]$ be the associative algebra defined by the relations

$$xy = \lambda yx + z,$$

$$zy = \mu yz,$$

$$xz = \mu zx, \qquad \lambda, \mu \in K.$$

1) A has a basis $\{x^i y^j z^k, i, j, k \in \mathbb{N}\}$.
2) Only in the case $\lambda = \mu$ the space $L = \langle x, y, z \rangle$ can be endowed with the structure of a colour Lie superalgebra and $A \cong U(L)$.
3) The following three properties are equivalent: a) A is a PI-algebra. b) If in addition K is algebraically closed then the dimensions of all simple A-modules are totally bounded. c) λ, μ are roots of unity and $\lambda \neq \mu$ for char $K = 0$.

Hints. As in 3.2.8 one has $J(A) = 0$ with the use of a natural filtration. Suppose that A is a PI-algebra. Consideration of $\text{gr}(A)$ as in 4.4.9, and Example 4.1.6 shows that λ, μ are roots of unity; by Proposition 5.1.4 the endomorphism field is algebraic. The claim $\lambda \neq \mu$ for char $K = 0$ follows from 4.4.2.

Let A satisfy the hypothesis c), in particular $\lambda^n = \mu^n = 1$ for some integer n. Suppose also that in the case char $K = p > 0$, p divides n. It is easy to verify that $B = K[x^n, y^n, z] \subset A$ is a commutative subalgebra (use that ${}_B A$ is of finite rank). We conclude that A is a PI-algebra.

4.2. Let $A = K[e, f, h]$ be the associative algebra defined by the relations

$$ef - \theta fe = h,$$

$$he - \lambda eh = e,$$

$$fh - \lambda hf = f, \qquad \lambda, \theta \in K.$$

1) A has a basis $\{e^i f^j h^k, i, j, k \in \mathbb{N}\}$.
2) The following three properties are equivalent: a) A is a PI-algebra. b) If in addition K is algebraically closed then the dimensions of all simple A-modules are totally bounded. c) λ, θ are roots of unity and $\theta \neq 1$, $\lambda \neq 1$, $\lambda \cdot \theta \neq 1$ if char $K = 0$.

Hints. To show that a) and b) are equivalent and that θ, λ must be roots of unity we apply the same argument as in the problem above.

Suppose that A is a PI-algebra and char $K = 0$. By using Example 4.4.1 for a subalgebra generated by h, e we obtain $\lambda \neq 1$. To show that $\theta \neq 1$ and $\lambda \cdot \theta \neq 1$ if char $K = 0$ one can construct infinite-dimensional representations

$$V = \langle v_i | i \in \mathbb{N} \rangle, \quad e \circ v_i = v_{i+1}, \quad f \circ v_i = \beta_i v_{i-1}, \quad h \circ v_i = \gamma_i v_i,$$

and find $\beta_i, \gamma_i \in K$ such that V is irreducible.

Suppose that A satisfies the hypotheses of c), in particular $\theta^n = \lambda^n = 1$ for some integer n (suppose that also p divides n if char $K = p > 0$). Set $\bar{h} = h + (1/\lambda - 1)$. From the given relations we deduce that $\bar{h}e = \lambda e\bar{h}$ and $f\bar{h} = \lambda \bar{h}f$. It is easy to check that $K[e^n, f^n, \bar{h}^n]$ is a central subalgebra in A. This proves that A is a PI-algebra.

Remark. Suppose that A is as in c) and K is algebraically closed. By Schur's Lemma for any irreducible representation $\rho: A \to \operatorname{End} V$ the elements $\rho(e^n)$, $\rho(f^n)$, $\rho(\bar{h}^n)$ act as some scalars α, β, γ, which are in K by Proposition 5.1.4. So, one can define "restricted enveloping algebras"

$$A(\alpha, \beta, \gamma) = A/(e^n - \alpha, f^n - \beta, \bar{h}^n - \gamma), \quad \dim A(\alpha, \beta, \gamma) = n^3, \quad \alpha, \beta, \gamma \in K,$$

and A resembles the universal enveloping algebra for $sl_2(K)$ over a modular field.

The algebras with $\lambda = \theta^2 = q^{-8}$, where q is not a root of unity, appear as subalgebras in the quantum group $U_q(sl_2(K))$, see e.g. [Montgomery, 1991].

Comments to Chapter 5

The theorem about the Jacobson radical in the case of a semi-direct product of Lie superalgebras (see 1.1) is due to [Bahturin, 1987b], [Bahturin, 1989a]. This result also follows from [Bergen, Montgomery, Passman, 1987]. Lemma 1.3 and the idea of 1.4 are taken from [Dixmier, 1974]; see also another approach in [McConnell, 1982].

The first result about the dimensions of simple modules for Lie algebras was obtained in [Bahturin, 1979]. Theorems 2.2, and 2.1 were obtained in [Bahturin, 1989b] and [Petrogradsky, 1991b], respectively. Claim 2 of Lemma 2.6 is an analogue of a well-known lemma for group algebras.

In Section 3 we study analogues of known results for group algebras. The methods are close to those used for group algebras [Passman, 1977] considering the reduction from the infinite-dimensional to the finite-dimensional case, but we need essentially Lie algebra methods in the finite-dimensional case. To exhibit the analogies we cite some results (see also the comments to Chapter 4):

Theorem [Villamayor, 1959]. *Let K be a field, char $K = p > 0$. Then a group ring $K[H]$ is regular if, and only if, H is a locally finite group not having elements of order p.* □

Theorem [Farkas, Snider, 1974], [Hartley, 1977]. *Let K be a field with char $K = p > 0$, and H a countable group. Then any simple $K[H]$-module is injective if, and only if, H is an abelian-by-finite locally finite group without elements of order p.* □

Theorem [Renault, 1971]. *A group ring is self-injective if, and only if, the group is finite.* □

See also the monograph [Passman, 1977].

Theorem 3.10 is a precise analogue of this result. This is not the case for Theorem 3.9 which in the particular case without condition 1 was obtained in [Petrogradsky, 1988b]. The idea of studying regularity was suggested by D.S. Passman. Theorem 3.9 is also an infinite-dimensional colour generalization of G. Hochschild's Theorem on semisimplicity of a restricted enveloping algebra for a finite-dimensional Lie p-algebra [Hochschild, 1954].

Lemmas 3.2, 3.3, and 3.8 are taken from [Passman, 1977], [Farkas, Snider, 1974], and [Curtis, Reiner, 1962], respectively; Lemma 3.4 for ordinary Lie p-algebras was proved in [Berkson, 1964].

Some similar results on Lie algebras have been independently, and by different methods, obtained by J. Feldvoss [Feldvoss, 1991].

Chapter 6

Finiteness conditions for colour Lie superalgebras with identities

The aim of the present chapter is to investigate connections between the identities of colour Lie superalgebras over a field K, char $K \neq 2$, and some finiteness conditions on its subalgebras. Our central result is the description of sufficient conditions for residual finiteness and matrix representability of colour Lie superalgebras in the language of identities. For a wide class of varieties of algebras these conditions are necessary as well.

§ 1. Various types of finiteness conditions. Examples

1.1. Maximal conditions and the Hopf property. One of the most important finiteness conditions on a ring is the condition to be Noetherian: a colour Lie superalgebra $L = \oplus L_g$ is called *Noetherian* if it satisfies the maximal condition for ideals, i.e. if every properly ascending chain of ideals $H_1 \subset H_2 \subset \cdots$ has finite length. Consider L as a left module over itself with the natural action $xv = \operatorname{ad} x(v) = [x, v]$. Then L is a Noetherian colour Lie superalgebra if and only if the maximal condition holds for the submodules of L.

Recall that the grading group G consists of even and odd elements, $G = G_+ \cup G_-$, and $L = L_+ \oplus L_-$ where

$$L_+ = \bigoplus_{g \in G_+} L_g, \quad L_- = \bigoplus_{g \in G_-} L_g.$$

The even component L_+ is a colour Lie superalgebra over K graded by the subgroup G_+ of the group G. Now we may consider L as a module over its subalgebra L_+. Another finiteness condition which we may impose on L is the maximal condition of L as L_+-module. Clearly, if L satisfies this condition then L is Noetherian.

We shall also consider superalgebras with the Hopf property. We say that a colour Lie superalgebra L possesses the *Hopf property* if any epimorphism of L (i.e. surjective homomorphism onto itself) is an automorphism. In other words, if φ is an endomorphism of L then the equality $\varphi(L) = L$ implies Ker $\varphi = 0$.

The infinite-dimensional Lie algebra L with zero multiplication is an example of a superalgebra without the Hopf property. If $w_1, w_2, \ldots$ is a basis of L then the mapping $\varphi: L \to L$, $\varphi(w_1) = 0$, $\varphi(w_i) = w_{i-1}$, $i \geq 2$, is an epimorphism and $\operatorname{Ker} \varphi \neq 0$.

Any finite-dimensional algebra is an algebra with the Hopf property. It is easy to verify that if L is a Noetherian superalgebra then it possesses the Hopf property.

Indeed, let φ be an endomorphism of L satisfying $\varphi(L) = L$ and let $H = \operatorname{Ker} \varphi$. If $H \neq 0$ then we may construct an infinite chain of ideals

$$H_0 \subset H_1 \subset \cdots \subset H_i \subset \cdots$$

where $H_0 = H$ and for any $i \geq 1$

$$H_i = \{x \in L \mid \varphi(x) \in H_{i-1}\}.$$

Since $\varphi(L) = L$ we have $L/H_i \cong L$ for every $i = 0, 1, \ldots$. Therefore $H \neq 0$ implies $H_{i+1} \neq H_i$ for any $i \geq 0$. On the other hand, L is a Noetherian superalgebra. This contradiction shows that we cannot find such an epimorphism of L with nonzero kernel. Hence L has the Hopf property.

The Hopf property is a weaker condition than the Noetherian property. For example, the free colour Lie superalgebra $L = L(X)$ over an infinite field K is not Noetherian, but it satisfies the Hopf condition whenever $|X| < \infty$. To verify this let us prove first that the free colour Lie superalgebra $L = L(X, \mathfrak{B})$ of an arbitrary variety $\mathfrak{B}$ (see Subsection 1.2.1) possesses the Hopf property if $|X| < \infty$ and the ground field is infinite.

Let φ be an endomorphism of L with $\varphi(L) = L$. If $X = \{x_1, \ldots, x_n\}$ then by the definition of $L(X, \mathfrak{B})$ any relation on $x_1, \ldots, x_n$ is an identity of $\mathfrak{B}$. Let $f = f(x_1, \ldots, x_n) = f_1 + \cdots + f_m = 0$ be an identity of $\mathfrak{B}$ where $\deg f_j = j$. Since K is an infinite field, the identifies $f_1 = \cdots = f_m = 0$ hold in $\mathfrak{B}$ (see Subsection 1.2.3). Using this fact we conclude that $x_1, \ldots, x_n$ are linearly independent modulo L^2 and $\dim L/L^2 = n$.

Any element $\varphi(x_i)$ can be written as a sum $\varphi(x_i) = y_i + t_i$ where y_i is a linear combination of $x_1, \ldots, x_n$ and $t_i \in L^2$. Since $\varphi(L) = L$ the vectors $y_1, \ldots, y_n$ are linearly independent. Hence the equations $\psi(x_i) = y_i$, $i = 1, \ldots, n$, define a non-degenerate linear map on $V = \langle x_1, \ldots, x_n \rangle$ with inverse map ψ^{-1}.

By the definition of free generators any mapping τ of X into L with $d(x_i) = d(\tau(x_i))$, $i = 1, \ldots, n$, extends to an endomorphism of L. Therefore we can extend both ψ and ψ^{-1} to some endomorphisms of L. Hence ψ is an automorphism of L and $y_1, \ldots, y_n$ are free generators of L.

If $f = f(x_1, \ldots, x_n) \in \operatorname{Ker} \varphi$ then $\varphi(f) = f(y_1 + t_1, \ldots, y_n + t_n) = 0$. We shall prove that $f(y_1, \ldots, y_n) = 0$ in L.

The elements $y_1, \ldots, y_n$ generate L as a superalgebra and $t_1, \ldots, t_n$ lie in L^2. It follows that any t_j can be written as a sum of monomials in $y_1, \ldots, y_n$ of degree greater than one. Let us write f in the form $f = f_0 + f'$ where f_0 is the sum of monomials of minimal degree in f and f' is the sum of monomials of degree greater than $\deg f_0$. Then

$$f(y_1 + t_1, \ldots, y_n + t_n) = f_0(y_1, \ldots, y_n) + f_1$$

where $f_1 = f_1(y_1, \ldots, y_n)$ is the sum of monomials in $y_1, \ldots, y_n$ of degree greater than $\deg f_0$. Since K is an infinite field we have $f_0 = f_1 = 0$. Continuing this process we get $f(y_1, \ldots, y_n) = 0$. Since $y_1, \ldots, y_n$ freely generate L it follows that $\operatorname{Ker} \varphi = 0$. Hence $L = L(X, \mathfrak{B})$ is a colour Lie superalgebra with the Hopf property.

Later we shall prove that a free colour Lie superalgebra $L(X)$ with $|X| > 1$ over an infinite field cannot be Noetherian (see Subsection 2.9).

1.2. Residual finiteness and matrix representability

Definition. A Lie superalgebra L over a field K is called *residually finite* if for any $x \neq 0$, $x \in L$ there exists a homomorphism $\varphi : L \to H$ onto a finite-dimensional superalgebra H over K such that $\varphi(x) \neq 0$.

It is easy to see that L is residually finite if and only if it can be embedded in a cartesian product $\prod_\alpha L_\alpha$ where the L_α are of finite dimension.

Definition. Let $L = \bigoplus_{g \in G} L_g$ be a colour Lie superalgebra over a field K graded by the group G with a given bilinear form ε. We say that L is *matrix representable* if there exist an extension $\tilde{K}$ of the ground field K and a finite-dimensional $\tilde{K}$-algebra $\tilde{L} = \bigoplus_{g \in G} \tilde{L}_g$ with the same G and ε such that L is embeddable in $\tilde{L}$ as a K-algebra.

It will be shown that these two properties are closely related. In particular any finitely generated representable algebra is residually finite.

Now let K be an infinite field and let L be a finitely generated free colour Lie superalgebra over K. We shall verify that the intersection of all terms of the lower central series of L is equal to zero.

Let x be an arbitrary element in $\bigcap_{m \geq 1} L^m$. We can write x as a sum of monomials of degree less than t for some t. On the other hand, x is contained in L^{t+1}. Hence, it may be written as a sum of monomials of degree greater than t. We obtain an equation $a_{i_1} + \cdots + a_{i_k} = a_{j_1} + \cdots + a_{j_n}$ where $\deg a_{i_r} = i_r$, $\deg a_{j_r} = j_r$, $i_1 < \cdots < i_k < j_1 < \cdots < j_n$. Since L is a free superalgebra and K is an infinite field it follows that $a_{i_1} = \cdots = a_{i_k} = a_{j_1} = \cdots = a_{j_n} = 0$ (see Subsection 1.2.3). Therefore $x = 0$.

For a finitely generated Lie superalgebra L every quotient algebra L/L^m is of finite dimension. Therefore, if all terms of the lower central series intersect trivially, then L is residually finite.

On the other hand, if L is a free Lie algebra with at least two generators then L cannot be represented on a finite-dimensional algebra. Indeed, if H is a finite-dimensional Lie algebra over some field, $\dim H = m$, then H satisfies the identity

$$\sum_{\sigma \in \text{Sym}(m+1)} (-1)^\sigma [x_{\sigma(1)}, \ldots, x_{\sigma(m+1)}, y] = 0$$

(see, for example, [Bahturin, 1987a], Chapter 2, Subsection 2.1.3). As L is a free Lie algebra it cannot be embedded in H. For colour Lie superalgebras the situation is just the same.

Clearly, all the above mentioned finiteness conditions are satisfied by any superalgebra of finite dimension.

1.3. Examples. The following examples illustrate the above definitions.

Let G be an arbitrary abelian group with alternating bilinear form ε and let g, h be some elements of G (not necessarily distinct). An example of a colour Lie superalgebra which is not residually finite is given by the Heisenberg superalgebra $\Gamma = \Gamma(g, h) = \bigoplus_{r \in G} \Gamma_r$. All components Γ_r, except Γ_g, Γ_h and Γ_{g+h}, are equal to zero. The basis of this superalgebra is formed by the set $\{a_i, b_i, c | i \in \mathbb{Z}\}$ where $a_i \in \Gamma_g$, $b_i \in \Gamma_h$, $c \in \Gamma_{g+h}$. The commutator is given by the formula

$$[a_i, b_i] = -\varepsilon(g, h)[b_i, a_i] = c, \qquad i \in \mathbb{Z},$$

with all the remaining commutators equal to zero. Γ is not residually finite since any ideal of finite codimension contains some nontrivial linear combination $\lambda_1 a_1 + \cdots + \lambda_n a_n$ and hence contains c.

To obtain an example of a finitely generated colour Lie superalgebra it is necessary to adjoin to $\Gamma(g, h)$ a binding derivation $\delta: \Gamma(g, h) \to \Gamma(g, h)$ and to place it into the zero component of the new algebra. As a result, we obtain a superalgebra $B = B(g, h)$ all of whose components, except the zero component B_0, are the same as in $\Gamma(g, h)$ and $B_0 = \Gamma_0 \oplus \langle \delta \rangle$ with commutators given by

$$[\delta, a_i] = \varepsilon(g, h)a_{i+1} - \varepsilon(h, g)a_{i-1},$$

$$[\delta, b_i] = \varepsilon(h, g)b_{i+1} - \varepsilon(g, h)b_{i-1}.$$

§1. Various types of finiteness conditions

We ask the reader to verify that each of the algebras $B(g, h)$ is indeed a colour Lie superalgebra and that it is finitely generated. It is easy to see that $B(g,h)$ is not residually finite since it contains $\Gamma(g, h)$.

Our next example relates to the Hopf property of colour Lie superalgebras. Let g and h be arbitrary elements of the group G. We denote by P_g, P_h, P_{g+h} the linear spaces spanned by $\{a_i\}$, $\{b_i\}$, $\{c_j\}$ respectively where $i \in \mathbb{Z}$ and $j = 0, 1, \ldots$. The space $P = P(g, h) = P_g \oplus P_h \oplus P_{g+h}$ becomes a colour Lie superalgebra if we put

$$[a_i, b_j] = -\varepsilon(g, h)[b_j, a_i] = \begin{cases} c_{i-j}, & \text{for } i \geq j \\ 0, & \text{for } i < j \end{cases}$$

with all other products equal to zero. Similar to the Heisenberg superalgebra, $P(g, h)$ has a homogeneous derivation $\delta: P(g, h) \to P(g, h)$ with $d(\delta) = 0$. The action of δ on P is given by the formulae:

$$\delta(a_i) = \varepsilon(g, h)a_{i+1} + \varepsilon(h, g)a_{i-1},$$

$$\delta(b_j) = -\varepsilon(g, h)b_{j-1} - \varepsilon(h, g)b_{j+1},$$

$$\delta(c_k) = 0.$$

Let us check that δ is a derivation of P with $d(\delta) = 0$. It is sufficient to verify the relation

$$\delta([x, y]) = [\delta(x), y] + [x, \delta(y)] \tag{1}$$

for all basis elements x, y of P. If $x = a_i$, $y = b_j$ then the left hand side of (1) is zero. The right hand side is equal to

$$[\delta(a_i), b_j] + [a_i, \delta(b_j)] = [\varepsilon(g, h)a_{i+1} + \varepsilon(h, g)a_{i-1}, b_i]$$

$$- [a_i, \varepsilon(g, h)b_{j-1} + \varepsilon(h, g)b_{j+1}]. \tag{2}$$

If $i \leq j - 2$ then (2) is zero. For $i = j - 1$ it equals

$$\varepsilon(g, h)[a_j, b_j] - \varepsilon(g, h)[a_{j-1}, b_{j-1}] = \varepsilon(g, h)(c_0 - c_0) = 0.$$

If $i = j$ then we have

$$[\varepsilon(g, h)a_{i+1} + \varepsilon(h, g)a_{i-1}, b_i] - [a_i, \varepsilon(g, h)b_{i-1} + \varepsilon(h, g)b_{i+1}]$$

$$= \varepsilon(g, h)c_1 - \varepsilon(g, h)c_1 = 0.$$

Finally, for $i \geq j + 1$ the value of (2) is

$$\varepsilon(g,h)c_{i-j+1} + \varepsilon(h,g)c_{i-j-1} - \varepsilon(h,g)c_{i-j-1} - \varepsilon(g,h)c_{i-j+1} = 0.$$

Thus the right hand side of (1) is also zero if $x = a_i$, $y = b_j$. For all other basis elements x, y the verification of (1) is trivial.

The above enables us to define a superalgebra $Q = Q(g,h) = P(g,h) \oplus \langle \delta \rangle$ as the semidirect product of $\langle \delta \rangle$ and P with $Q_0 = P_0 \oplus \langle \delta \rangle$.

Let φ be the linear map on Q defined by the equations $\varphi(\delta) = \delta$, $\varphi(a_i) = a_{i-1}$, $\varphi(b_j) = b_j$, $\varphi(c_k) = c_k$ for $k \geq 1$ and $\varphi(c_0) = 0$. One can easily prove that $\varphi([x,y]) = [\varphi(x), \varphi(y)]$, i.e. φ is an endomorphism of the colour Lie superalgebra Q. Clearly $\varphi(Q) = Q$ and Ker $\varphi \neq 0$. Therefore Q is a finitely generated superalgebra without Hopf property. Note that each of the algebras $B(g,h)$, $Q(g,h)$ is a centre-by-metabelian superalgebra in the sense that it satisfies the identity

$$[x,[y,z],u,v] = 0.$$

§2. Maximal condition and Hopf property

Most results of this section concern soluble colour Lie superalgebras having finitely many generators. Recall that any colour Lie superalgebra L is the sum of its even and odd components, $L = L_+ \oplus L_-$. We use the notation L_+^m for the superalgebra $(L_+)^m$ (not for $(L^m)_+$!) where $H^1 = H$, $H^m = [H, H^{m-1}]$ for any superalgebra H.

2.1. Identities of representations of colour Lie superalgebras. Let $L = \bigoplus_{g \in G} L_g$ be a colour Lie superalgebra over a field K. If $M = \bigoplus_{g \in G} M_g$ is an L-module then there exists a homomorphism ρ of L into the G-graded associative algebra End M. Although ρ may not be a monomorphism in general, we shall identify L with its image in End M if we speak about the action of elements of L on M. In this case we denote the result of the action of x on v ($x \in L, v \in M$) by xv instead of $\rho(x)v$.

Let $X = \bigcup X_g$ be a G-graded set and $f = f(x_1, \ldots, x_n)$ an associative monomial in $x_1, \ldots, x_n \in X$. Then the relation

$$f(x_1, \ldots, x_n) = 0$$

is called an identity of the representation of L on M if $f(a_1, \ldots, a_m)$ is a zero linear mapping on M for any $a_1, \ldots, a_m \in L$ with $d(a_i) = d(x_i), i = 1, \ldots, n$.

For example, we say that an algebra L acts on M as a *nilpotent space of transformations* if the family of identities

$$x_1 \ldots x_n = 0, \qquad d(x_i) = g_i, \qquad i = 1, \ldots, n,$$

holds in the representation of L on M for all $g_1, \ldots, g_n$ in G.

We shall also consider the identities on one of the homogeneous components of M. Such an identity is a relation of the form

$$f(x_1, \ldots, x_n)u = 0, \qquad d(u) = g,$$

and we only require that the restriction on M_g of an operator $f(a_1, \ldots, a_n)$ is zero if $a_i \in L$, $d(a_i) = d(x_i)$, $i = 1, \ldots, n$.

Sometimes, we shall use non-associative expressions with square brackets to denote the identities of the action of L on M. In this case $[a, b]$ with $a, b \in L$ denotes the product of a and b in the Lie superalgebra $[\operatorname{End} M]$. This agreement is justified because ρ is a homomorphism of colour Lie superalgebras, $\rho\colon L \to \operatorname{End} M$.

2.2. Separating variables in identities. Consideration of identities with separated variables proves to be rather advantageous. By such an identity we mean a relation of the form

$$f(\operatorname{ad} x_1, \ldots, \operatorname{ad} x_n)(w) = 0$$

where w is a Lie polynomial in variables $y_1, \ldots, y_m$ different from $x_1, \ldots, x_m$. The advantage of using identities with separated variables was demonstrated by P.J. Higgins [Higgins, 1954]. We now describe the method which enables us to separate variables in the identities of graded superalgebras, in some cases.

Let G be an additive commutative semigroup with zero element and let $L = \bigoplus L_g$ be a G-graded algebra over an infinite field K. Suppose that for any g, h, r in G the algebra L satisfies graded identities of the form

$$[x, y^{(n)}, z] = \sum_{j=1}^{n} \alpha_j [y^{(j)}, x, y^{(n-j)}, z], \qquad d(x) = d(y) = g, \qquad d(z) = h, \quad (1)$$

$$[x, y^{(n)}, z] = \sum_{j=1}^{n} \beta_j [y^{(j)}, x, y^{(n-j)}, z], \qquad d(x) = r, \qquad d(y) = g, \qquad d(z) = h, \tag{2}$$

if g is of finite order. In (1) and (2) all α_j, β_j are in K, $j = 1, \ldots, n$, and at least one β_j is nonzero. Recall that we use the notation $[y^{(j)}, x]$ to denote the product $[y,\ldots,y,x]$ where y occurs j times. In this subsection we do not assume that L is a colour Lie superalgebra. Nevertheless, we maintain the notation $[a,b,\ldots,c,f]$ for the right-normed product $a(b(\ldots(cf)\ldots))$ in a non-associative algebra L.

We show that, for some k and m, L satisfies the identity

$$[x, t^{(k)}, Z] = \sum_{j=1}^{k} \beta_j [t^{(j)}, x, t^{(k-j)}, Z], \qquad d(x) = r, \qquad d(t) = g, \qquad (3)$$

where $\beta_k \neq 0$, $Z = [y^{(m)}, z]$, $d(y) = g$ and $d(z) = d(Z) = h$.

Let k be maximal with respect to $\beta_k \neq 0$ in (2). Denote by A, B, C the left multiplications with x, y, z, respectively, in the algebra L where $d(t) = g$. Then (2) can be written in the form $f(A, B)(z_1) = 0$ where

$$f(A, B) = \beta_k B^k A + \cdots + \beta_1 BAB^{k-1} - AB^k$$

and $z_1 = (y^{(n-k)}, z]$. Let s be an integer such that $sg = 0$ in G. Replacing y and z in (2) by $y + t$ and $[y^{(s)}, z]$, respectively, we obtain

$$f(A, B + C)([(y + t)^{(n-k)}, y^{(s)}, z]) = 0. \qquad (4)$$

Denote by $f_j(X_1, X_2, X_3)$ the sum of the monomials of degree j in X_3 in the associative polynomial $f(X_1, X_2 + X_3)$. Using induction on j we will prove that one can choose numbers s_j, $j = 1, \ldots, k$, such that

$$f_j(A, B, C)([y^{(s_j)}, z]) = 0 \qquad (5)$$

in L and $s_j g = 0$ in G. If $j = 0$ then (5) holds for $s_0 = n - k$. Suppose that the required numbers $s_0 \leq s_1 \leq \cdots \leq s_j$ have been already obtained. Let D be the sum of the monomials of degree $j + 1$ in t on the left hand side of (4). The number s in (4) will be chosen later. Since K is an infinite field, D is equal to zero (see Chapter 1, Subsection 2.3). On the other hand,

$$D = f_{j+1}(A, B, C)(z_1) + \sum_{i=0}^{j} f_i(A, B, C)(w_i), \qquad (6)$$

where $z_1 = [y^{(s)}, z]$ and w_i is the sum of the monomials of degree $j + 1 - i$ in t in the commutator $[(y + t)^{(s)}, z]$.

In the identity (1) replace x by t. This enables us to rewrite w_i as a sum of right-normed commutators of the form $[y,\ldots,y,t,y,\ldots,y,t,y,\ldots,z]$ where not more than $n - 1$ letters y occur between any two neighbouring variables t. Since the degree of any w_i in t is not larger than $j + 1$ for $s \geq (j + 1)(n - 1) + s_j$ we obtain the expression $w_i = [y^{(s_j)}, v_i]$ with some $v_i \in L$, $d(v_i) =$

$d(z) = h$. By the inductive hypothesis it follows that every term under the summation sign in (6) is zero. Hence,

$$f_{j+1}(A, B, C)([y^{(s_{j+1})}, z]) = 0$$

if $s_{j+1} \geq s_j + (j+1)(n-1)$ and $s_{j+1}g = 0$ in G.

For $j = k$, $m = s_k$ we get $f(A, C)([y^{(m)}, z]) = 0$. This is precisely the desired identity (3).

Note that (3) is similar to (2), but the variables in (3) are separated since Z in (3) is not dependent on x, t.

The following two results will be used to perform calculations.

2.3. Lemma. *Let L be a finite-dimensional colour Lie superalgebra over a field K with basis $\{e_1, \ldots, e_m\}$, M an L-module and let v be an element of M. If the sequence $i_1, \ldots, i_q$ consists of numbers $1, \ldots, m$ and includes j exactly t_j times ($j = 1, \ldots, m$) then the difference*

$$e_{i_1} \ldots e_{i_q} v - e_1^{t_1} \ldots e_m^{t_m} v$$

can be written in M as a linear combination of the elements

$$e_1^{r_1} \ldots e_m^{r_m} v, \qquad r_1 + \cdots + r_m \leq q - 1.$$

Proof. The result follows by induction on q. □

We shall demonstrate the advantages of identities with separated variables using usual Lie algebras as an example.

2.4. Lemma. *Let L be a finitely generated soluble Lie algebra over an infinite field K. Let M be an L-module, and assume that the following identities hold for the representation of L on M:*

$$[y^{(n)}, z] = \sum_{j=1}^{n} \lambda_j y^j [y^{(n-j)}, z], \qquad (7)$$

$$[x, y^{(n)}, z] = \sum_{j=1}^{n} \alpha_j [y^{(j)}, x, y^{(n-j)}, z], \qquad (8)$$

with some α_j, λ_j in K, $j = 1, \ldots, n$. Then there exists q such that L^q acts on M as a nilpotent space of transformations.

Proof. We construct the semidirect product $H = L \curlywedge M$ where M is an ideal of H with zero multiplication. The Lie algebra H may be considered as graded by the group $\mathbb{Z}_2$, where $H_0 = L$, $H_1 = M$. Now (7) can be written in H in the form

$$[x, y^{(n)}, z] = \sum_{j=1}^{n} \beta_j [y^{(j)}, x, y^{(n-j)}, z],$$

$$d(x) = 1, \qquad d(y) = d(z) = 0,$$
(9)

where $\beta_j = \pm \lambda_j$, since the commutator $[b, x]$ in H of elements $x \in M$, $b \in L$ is equal to bx by the definition of the semidirect product (see Subsection 1.1.6).

If all λ_j in (7) are zero then $(\operatorname{ad} y)^n = 0$ in L and $L^q = 0$ for some q (see, for example, [Bahturin, 1987a], Theorem 4.7.2). Hence, we may assume that some β_j in (9) is different from zero, say $\beta_k \neq 0$.

We apply the method of separating variables to the identity (9) assuming that $r = 1$, $g = h = 0$ in the relations (1), (2) where 0, 1 are the elements of the group $\mathbb{Z}_2$ which grades H. By Subsection 2.2 an identity of type (3) holds in H with t, y, $z \in L$, $x \in M$. Since $\beta_k \neq 0$, replacing t and Z in (3) by $[y^{(m)}, z]$ yields the identity

$$([y^{(m)}, z])^k = 0 \tag{10}$$

of the representation of L on M.

If $f(x_1, \ldots, x_n)$ is a Lie polynomial and L is a Lie algebra over an infinite field K then the linear span of the set

$$\{f(a_1, \ldots, a_n) | a_i \in L, i = 1, \ldots, n\}$$

is an ideal of L (see [Bahturin, 1987a], Theorem 4.2.9). It follows that the linear span P of all elements $[a^{(m)}, b]$, $a, b \in L$, is an ideal of the finitely generated soluble Lie algebra L and $(\operatorname{ad} x)^m = 0$ in L/P. In this case L/P is nilpotent, i.e. $L^q \subset P$ for some q.

It follows by (3), (10) that $b^k = 0$ on M if b is contained in L^q. By hypothesis L is finitely generated, hence $\dim L/L^q = s < \infty$. As an ideal, L^q is generated by all commutators of degree q in the generating set of L. Denote by $a_1, \ldots, a_r$ all commutators of degrees q, $q + 1$, $\ldots$, $2q - 2$ in the generating set of L. Then also $a_1, \ldots, a_r$ generates L^q as an ideal of L. Let $e_1, \ldots, e_s$ be a basis of L modulo L^q where every e_i is a commutator of generators of L. The reader can easily check that L^q, as a Lie algebra, is generated by all products

$$[e_1^{(t_1)}, \ldots, e_s^{(t_s)}, a_i], \qquad t_1, \ldots, t_s \geq 0, \qquad i = 1, \ldots, r. \tag{11}$$

Let Q be the subalgebra of L^q generated by all elements of the form (11), with $t_1, \ldots, t_s \leq n - 1$. By Engel's Theorem Q is a finite-dimensional Lie algebra over K. If $w_1, \ldots, w_p$ is a basis of Q and v is in M then the Q-

submodule W, generated by v, is the linear span of the elements

$$w_1^{r_1} \ldots w_p^{r_p} v.$$

For any $a \in L^q$, $w \in W$ we have $a^k w = 0$. It follows that $\dim W \le N = p(s+1)$. It is known that Q may be represented on W by upper triangular matrices (see, for example, [Bahturin, 1987a], Subsection 1.7.5). Therefore, $h_1 \ldots h_N v = 0$ for any $h_1, \ldots, h_N \in Q$.

For the completion of the proof of the lemma it is sufficient to verify that $b_1 \ldots b_N M = 0$ for any $b_1, \ldots, b_N$ of the form (11) which generate L^q.

We introduce a partial order on the set of elements (11) by setting

$$[e_1^{(t_1)}, \ldots, e_s^{(t_s)}, a_i] < [e_1^{(u_1)}, \ldots, e_s^{(t_s)}, a_j]$$

if $t_1 + \cdots + t_s < u_1 + \cdots + u_s$ or $t_1 + \cdots + t_s = u_1 + \cdots + u_s$ and the tuple $(t_1, \ldots, t_s)$ precedes lexicographically the tuple $(u_1, \ldots, u_s)$ if we compare the components from the right to the left. Analogously, we can order all tuples $(b_1, \ldots, b_N)$ where any b_i is of type (11) by comparing their components from the right to the left.

Suppose that $b_1, \ldots, b_N$ are generators of Q. Then $b_1 \ldots b_N v = 0$ for any $v \in M$. Now let b_i, for some i, be a commutator

$$[e_1^{(t_1)}, \ldots, e_s^{(t_s)}, a_j]$$

and suppose that at least one of the exponents, say t_s, is greater than $n-1$. Set

$$c = [e_s^{(t_s)}, e_1^{(t_1)}, \ldots, e_{s-1}^{(t_{s-1})}, a_j].$$

Using the Jacobi identity we can move all the letters e_s in b_i from the right to the left. Hence we may write $b_i - c$ as a linear combination of commutators which are similar to b_i with the degree in e_s strictly less than $n-1$, and with replacing one of the entries of some basis element e_{i_1} ($i_1 < s$) by the product $[e_{i_1}, e_{i_2}]$. The basis of L modulo L^q has been chosen in such a way, that any commutator $[e_{i_1}, e_{i_2}]$ is either one of the $a_1, \ldots, a_r$ or a linear combination of $e_1, \ldots, e_s$. Note that, if f is of the form (11) then substituting some a_p in place of one of the e_j in f yields an element f' which can be written as a linear combination of products $[f_1, f_2]$ where f_1, f_2 are of the form (11) and $f_1, f_2 < f$. Applying Lemma 2.3 to the module $L^q/(L^q)^2$ over the Lie algebra L/L^q we obtain

$$b_i - c = A + B$$

where A is a linear combination of some elements f_1 of the form (11) and B is a linear combination of products $[f_2, f_3]$ where f_2, f_3 are of type (11). Moreover, all f_1, f_2 and f_3 are strictly less than b_i.

It follows that $b_1 \ldots b_{i-1}(b_i - c)b_{i+1} \ldots b_N v$ can be expressed as a linear combination of the vectors $b_1 \ldots b_{i-1} f_1 f_2 b_{i+1} \ldots b_N v$, $b_1 \ldots b_{i-1} f b_{i+1} \ldots b_N v$ where $f, f_1, f_2 < b_i$. Using induction on the partial order on the set of tuples $(b_1, \ldots, b_N)$ we may assume that $b_1 \ldots b_{i-1}(b_i - c)b_{i+1} \ldots b_N v = 0$.

Now set $w = b_{i+1} \ldots b_N v$. By (7) cw is a linear combination of the elements $e_s^t c_t w$ where

$$c_t = [e_s^{(t_s - t)}, e_1^{(t_1)}, \ldots, e_{s-1}^{(t_{s-1})}, a_j], \qquad t > 0.$$

Obviously, $b_1 \ldots b_{i-1} e_s^t c_t w$ is contained in the L-module generated by the elements of the form $f_1 \ldots f_{i-1} c_t w$ where $f_l = [e_s^{(p)}, b_l]$ for some $p \geq 1$ ($l = 1, \ldots, i - 1$). For any j the commutator $[e_j, b_l]$ is a linear combination of some h_1, $[h_2, h_3]$ with h_1, h_2, h_3 of type (11). It follows that $b_1 \ldots b_{i-1} e_s^t c_t w$ lies in the module generated by the elements $g_1 \ldots g_l c_t w$ where $g_1, \ldots, g_l$ are of the form (11) and $l \geq i - 1$.

The commutator c_t is not of the form (11), but it may be represented as a linear combination of some h_1, $[h_2, h_3]$ with h_1, h_2, h_3 of type (11) and smaller than b_i. Consequently, $b_1 \ldots b_{i-1} e_s^t c_t w$ is contained in the L-module generated by elements $f_1 \ldots f_l b_i' w$, where $f_1, \ldots, f_l, b_i'$ are of the form (11), $l \geq i - 1$ and $b_i' < b_i$.

Thus we have verified that $b_1 \ldots b_N v$ belongs to the L-module generated by elements of the form $C = f_1 \ldots f_l b_i' b_{i+1} \ldots b_N v$. Since $b_i' < b_i$ and $l \geq i - 1$, the inductive hypothesis allows us to conclude that $C = 0$. This completes the proof of Lemma 2.4. □

For a colour Lie superalgebra the nilpotence of the action L^q on an L-module M with some additional restrictions implies the maximal condition on M.

2.5. Theorem. *Let $L = \bigoplus_{g \in G} L_g$ be a finitely generated colour Lie superalgebra over a field K, char $K \neq 2$, graded by a finite group G. Suppose that $M = \bigoplus M_g$ is a finitely generated L-module such that the representation of L on M, for any homogeneous component M_h and $g_1, g_2 \in G$, satisfies an identity of the form*

$$[y^{(n)}, z]v = \sum_{j=1}^{n} \lambda_j y^j [y^{(n-j)}, z]v,$$

$$d(z) = g_1, \qquad d(y) = g_2, \qquad v \in M_h, \qquad \lambda_1, \ldots, \lambda_n \in K.$$

(12)

If L^m acts on M as a nilpotent space of transformations for some m then M is a Noetherian L-module.

Proof. Consider the chain of L-submodules of M

$$M = M_0 \supset M_1 \supset \cdots \supset M_N \supset M_{N+1} = 0$$

where $M_i = L^m M_{i-1}$ for $i \geq 1$. Obviously it is sufficient to prove the maximal condition for every quotient $V_i = M_i/M_{i-1}$, $i = 1, \ldots, N$. Note that V_i is a module over the finite-dimensional superalgebra L/L^m. By Proposition 3.2.8 the universal enveloping algebra $U(L/L^m)$ is Noetherian and therefore the theorem will be proved if we verify that V_i is a finitely generated L-module.

The codimension of L^m in L is finite, because L is finitely generated. Furthermore, as an ideal of L, L^m is generated by a finite set of elements, e.g. by all products $[x_1, \ldots, x_m]$ where $x_1, \ldots, x_m$ are in a finite set of generators of L.

The following lemma implies Theorem 2.5.

2.6. Lemma. *Let L be a finitely generated G-gradèd colour Lie superalgebra over a field K, and let $\{e_1, \ldots, e_q\}$ be a homogeneous basis of L modulo L^m. Assume the homogeneous set $\{a_1, \ldots, a_r\}$ generates L^m as an ideal. Let M be a finitely generated L-module, and, for the representation of L on M, suppose that every homogeneous component M_h satisfies a family of identities of the form (12) with $d(z) = g_1$, $d(y) = g_2$, where h, g_1, g_2 are arbitrary elements in G, and $\lambda_1, \ldots, \lambda_n$ depend only on h, g_1, g_2. Put $M_0 = M$ and $M_j = L^m M_{j-1}$ if $j \geq 1$. Then for any $k \geq 0$ we have the equation $M_k = M_{k+1} + P_k$ where P_k is the L-submodule generated by all elements $b_1 \ldots b_k v$ with b_i of the form*

$$[e_1^{(t_1)}, \ldots, e_q^{(t_q)}, a_i], \quad 0 \leq t_1, \ldots, t_q \leq n-1, \quad 1 \leq i \leq r, \qquad (13)$$

and the vector v is one of the generators of M as an L-module.

Proof. We shall prove the relation $M_k = M_{k+1} + P_k$ by induction on k.

For $k = 0$ the statement is trivial. Suppose that $k \geq 1$ and that the relation $M_{k-1} = M_k + P_{k-1}$ is already proved. Then $M_k = M_{k+1} + H$ where H is the linear span of elements of the form

$$f e_1^{i_1} \ldots e_q^{i_q} b_2 \ldots b_k v$$

where $b_2, \ldots, b_k$ are of the form (13) and $f \in L^m$. It follows that M_k, as an L-module, is generated by M_{k+1} and some elements $f b_2 \ldots b_k v$ where v is one of the generators of M and $f \in L^m$. Moreover, we may assume that

$$f = [e_1^{(j_1)}, \ldots, e_q^{(j_q)}, a_i].$$

If $j_1, \ldots, j_q \leq n - 1$ then f is of the form (13), as required. Otherwise, one of the $j_1, \ldots, j_n$ is larger than $n - 1$. For example, let $j_q \geq n$. Set $B = b_2 \ldots b_k v$,

$$C = [e_q^{(j_q)}, e_1^{(j_1)}, \ldots, e_{q-1}^{(j_{q-1})}, a_i].$$

Using the identity (12) we obtain

$$CB = \sum_{t \geq 1} \lambda_t e_q^t C_t B$$

where

$$C_t = [e_q^{(j_q - t)}, e_1^{(j_1)}, \ldots, e_{q-1}^{(j_{q-1})}, a_i].$$

By Lemma 2.3 both C_t and $f - C$ can be written modulo $(L^m)^2$ as a linear combination of commutators $[e_1^{(s_1)}, \ldots, e_q^{(s_q)}, a_r]$ with $s_1 + \cdots + s_q < j_1 + \cdots + j_q$. Therefore, lowering the total degree $j_1 + \cdots + j_q$ we obtain $fB \in M_{k+1} + P$. Now the proof of Lemma 2.6 is complete, proving also Theorem 2.5. □

Later we will show that the identities of the form (2) of a colour Lie superalgebra L lead to identities of the form (12) of the adjoint representation of L. Hence Theorem 2.5 plays a significant role for the investigation of maximal conditions of colour Lie superalgebras.

We are now able to prove that a finitely generated soluble colour Lie superalgebra with identities of type (2) is a Noetherian module over itself. However, we aspire to a stronger result. The point is that L is not only an L-module but also an L_+- and L_0-module. Thus the question arises under which conditions does the Noetherian property hold for L as L_+- or L_0-module? The main obstacle in applying Theorem 2.5 to L as L_+- or L_0-module is the requirement of finite generation.

For a finitely generated superalgebra L the subalgebra L_+ is in general not finitely generated. Theorem 3.28 in Chapter 2 illustrates this fact. The following result is therefore of major importance for the purpose of the present chapter.

2.7. Lemma. *Let $L = \bigoplus_{g \in G} L_g$ be a finitely generated soluble colour Lie superalgebra over a field K such that for any $g_1, g_2 \in G$, $h \in G_+$ we have an identity of the form*

$$[x, y^{(n)}, z] = \sum_{j=1}^n \alpha_j [y^{(j)}, x, y^{(n-j)}, z], \quad d(x) = g_1, \quad d(z) = g_2, \quad d(y) = h, \quad (14)$$

with $\alpha_1, \ldots, \alpha_k \in K$. Then some finite set of elements generates L_+ as a colour Lie superalgebra. Furthermore L_- is a finitely generated L_+-module.

§2. Maximal condition and Hopf property

Proof. We shall construct a finitely generated subalgebra B in L_+ such that L_+ and L_- are modules of finite type over B. Given such a subalgebra B it is sufficient to check that the superalgebra L^2 is a finitely generated B-module, because $\dim L/L^2 < \infty$.

Suppose that the homogeneous elements $x_1, \ldots, x_m, y_1, \ldots, y_m$ generate L and $y_i \in L_+$, $x_i \in L_-$, $i = 1, \ldots, m$. It is easy to see that the colour Lie superalgebra L^2 is generated by

$$[y_1^{(t_1)}, \ldots, y_m^{(t_m)}, x_1^{(q_1)}, \ldots, x_m^{(q_m)}, u], \quad t_1 + \cdots + t_m + q_1 + \cdots + q_m \geq 1, \quad (15)$$

where $u \in \{x_1, \ldots, x_m, y_1, \ldots, y_m\}$.

We denote by H the subalgebra of L_+ generated by the elements y_i, x_i^2, $i = 1, \ldots, m$ and by M the H submodule in L generated by all commutators $[a_1, \ldots, a_r]$, $r \geq 1$, where $a_1, \ldots, a_r$ are of the form (15) and $0 \leq t_1, \ldots, t_m < n$, $0 \leq q_1, \ldots, q_m < 2n$.

We claim that $L^2 = M$. It is sufficient to verify the inclusion $[a_1, \ldots, a_r] \in M$ for all $a_1, \ldots, a_r$ of the form (15) without restrictions on the exponents $t_1, \ldots, t_m, q_1, \ldots, q_m$.

Introduce a partial order on the set of elements of type (15). We say that $a < a'$ if $t_1 + \cdots + t_m + q_1 + \cdots + q_m < t'_1 + \cdots + t'_m + q'_1 + \cdots + q'_m$ where the tuple $(t_1, \ldots, t_m, q_1, \ldots, q_m)$ defines a and the tuple $(t'_1, \ldots, t'_m, q'_1, \ldots, q'_m)$ defines a'. Extend this partial order to the set of tuples $(a_1, \ldots, a_r)$ with $a_1, \ldots, a_r$ of type (15) by comparing two tuples lexicographically from the right to the left.

Suppose that $[a_1, \ldots, a_r]$ is a commutator of elements of type (15) such that for any a_i we have $t_j \leq n - 1$, $q_j \leq 2n - 1$. Then $[a_1, \ldots, a_r]$ is contained in M by construction. This enables us to use induction on the partial order on the set of tuples $(a_1, \ldots, a_r)$. Now suppose that

$$a_i = [y_1^{(t_1)}, \ldots, y_m^{(t_m)}, x_1^{(q_1)}, \ldots, x_m^{(q_m)}, u]$$

and one of the exponents, say t_m, is greater than $n - 1$.

Set

$$c_i = [y_1^{(t_1)}, \ldots, y_{m-1}^{(t_{m-1})}, x_1^{(q_1)}, \ldots, x_m^{(q_m)}, u].$$

Using the Jacobi identity and Lemma 2.3 we can write the difference $a_i - [y_m^{(t_m)}, c_i]$ in the form of a linear combination of commutators $[b_1, b_2]$ where both b_1 and b_2 have the form (15) and are strictly less than a_i. Using the identities (14) one can express the commutator $[[y_m^{(t_m)}, c_i], f]$, f being arbitrary homogeneous, as a linear combination of commutators of the form

$$[y_m^{(j)}, [y_m^{(t_m - j)}, c_i], f], \quad j \geq 1.$$

Hence the element

$$[a_1,\ldots,a_{i-1},[y_m^{(t_m)},c_i],a_{i+1},\ldots,a_r]$$

is a linear combination of the products

$$[a_1,\ldots,a_{i-1},y_m^{(j)},c_{ij},a_{i+1},\ldots,a_r] \qquad (16)$$

where $c_{ij} = [y_m^{(t_m-j)}, c_i]$ and $j \geq 1$. The Jacobi identity allows us to move y_m in (16) from the right to the left. As a result of this procedure we obtain factors of the form $[y_m^{(l)}, a_k]$ in (16) instead of a_k, $k \leq i - 1$. If we express c_{ij} and $[y_m^{(l)}, a_k]$ as polynomials in the elements (15) then we can write (16) as a linear combination of the products $[y_m^{(t)}, \ldots, b_1, \ldots, b_s, a_{i+1}, \ldots, a_r]$. Since y_m is in H and taking into account the inductive hypothesis we obtain that $[a_1, \ldots, a_r]$ lies in $M + A$ where A is the H-module generated by elements $[b_1, \ldots, b_s, a_{i+1}, \ldots, a_r]$ where $s \geq i$, $b_1, \ldots, b_s$ are products of type (15) and $b_s < a_i$. Now the tuple $(b_1, \ldots, b_s, a_{i+1}, \ldots, a_r)$ is strictly less than $(a_1, \ldots, a_r)$, hence, by the inductive hypothesis, $A \subset M$ and $[a_1, \ldots, a_r] \in M$.

Arguing similarly in the case where one of the exponents $t_1, \ldots, t_{m-1}$ is greater than $n - 1$ or one of the $q_1, \ldots, q_m$ is greater than $2n - 1$ and moving y_i or x_i^2 to the left we get $[a_1, \ldots, a_r] \in M$ in the general case.

Now the equality $L^2 = M$ has been proved, and we are going to construct a finitely generated subalgebra $B \subset L_+$ such that L_+ and L_- are finitely generated B-modules. Recall that L is a soluble superalgebra. If $L^2 = 0$ then $B = L_+ = \langle y_1, \ldots, y_m \rangle$. For the general case denote by Q the subalgebra of L generated by the elements of type (15) with $t_i < n$, $q_i < 2n$, $i = i, \ldots, m$. Then Q is a finitely generated colour Lie superalgebra and the solubility length of Q is strictly smaller than that of L. Hence it is possible to assume that Q_+ contains a finitely generated subalgebra C such that Q_+ and Q_- are C-modules generated by sets of homogeneous elements of the form $\{Y_1, \ldots, Y_N\}$, $\{X_1, \ldots, X_N\}$ respectively. Set $B = \text{alg}\{C, y_1, \ldots, y_m, x_1^2, \ldots, x_m^2\}$. Then the B-module generated by $Y_1, \ldots, Y_N, X_1, \ldots, X_N$ contains Q since $B \supset C$. On the other hand, B contains H, therefore the B-module generated by $Y_1, \ldots, Y_N, X_1, \ldots, X_N$ contains L^2. This follows from $L^2 = M$. Consequently the elements $y_1, \ldots, y_m, Y_1, \ldots, Y_N$ generate L_+ and the elements $x_1, \ldots, x_m, X_1, \ldots, X_N$ generate L_- as a B-module, and the lemma is proved. □

2.8. Corollary. *Let L be a finitely generated soluble colour Lie superalgebra over a field K, char $K \neq 2$, which is graded by a finite group G. If for any $g_1, g_2 \in G$, $h \in G_+$ an identity of the form (14) holds in L and if L_+^m, for some $m \geq 1$, acts on L as a nilpotent space of transformations then L is a Noetherian L_+-module.*

§ 2. Maximal condition and Hopf property

Proof. By Lemma 2.7 L_+ is a finitely generated superalgebra and L_- is a finitely generated L_+-module. If $x \in L$, $y, z \in L_+$, $d(x) = g_1$, $d(y) = h$, $d(z) = g_2$ then the identity (14) can be rewritten in the form

$$[[y^{(n)}, z], x] = \sum_{j=1}^{n} \lambda_j [y^{(j)}, [y^{(n-j)}, z], x], \qquad (17)$$

where

$$\lambda_j = \alpha_j \varepsilon(g_2 + nh, g_1) \cdot \varepsilon(g_1, g_2 + jh).$$

The superalgebra L_+ can be viewed as a G-graded colour Lie superalgebra with zero components $(L_+)_g$ for every $g \in G_-$. Now the family of identities (17) may be written as a system of identities of type (12) for L_+ and its representation on L. Therefore one can apply Theorem 2.5 to L_+ and its module L to conclude the proof. □

2.9. Necessary conditions for Noetherian and Hopf properties in varieties of superalgebras. It was shown that the family of identities of the form (14) together with some additional conditions implies that a colour Lie superalgebra is Noetherian. Let $\mathfrak{B}$ be a variety of colour Lie superalgebras over an infinite field K and assume that every finitely generated superalgebra in $\mathfrak{B}$ is Noetherian. It will be shown that a part of the family of identities (14) holds in $\mathfrak{B}$.

In the first section we constructed a series of colour Lie superalgebras $Q(g, h)$. Every $Q(g, h)$ contains a central subalgebra of infinite dimension and therefore it is not Noetherian. Moreover, the $Q(g, h)$ do not satisfy the (weaker) Hopf property. Hence, none of these algebras lies in $\mathfrak{B}$.

Proposition. *Let $\mathfrak{B}$ be a variety of G-graded colour Lie superalgebras over an infinite field K, char $K \neq 2$. Assume $\mathfrak{B}$ does not contain all algebras $Q(g, h)$, $g, h \in G$. Then for every $g, h \in G$ an identity of the form*

$$[x, y^{(n)}, z] = \sum_{j=1}^{n} \alpha_j [y^{(j)}, x, y^{(n-j)}, z]$$

is satisfied in $\mathfrak{B}$ with $d(x) = g$, $d(z) = h$, $d(y) = 0$.

For the proof of this proposition we need two lemmas about (usual) Lie algebras.

Let L be a free Lie algebra of infinite rank over K and $C = [L, L^2, L^2]$.

2.10. Lemma. *Suppose that V is an ideal of identities in the algebra L and let x, y, z be free generators of L.*

1) *If the ideal $V + C$ contains the element*

$$[[x, y], x^{(k)}, y^{(l)}, z^{(m)}, x, z]$$

then it also contains the element

$$[[x, y], y^{(k+l+m+1)}, z].$$

2) *If the ideal $V + C$ contains the element*

$$[[x, y], x^{(k)}, y^{(l)}, x, y]$$

of odd degree then it contains the element

$$[[x, y], y^{(k+l+1)}, z].$$

Proof. Note first that any element of the type

$$[[x, y], z_1, \ldots, z_k, t, z] - [[x, y], z_{i_1}, \ldots, z_{i_k}, t, z]$$

lies in C if $\{i_1, \ldots, i_k\} = \{1, \ldots, k\}$. Furthermore

$$[[z_1, \ldots, z_k, x, y], z, t] \equiv (-1)^k [[x, y], z_1, \ldots, z_k, z, t] \pmod{C}.$$

This relations will be frequently used in what follows.

Let us prove the first statement of the lemma by induction on m. Assume $m = 0$. If $l = 0$ then it is sufficient to make the substitution $x \mapsto y$, $y \mapsto x$. Now let $l \geq 1$, $f = f(x, y, z) = [[x, y], x^{(k)}, y^{(l)}, x, z]$ and let f_1 be a partial linearization of f in y of degree 1. Recall that f_1 is the sum of monomials of degree 1 in t in the polynomial $f(x, y + t, z)$. Since K is an infinite field and $V + C$ is an ideal of identities in L, $f_1 \in V + C$ (see Subsection 1.2.3). On the other hand,

$$f_1 = [[x, t], x^{(k)}, y^{(l)}, x, z] + l[[x, y], x^{(k)}, y^{(l-1)}, t, x, z].$$

Denote by φ the endomorphism of L replacing x by y, t by x and preserving all other generators. Then $\varphi(f_1) \in V + C$ and $\varphi(f_1) = -[[x, y], y^{(k+l+1)}, z]$.

Now let $m > 0$. A partial linearization of $f = [[x, y], x^{(k)}, y^{(l)}, z^{(m)}, x, z]$ by z yields the polynomial

$$f = m[[x, y], x^{(k)}, y^{(l)}, z^{(m-1)}, t, x, z] + [[x, y], x^{(k)}, y^{(l)}, z^{(m)}, x, t]. \quad (18)$$

If $m \equiv 0 \pmod{p}$ where $p = \operatorname{char} K$ then we substitute in (18) y, z for z, t, respectively. As a result we obtain the element $[[x, y], x^{(k)}, y^{(l+m)}, x, z]$ which belongs to $V + C$. As before (where $m = 0$) the ideal $V + C$ contains the required element. If the characteristic of K does not divide m then we replace x by t in (18). As a result we get $m[[x, y], x^{(k+1)}, y^{(l)}, z^{(m-1)}, x, z]$. Induction on m completes the proof of the first statement.

Now let $V + C$ contain the element $f = [[x, y], x^{(k)}, y^{(l)}, x, y]$. If $l = 0$ then $f = [[x, y], x^{(k+1)}, y]$. Linearizing f in y we obtain

$$f_1 = [[x, y], x^{(k+1)}, z] + [[x, z], x^{(k+1)}, y].$$

Since k is odd, the first and the second summand of f_1 coincide. This completes the proof for $l = 0$ because char $K \neq 2$.

Suppose that $l > 0$. Linearization of f in y and reduction modulo C results in

$$f_1 = 2[[x, y], x^{(k)}, y^{(l)}, x, z] + l[[x, y], x^{(k)}, y^{(l-1)}, z, x, y].$$

If char K does not divide l then replacing z by x in f_1 we obtain $l[[x, y], x^{(k+1)}, y^{(l-1)}, x, y]$ and one can use induction on l. If $l \equiv 0 \pmod{\text{char } K}$ then $f_1 = 2[[x, y], x^{(k)}, y^{(l)}, x, z]$, and we can apply the first part of the lemma, completing the proof of the second statement. □

2.11. Lemma. *Let V be an ideal of identities in L. Assume L is contained in a variety of Lie algebras $\mathfrak{B}$ which does not include the algebra $Q(0, 0)$. Then the ideal $V + C$ contains the element $[[x, y], y^{(k)}, z]$ for some $k \geq 1$, where x, y, z are free generators of L.*

Proof. Recall that $Q = Q(0, 0) = \langle a_i, b_i, c_j, \delta \rangle$, where $i, j \in \mathbb{Z}, j \geq 0, [a_i, b_j] = c_{i-j}$ if $i \geq j$, $[\delta, a_i] = a_{i-1} + a_{i+1}$, $[\delta, b_j] = -b_{j-1} - b_{j+1}$, and all other products of basis elements are equal to zero. It is not difficult to check that $[Q, Q^2, Q^2] = 0$.

By the hypotheses of the lemma there exists an identity $f = 0$ for $\mathfrak{B}$ which does not hold in Q. Since K is infinite, f may be chosen homogeneous (see Subsection 1.2.3).

First let $f = 0$ be a multilinear identity depending on variables $x_1, \ldots, x_n$. We can assume $n > 3$, because any nontrivial multilinear identity $f = 0$ of degree $n \leq 3$ implies $[x_1, \ldots, x_n] = 0$. All elements of C become zero if we replace the variables by elements of Q. Therefore $f \in C$. We can represent f modulo $(L^2)^2$ as

$$f \equiv \sum_{i=2}^{n} \beta_i [x_n, x_{n-1}, \ldots, x_{i+1}, x_{i-1}, \ldots, x_2, x_i, x_1] \pmod{(L^2)^2}. \quad (19)$$

If one of the β_i is nonzero, for instance, $\beta_n \neq 0$, then the substitution $x_n \mapsto z$, $x_j \mapsto y$, $j \neq n$, yields the element $[y^{(n-1)}, z]$ on the right hand side of (19). A homogeneous element $g = g(y, z)$ in $(L^2)^2$ is zero if the degrees of g in y and z are $n - 1$ and 1, respectively. Hence there is an element $[y^{(n-1)}, z]$ in $V + C$ and multiplying it by $[x, z]$ we obtain

$$[[x, y], y^{(k)}, z] \in V + C.$$

Now suppose that f is in $(L^2)^2$. Modulo C, the element f is equal to

$$\sum_{i,j=1}^{n} \alpha_{ij}[[\ldots x_i, x_n], x_j, x_{n-1}] \qquad (20)$$

where $\alpha_{ij} \in K$. We use the notation $[\ldots x_i, x_n]$ for the commutator $[x_{k_1}, \ldots, x_{k_{n-4}}, x_i, x_n]$ where $k_1 > \cdots > k_{n-4}$ and $\{k_1, \ldots, k_{n-4}\} = \{1, \ldots, n-2\} \setminus \{i, j\}$ since the set of variables $\{x_{k_1}, \ldots, x_{k_{n-4}}\}$ and their order are defined entirely by i, j.

Suppose $\sum_j \alpha_{ij} \neq 0$ for some i. In (20) we substitute x for x_{n-1}, z for x_i and y for all the other x_k and obtain the product $[[x, y], y^{(n-3)}, z]$ with nonzero coefficient $\alpha_{i_1} + \cdots + \alpha_{i,n-2}$.

In case $\sum_i \alpha_{ij} \neq 0$ for some j, the same result is achieved if we write x, z in place of x_n, x_j, respectively, and y in place of all other variables in (20).

Now suppose that for some i, j the scalar $\alpha_{ij} - (-1)^n \alpha_{ji}$ is nonzero. Replace, in (20), x_i by z, x_j by x and all the other x_k by y. Then

$$[[\ldots x_i, x_n], x_j, x_{n-1}] = -[[y^{(n-3)}, z], x, y]$$

and

$$[[\ldots x_j, x_n], x_i, x_{n-1}] = [[y^{(n-3)}, x], y, z]$$
$$\equiv (-1)^{n-4}[[y, x], y^{(n-3)}, z] \pmod{C}.$$

Hence, modulo C, (20) is equal to

$$(\alpha_{ij} - (-1)^n \alpha_{ji})[[x, y], y^{(n-3)}, z].$$

To complete the proof for a multilinear polynomial f we must verify that (20) is equal to zero for any $x_1, \ldots, x_n$ in the Lie algebra $Q(0, 0)$ provided that

$$\left.\begin{aligned} \sum_{j=1}^{n-2} \alpha_{ij} &= 0, & i &= 1, \ldots, n-2 \\ \sum_{i=1}^{n-2} \alpha_{ij} &= 0, & j &= 1, \ldots, n-2 \\ \alpha_{ij} - (-1)^n \alpha_{ji} &= 0, & i, j &= 1, \ldots, n-2. \end{aligned}\right\} \qquad (21)$$

Since f is a multilinear polynomial it is sufficient to check the equality $f(x_1,\ldots,x_n) = 0$ in Q only for variables $x_1, \ldots, x_n$ taken from the set $\{a_i, b_j, c_k, \delta\}$. Let P be the linear span of all a_i, b_j, c_k in Q. Then $P^3 = 0$. Therefore not more than two elements from P can be substituted into (20).

If only one of the variables $x_1, \ldots, x_n$ is in P then f equals zero since $\dim Q/P = 1$.

Replace x_i, x_j in (20) by $A, B \in P$, where $i, j \leq n - 2$ and substitute δ for all other variables. Then

$$f_{ij} = [[\ldots x_i, x_n], x_j, x_{n-1}] = [[\delta^{(n-3)}, A], \delta, B] = A_0,$$

$$f_{ji} = [[\ldots x_j, x_n], x_i, x_{n-1}] = [[\delta^{(n-3)}, B], \delta, A]$$

$$= (-1)^{n-4}[[\delta, B], \delta^{(n-3)}, A] = (-1)^{n+1} A_0.$$

It follows that $f = (\alpha_{ij} + (-1)^{n+1}\alpha_{ji})A_0 = 0$. Replace x_n by A, x_j by B and x_i by δ for any $i \neq j, n$. Then $f_{ij} = -[[\delta^{(n-3)}, A], \delta, B]$ if $1 \leq i \leq n - 2$, $i \neq j$. After this substitution f becomes zero since $\alpha_{1j} + \cdots + \alpha_{n-2,j} = 0$. A similar result is achieved by the substitution $x_{n-1} \mapsto B$, $x_i \mapsto A$, $x_j \mapsto \delta$ for all $j = 1, \ldots, n - 2, j \neq i$, using the equation $\alpha_{i1} + \cdots + \alpha_{i,n-2} = 0$.

Finally, if we replace x_n by A, x_{n-1} by B and all x_i by δ, $i = 1, \ldots, n - 2$, in (20) then

$$f = \left(\sum_{i,j} \alpha_{ij}\right)[[\delta^{(n-3)}, A], \delta, B].$$

This element is zero since the sum all of α_{ij} is zero, as follows from (21). Hence if a multilinear identity $f = 0$ holds in $\mathfrak{B}$ but not in Q our lemma is proved.

Now let $f = f(x_1,\ldots,x_n) = 0$ be a multihomogeneous identity holding in $\mathfrak{B}$ but not in Q and let the degree of f in one of the variables, say x_n, be larger than 1. Then, modulo C, f is equal to

$$h = \sum_{1 \leq i \leq j \leq n-1} \gamma_{ij} h_{ij} \qquad (22)$$

where $h_{ij} = [[\ldots x_i, x_n], x_j, x_n]$.

Suppose first that $\gamma_{ii} = 0$ for all $i = 1, \ldots, n - 1$. If $\gamma_{ij} \neq 0$ for some i, j then after replacing in (22) x_i, x_j by z, y, respectively, and all other x_r by x, we obtain

$$[[x^{(k)}, x^{(l)}, z^{(m)}, z, x], y, x]$$

with a nonzero coefficient γ_{ij}. In this case Lemma 2.10 completes the proof.

Now let γ_{ii} be nonzero for some i. One may assume that h is of odd degree since the element h_{ii} of even degree lies in C. Substitute x for x_i and y for the other x_j, $1 \le j \le n, j \ne i$, in (22). From this substitution we obtain for h_{ii} the commutator $[[x^{(k)}, y^{(l)}, x, y], x, y]$ and zero for the other h_{jr}. The second part of Lemma 2.10 now completes the proof. □

2.12. Proof of Proposition 2.9. Since K is an infinite field we assume that all identities are multihomogeneous.

We start with considering the condition $\mathfrak{B} \not\supseteq Q(0,0)$. By Lemma 2.11 the ideal $V + C$ contains the element $f = [[x, y], y^{(k)}, z]$ where V is the ideal of identities of $\mathfrak{B}$ in a free colour Lie superalgebra L, $C = [L, L^2, L^2]$ and $d(x) = d(y) = d(z) = 0$. Any multihomogeneous element $b = b(x, y, z)$ in C of degrees 1, $k + 1$, 1 in x, y, z, respectively, belongs to the linear span of the products

$$[y^{(i)}, [y^{(j)}, x], y^{(k-i-j+1)}, z], \qquad i \ge 1.$$

Since $f + b \in V$ for some $b \in C$ we have

$$[x, y^{(k+1)}, z] = [y, x, y^{(k)}, z] + b, \qquad b \in C. \tag{23}$$

Taking into account the form of b we conclude that the relation (23) is an identity of required type with $d(x) = d(y) = 0$.

Now suppose $d(x) = d(y) = 0$, $d(z) = h \ne 0$. Consider the algebra $Q = Q(0, h)$. We have

$$Q_0 = \langle \delta, a_i | i \in \mathbb{Z} \rangle; \qquad Q_h = \langle b_i, c_j | i \in \mathbb{Z}, j \ge 0 \rangle$$

and $Q_g = 0$ if $g \ne 0, h$. If $X = \bigcup_{g \in G} X_g$ is a graded set of free generators of the colour Lie superalgebra $L(X)$ then any multihomogeneous identity $f = f(x_1, \ldots, x_n)$ holds in Q provided that some variable x_i is in $X \setminus (X_0 \cup X_h)$. Consequently, the condition $\mathfrak{B} \not\supseteq Q(0, h)$ implies the existence of a multihomogeneous identity in $\mathfrak{B}$ which does not hold in Q and depends only on variables from $X_0 \cup X_h$. Since Q_h is an abelian ideal, Q satisfies any multihomogeneous identity depending on more than one variable in X_h.

Assume first that there is no identity with one variable in X_h which is satisfied in $\mathfrak{B}$ but not in Q. Now since Q_0 is metabelian, any multihomogeneous identity in variables of X_0 valid in $\mathfrak{B}$ but not in Q can be written in the form

$$\sum_{k=1}^{n} \beta_k [x_1, \ldots, x_{k-1}, x_{k+1}, \ldots, x_n, x_k, x_0] = 0 \tag{24}$$

modulo the metabelian identity, where alike terms are reduced. (Note that this identity as well as some others to follow are not assumed to be multilinear.) One may assume that the set X is linearly ordered and $x_0 \geq x_1 \geq \cdots \geq x_n$ in (24). If not all β_k equal zero, e.g. $\beta_1 \neq 0$, then setting in (24) $x_1 = x + y$, $x_i = y$ for $x_i \neq x_1$ with $d(x) = d(y) = 0$ and taking the sum of monomials of degree 1 in x we get $[y^{(n)}, x] = 0$. Since K is an infinite field, the identity $[y^{(n)}, x] = 0$, $d(x) = d(y) = 0$, follows from (24). It is obvious that the same substitution in each of the consequences of $[[x, y], z, u] = 0$ gives zero. Multiplying by $z \in X_h$ we obtain, from our assumption, that $\mathfrak{B}$ satisfies the identity $[z, y^{(n)}, x] = 0$, $d(x) = d(y) = 0$, $d(z) = h$.

Note that for any $k \geq 0$ we have the relation

$$[z, y^{(k)}, x] = (-1)^{k+1}\varepsilon(2h, 0)[x, y^{(k)}, z] + \sum_{j=1}^{k} \gamma_j [y^{(j)}, x, y^{(k-j)}, z] \qquad (25)$$

with $\gamma_1, \ldots, \gamma_k \in K$. For $k = 1$ this follows from the Jacobi identity for colour Lie superalgebras. If $k > 1$ then we write

$$[z, y^{(k)}, x] = [[z, y], y^{(k-1)}, x] + \varepsilon(h, 0)[y, z, y^{(k-1)}, x]. \qquad (26)$$

By induction on k we may express the second summand in (26) as

$$\sum_{j \geq 1} \beta_j [y^{(j)}, x, y^{(k-j)}, z].$$

The first summand in (26) is

$$-\varepsilon(h, 0)[[y^{(k-1)}, x], z, y] = \varepsilon(2h, 0)[[y^{(k-1)}, x], y, z]$$

$$= \varepsilon(2h, 0) \sum_{j=0}^{k} (-1)^j \binom{k-1}{j} [y^{(k-1-j)}, x, y^{(j)}, y, z].$$

This relation proves (25).

Since the identity $[z, y^{(n)}, x] = 0$, $d(x) = d(y) = 0$, $d(z) = h$, holds in $\mathfrak{B}$, (25) gives us the required identity

$$[x, y^{(n)}, z] = \sum_{j=1}^{n} \alpha_j [y^{(j)}, x, y^{(n-j)}, z], \qquad d(x) = d(y) = 0, \qquad d(z) = h.$$

Now suppose $\mathfrak{B}$ satisfies an identity $f = 0$ with one variable in X_h which is not satisfied in $Q(0, h)$. Since $Q(0, h)$ satisfies

$$[x_1, [x_2, x_3], z] = 0, \qquad x_i \in X_0, \qquad z \in X_h, \qquad (27)$$

the additional identity f is not a consequence of (27). We reduce it modulo (27) to the form

$$\sum_{k=0}^{n} \gamma_k [x_k, x_0, \ldots, x_{k-1}, x_{k+1}, \ldots, x_n, z] = 0 \qquad (28)$$

with $d(z) = h$, $d(x_0) = \cdots = d(x_n) = 0$ and $x_0 \geq \cdots \geq x_n$. Let $\gamma_n \neq 0$. Then, setting $x_n = x + y$ and $x_i = y$ for $x_i \neq x_n$ in (28) and taking the sum of monomials of degree 1 in x yields an identity of the form

$$\gamma_n [x, y^{(n)}, z] + \beta [y, x, y^{(n-1)}, z] = 0.$$

On the other hand, if we apply the same substitution to the consequences of (27) we obtain terms of the form $[y^{(i)}, x, y^{(j)}, z]$, $i \geq 1$. Hence $\mathfrak{B}$ satisfies an identity of the required form.

We now consider the general case $Q = Q(g, h)$ with $g, k \neq 0$ and assume that $Q \notin \mathfrak{B}$. As before, $\mathfrak{B}$ must satisfy an identity in variables from $X_0 \cup X_g \cup X_h \cup X_{g+h}$.

If $g + h \neq 0$ then $Q_{g+h} = \langle c_1, c_2, \ldots \rangle$, and the only identity including a variable $u \in X_{g+h}$ which does not hold in Q is $u = 0$. However, it follows from this identity that $[x, z] = 0$ with $x \in X_g$, $z \in X_h$, and this implies $[x, y, z] = 0$ with $x \in X_g$, $y \in X_0$, $z \in X_h$, i.e. the required identity. Hence, we may assume that the identites of $\mathfrak{B}$ which do not hold in Q include only variables in $X_0 \cup X_g \cup X_h$.

It is quite clear that an "additional" identity cannot contain more than two variables in $X_g \cup X_h$ since $Q_g + Q_h + Q_{g+h}$ is a 2-step nilpotent ideal in Q. If this identity depends only on variables in X_0 then it must be $y = 0$ with $y \in X_0$ since Q is abelian.

If the additional identity contains only one variable which is not in X_0, say $z \in X_h$, then since $[y_1, y_2] = 0$ in Q_0 for $y_1, y_2 \in X_0$ this multihomogeneous identity is equivalent to $[y_1, \ldots, y_n, z] = 0$, $y_1 \geq \cdots \geq y_n$.

We identify all $y_1, \ldots, y_n$ with some y and multiply the result by x to get $[x, y^{(n)}, z] = 0$, as required.

Now let the additional identity contain two variables $x \in X_g$, $z \in X_h$ and the remaining variables in X_0. Since $Q = Q(g, h)$ satisfies

$$[y, x, z] = 0, \quad d(y) = 0, \quad d(x) = g, \quad d(z) = h,$$
$$[y_1, y_2] = 0, \quad d(y_1) = d(y_2) = 0, \qquad (29)$$

our additional identity reduces, modulo (29), to the form $[x, y_1, \ldots, y_n, z] = 0$. If we identify $y_1, \ldots, y_n$ then we obtain $[x, y^{(n)}, z] = 0$. As for the consequences of the first identity in (29), either they are zero or they have the form of a linear combination of monomials of the form

$$[y^{(i)},[y^{(k)},x],y^{(n-i-k)},z].$$

If $g + h = 0$, the consequences of the second identity in (29) have the same form, and if $g + h \neq 0$ then all the consequences of the second identity are zero. The proof is complete. □

2.13. Local properties of varieties. A variety of colour Lie superalgebras is defined by a family $\{f_i = 0, i \in I\}$ of graded identities. Nevertheless sometimes we shall be using nongraded identities of the form $f(x_1,\ldots,x_n) = 0$ where f is a multilinear polynomial. If the inclusions $x_i \in L_g$, $i = 1,\ldots,n$ are not indicated then $f(a_1,\ldots,a_n)$ must be zero for any $a_1,\ldots,a_n \in L$. On the other hand, $f = 0$ may be considered as a family of graded identities

$$f(x_1,\ldots,x_n) = 0, \qquad d(x_1) = g_1, \ldots, d(x_n) = g_n \qquad (30)$$

for all $g_1,\ldots,g_n \in G$. This means that a class of superalgebras satisfying the nongraded identity $f = 0$ is the variety of graded algebras defined by the family (30). For a finite group G the set of identities (30) is finite. For example, all soluble superalgebras with solubility length less than m form a variety of colour Lie superalgebras.

Denote by $\mathscr{P}$ some property for colour Lie superalgebras over a field K. We shall write $L \in \mathscr{P}$ if this property holds in L. For example, if $\mathscr{P}$ stands for finiteness of dimension, then $L \in \mathscr{P}$ simply means $\dim L < \infty$.

We say that $\mathscr{P}$ holds in a variety $\mathfrak{B}$ locally if $L \in \mathscr{P}$ for every finitely generated superalgebra L in $\mathfrak{B}$. For example, let $\mathscr{P}$ be the solubility of colour Lie superalgebras. Then $\mathfrak{B}$ is called *locally soluble* if for any finitely generated $L \in \mathfrak{B}$ there exists a number m such that $L^{(m)} = 0$, where $L^{(0)} = L$, $L^{(i+1)} = [L^{(i)}, L^{(i)}]$ for $i \geq 1$.

Now we can define a *locally Noetherian* variety $\mathfrak{B}$ of colour Lie superalgebras as one in which any algebra with finitely many generators satisfies the maximal condition for ideals. Similarly, $\mathfrak{B}$ is said to be *locally Hopf* is every finitely generated superalgebra in $\mathfrak{B}$ possesses the Hopf property. Corollary 2.8 implies that any locally soluble variety $\mathfrak{B}$ of colour Lie superalgebras defined by the set of identities (14) with homogeneous $x, y, z \in L_+$, is locally Noetherian provided that it has the additional system of identities

$$[[x_1,\ldots,x_m],[y_1,\ldots,y_m],\ldots,[z_1,\ldots,z_m],v] = 0$$

with homogeneous $x_i, y_i, z_i \in L_+$, $i = 1,\ldots,m$. As it is shown in Subsection 1.1, $\mathfrak{B}$ is also locally Hopf.

For usual $\mathbb{Z}_2$-graded Lie superalgebras over an infinite field we give a complete classification of the varieties with maximal condition.

2.14. Theorem. *For a locally soluble variety $\mathfrak{B}$ of Lie superalgebras over an infinite field K the following conditions are equivalent.*
a) *$\mathfrak{B}$ is a locally Noetherian variety.*
b) *Any finitely generated Lie superalgebra $L = L_0 \oplus L_1$ in $\mathfrak{B}$ is a Noetherian L_0-module.*
c) *$\mathfrak{B}$ is a locally Hopf variety.*
d) *$\mathfrak{B}$ satisfies for some n the following three identities*

$$[x, y^{(n)}, z] = \sum_{j=1}^{n} \alpha_j^i [y^{(j)}, x, y^{(n-j)}, z] \tag{31}$$

with $\alpha_j^i \in K$, $i = 1, 2, 3$, $d(y) = 0$ where $d(x) = d(z) = 0$ for $i = 1$, $d(x) = 0$, $d(z) = 1$ for $i = 2$ and $d(x) = d(z) = 1$ if $i = 3$.

Proof. Obviously, b) implies a). As it has been shown earlier (see Subsection 1.1) any Noetherian superalgebra possesses the Hopf property, hence c) follows from a). If $\mathfrak{B}$ satisfies c) then it does not contain any of the superalgebras $Q(g, h)$, $g, h \in \mathbb{Z}_2$ (see Subsection 1.2). By Proposition 2.9 $\mathfrak{B}$ satisfies the identities (31). Thus, c) implies d).

To prove the remaining implication, d) $\Rightarrow$ b), let $\mathfrak{B}$ satisfy the identities (31). First prove that a similar identity holds in $\mathfrak{B}$ for $d(x) = 1$, $d(z) = 0$. Consider (31) with $i = 2$. As in the proof of Proposition 2.9 we have a relation similar to (25):

$$[x, y^{(k+1)}, z] = (-1)^k [z, y^{(k+1)}, x] + \sum_{i=0}^{k} \gamma_i [y^{(1+i)}, z, y^{(k-i)}, x]$$

for any $k \geq 0$.

Now let $L = L_0 \oplus L_1$ be a finitely generated Lie superalgebra in $\mathfrak{B}$. According to the hypotheses of the theorem L is soluble. By Lemma 2.7 L_0 is finitely generated and so is L, as an L_0-module. It is easy to check (cf. the proof of Corollary 2.8) that the representation of L_0 both on L_0 and L_1 satisfies an identity of the form

$$[y^{(n)}, z] = \sum_{j=1}^{n} \lambda_j y^j [y^{(n-j)}, z].$$

By Lemma 2.4 there exists a number m such that L_0^m acts on L as a nilpotent space of transformations. By Corollary 2.8 L is a Noetherian L_0-module. □

§3. Sufficient conditions for residual finiteness

3.1. Residual finiteness and identities. Recall that an algebra L is residually an algebra in the family $\{L_i\}$, $i \in I$, if for any nonzero $x \in L$ there exists a homomorphism φ such that $\varphi(L) \in \{L_i\}$ and $\varphi(x) \neq 0$. Suppose that the family $\{L_i\}$ contains with L_i every subalgebra $H_i \subset L_i$ as well. Then L is residually in $\{L_i\}$, $i \in I$, if and only if L can be embedded in the cartesian product $\prod_{i \in I} L_i$. In this section we prove the residual finiteness of finitely generated colour Lie superalgebras satisfying identities of some special type. Later it will be shown that the results of this section can be extended and that a colour Lie superalgebra satisfying these identities is residually an n-dimensional superalgebra, for some $n \in \mathbb{N}$. Residual finiteness is an important property in the study of a number of algorithmic problems. It is also connected with matrix representations of finitely generated algebras as will be shown in the next section.

Our present interest is to find identities which imply residual finiteness. The example of the Heisenberg superalgebra $\Gamma(g,h)$ (see Section 1) shows that even the identity $[x_1, x_2, x_3] = 0$ is not sufficient for residual finiteness. One can easily prove that, for an infinite field K, there is a unique variety consisting of residually finite colour Lie superalgebras; it is defined by $[x, y] = 0$, $d(x) = g$, $d(y) = h$ for all $g, h \in G$. If we assume residual finiteness only for finitely generated superalgebras in $\mathfrak{B}$ then, following the terminology of Subsection 2.13, $\mathfrak{B}$ is called locally residually finite.

The family of superalgebras $B(g, h)$ constructed in Section 1 shows that the center-by-metabelian variety is not locally residually finite. However, the identity

$$[[x, y], z, t] = 0 \tag{1}$$

implies residual finiteness for a colour Lie superalgebra with finitely many generators. Recall that an algebra with identity (1) is called metabelian.

3.2. Theorem. *A finitely generated metabelian colour Lie superalgebra with finite grading group is residually finite.* □

In fact Theorem 3.2 is a particular case of a more general result which will be proved later.

We need some auxiliary results.

3.3. Lemma. *Let L be finitely generated soluble colour Lie superalgebra with grading group G. Suppose that for any $g, h \in G$, $r \in G_+$ an identity of the following form holds in L:*

$$[x, y^{(n)}, z] = \sum_{j=1}^{n} \alpha_j [y^{(j)}, x, y^{(n-j)}, z], \qquad d(x) = g, \qquad d(y) = r, \qquad d(z) = h. \tag{2}$$

Let M be a finitely generated L-module, and assume that for every homogeneous component M_g we have an identity of the form

$$[y^{(n)}, z]v = \sum_{j=1}^{n} \lambda_j y^j [y^{(n-j)}, z]v,$$
(3)

$v \in M_g$, $d(z) = h$, $d(y) = r$; $g, h \in G$, $r \in G_+$.

Then, for any $m \geq 1$, M is a finitely generated module over the Lie superalgebra $H = H_+ \oplus H_-$, where $H_+ = L_+$, $H_- = [L_+^m, L_-]$.

Proof. By Lemma 2.7 L_+ is a finitely generated superalgebra and L_- is a finitely generated L_+-module with homogeneous generating set $x_1, \ldots, x_q$. If $e_1, \ldots, e_k$ is a homogeneous basis for L_+ modulo L_+^m then L_- modulo $[L_+^m, L_-]$ is the linear span of the commutators

$$[e_1^{(t_1)}, \ldots, e_k^{(t_k)}, x_i], \quad 1 \leq i \leq q, \quad t_1 + \cdots + t_k \geq 0. \tag{4}$$

We introduce a linear order on the set of products (4) by setting $b < b'$ if $t_1 + \cdots + t_k < t_1' + \cdots + t_k'$ or else if $t_1 + \cdots + t_k = t_1' + \cdots + t_k'$ and $(t_1, \ldots, t_k, i)$ is lexicographically smaller than $(t_1', \ldots, t_k', i')$ where the components are compared from the right to the left. If $\{u_j\}$ is a finite set of homogeneous generators of M as an L-module then by the Poincaré-Birkhoff-Witt Theorem (Subsections 3.2.2, 3.2.3) all the elements

$$b_1 \ldots b_r u_j, \quad b_1 > \cdots > b_r, \quad r \geq 0, \tag{5}$$

generate M as an H-module where $b_1, \ldots, b_r$ are of the form (4).

Next we consider the H-submodule T in M generated by those elements of type (5) with $t_1, \ldots, t_k < n$ for all b_i. To prove the lemma it is sufficient to verify that all the elements of type (5) belong to T.

Actually, we prove a more general statement: for any $c_1, \ldots, c_r \in L$ the element $c_1 \ldots c_r u_j$ lies in the H-module generated by the elements $b_1 \ldots b_{r'} u_j$, where $r' \leq r$, the b_i are of type (4) with $t_1 + \cdots + t_k \leq n - 1$, and $b_1 > \cdots > b_{r'}$.

For $r = 0$ this is trivial. Suppose that it is true for all numbers smaller than r. Then $c_1 \ldots c_r u_j$ is a linear combination of some

$$c_1 a_1 \ldots a_i b_2 \ldots b_r u_j \tag{6}$$

where $a_1, \ldots, a_i \in H$, $i \geq 0$, and $b_2, \ldots, b_r$ are of the form (4) with the required restrictions. By the Jacobi identity any vector (6) is in the H-module generated by all $f = b b_2 \ldots b_r u_j$ with $b \in L$. If $b \in H$ then f lies in T by construction. If $b \notin H$ then one may assume that b is of the form (4).

Thus it is sufficient to prove that T contains any vector $b_1 \ldots b_r u_j$ provided that all b_i are of the form (4) and $b_2, \ldots, b_r$ satisfy the restriction $t_1, \ldots, t_k \leq n - 1$. Moreover, $b_2 > \cdots > b_r$.

Now let $b = [e_1^{(q_1)}, \ldots, e_k^{(q_k)}, x_i]$. First suppose that $q_1, \ldots, q_k \leq n - 1$. If $b_1 > b_2$ then, by definition, $b_1 \ldots b_r u_j$ is contained in T. If $b_i \geq b_1 \geq b_{i+1}$ then, using induction on r and the Jacobi identity, we conclude that $b_1 \ldots b_r u_j$ modulo T is equal to $B = b_2 \ldots b_i b_1 b_{i+1} \ldots b_r u_j$. For $b_1 \neq b_{i+1}$ the element B is in T by construction. If $b_1 = b_{i+1}$ then $b_1^2 \in L_+ \subset H$ and $(\operatorname{ad} b_1)^2 = \frac{1}{2} \operatorname{ad} b_1^2$. Moving b_1^2 to the left by the Jacobi identity, we express B by $(b_1)^2 b_2 \ldots b_{i-1} b_{i+1} \ldots b_r u_j$ together with a combination of elements of the form $B' = b_2' \ldots b_{i-1}' b_{i+1} \ldots b_r u_j$ with $b_2', \ldots, b_r' \in L_-$. Using the induction hypothesis, one derives $B' \in T$. It follows then that $B \in T$.

Now let the exponents $q_1, \ldots, q_k$ in the expression for b_1 be arbitrary. Set $v = b_2 \ldots b_r u_j$. If $q_1 \geq n$ then using (3) we can write $b_1 v$ as a linear combination of $e_1^j c_j v$ where c_j is a commutator of type (4) with the same $t_2, \ldots, t_k$ as in b_1 and with t_1 smaller than n. Suppose $q_s \geq n$ for some $s > 1$. Set

$$c = [e_s^{(q_s)}, e_1^{(q_1)}, \ldots, e_{s-1}^{(q_{s-1})}, e_{s+1}^{(q_{s+1})}, \ldots, e_k^{(q_k)}, x_i].$$

Using Lemma 2.3 from Section 2 we can express $b_1 - c$ modulo H as a linear combination of elements (4) which are strictly smaller than b_1. Using induction on the order of elements in (4) we may assume $b_1 - c \in T$. Now as in the case $s = 1$, lowering the degree of c in e_s and using (3) we obtain $cb_2 \ldots b_r u_j \in T$. □

3.4. Lemma. *Let L, M and H be as in Lemma 3.3 and let Q be an H-submodule in M of codimension t. Then Q contains some L-submodule $\tilde{Q}$ such that the codimension of $\tilde{Q}$ in M is restricted by some function of t.*

Proof. As in the previous lemma we consider the set of commutators

$$[e_1^{(t_1)}, \ldots, e_k^{(t_k)}, x_i]; \quad 1 \leq i \leq q; \quad t_1 + \cdots + t_k \geq 0, \qquad (7)$$

where $x_1, \ldots, x_q$ are homogeneous generators of L_- as an L_+-module and $e_1, \ldots, e_k$ is a homogeneous basis of L_+ modulo L_+^m. The linear span of the elements (7) coincides modulo $[L_m^+, L_-]$ with L_-. Denote by $z_1, \ldots, z_N$ all products (7) with $t_1 + \cdots + t_k \leq n - 1$.

Denote by $A_1, \ldots, A_T$ all linear maps of M into itself of the form $z_{i_1} \ldots z_{i_s}$, $N \geq i_1 > \cdots > i_s \geq 1$. There exists a subspace W_1 in Q of codimension at most t such that $A_1 W_1 \subset Q$. Choose a subspace W_2 in W_1 such that $A_2 W_2 \subset Q$ and $\dim W_1/W_2 \leq t$. By induction we obtain a subspace V in Q of codimension at most tT, and $A_i V$ lies in Q for any $i = 1, \ldots, T$.

Now let $U(L)$ be the universal enveloping algebra of L and let P be the subspace in $U(L)$ spanned by fA_i, $i = 0, \ldots, T$ where $A_0 = 1$ and $f \in U(H)$. Then $V \subset PV \subset Q$. The codimension of V in Q does not depend on M. Hence, to prove our lemma it is sufficient to check that PV is an L-module.

Consider some $a \in L$, $v \in V$ and show that $aPv \subset PV$. Since L, modulo H, is the linear span of the elements of the form (7) we may assume that a is one of the commutators (7). Now the family of identities (3) implies the relation

$$aw \equiv \sum h_j z_j w \qquad (\text{mod } V(H)w) \qquad (8)$$

where $w \in M$ and all h_j are in $U(H)$. Therefore, for proving $aPv \subset PV$, it is sufficient to verify $z_j A_i v \in PV$.

Let $A_i = z_{j_1} \ldots z_{j_s}$. If $s = 1$ then $[z_j, A_i] \in L_+ \subset H$. For $j = j_1$ the element $z_j A_i v$ is equal to $\frac{1}{2}[z_j, z_j]v$, which is contained in PV since $[z_j, z_j] \in L_+$. If $j \neq j_1$ then either $z_j A_i$ or $A_i z_j$ is an element of the set $\{A_1, \ldots, A_T\}$. Hence, either $z_j A_i v$ or $A_i z_j v$ lies in PV by the definition of P. On the other hand, $[A_i, z_j]v \in U(H)v \subset PV$ and, for $s = 1$, the desired inclusion $z_j a_i v \in PV$ holds.

Assume $s > 1$. If $j > j_1$ then the product $z_j A_i$ is one of $A_1, \ldots, A_T$, hence, $z_j A_i \in PV$. Suppose that $j_{l-1} > j > j_l$ for some l. Move z_j to the right in $z_j A_i v$. Then we express $z_j A_i v$ as a sum of

$$u = z_{j_1} \ldots z_{j_{l-1}} z_j z_{j_l} \ldots z_{j_s} v$$

and the vectors

$$z_{j_1} \ldots z_{j_r} g_r z_{j_{r+2}} \ldots z_{j_s} v \qquad (9)$$

where $g_r \in L_+$, $r \leq l - 1$. If $j > j_l$ then u is one of $A_1 v, \ldots, A_T v$, therefore it belongs to PV. If $j = j_l$ then u is of the same type as the element (9) since $z_j^2 = \frac{1}{2}[z_j, z_j] \in L_+$. Moving g_r to the left we get an expression of (9) as a sum of elements $f_b z_{i_b} \ldots z_{i_s} v$ where $i_b > i_{b+1} > \cdots > i_s$, $2 \leq b \leq s$ and $f_b \in L$. By (8) $z_j A_i v$, modulo PV, is the sum of elements $h_t z_t A_u v$ where $h_t \in U(H)$, $A_u = z_{q_1} \ldots z_{q_r}$ and $r \leq s - 1$. Using induction on s we derive that all these elements are in PV, proving Lemma 3.4. $\square$

Now we shall prove the first in a series of results related to residual finiteness.

3.5. Theorem. *Let L be a finitely generated colour Lie superalgebra over a field K, char $K \neq 2$, with finite grading group G. Let A be an abelian ideal of L such that L/A is finite-dimensional. If, in addition, char $F = 0$, we assume that $L_+^2 \subset A$. Then L is residually finite.*

§ 3. Sufficient conditions for residual finiteness

Proof. Let $h_1, \ldots, h_t$ be a homogeneous generating system of L such that $h_1 + A, \ldots, h_t + A$ contains a basis of $H = L/A$. By a linear change of variables we may assume that $\{h_1 + A, \ldots, h_s + A\}$ is a basis in L/A and $h_{s+1}, \ldots, h_t \in A$. Let $\{c_{ij}^k\}$ be the set of structure constants with respect to the given basis of H. For any i, j, $1 \le i, j \le s$, we set

$$r_{ij} = [h_i, h_j] - \sum_k c_{ij}^k h_k.$$

It is obvious that $r_{ij} \in A$. If $J = \mathrm{id}_L \{r_{ij}, h_{s+1}, \ldots, h_t\}$ then $A \supset J$. On the other hand, $\dim L/J \le s = \dim L/A$. Hence $A = J$, i.e. A is the ideal generated by $\{r_{ij}, h_{s+1}, \ldots, h_t | 1 \le i, j \le s\}$.

Thus A is a finitely generated L-module with respect to the adjoint action. Since $A^2 = 0$, A is a finitely generated H-module. According to Proposition 3.2.8, $U(H)$ is a Noetherian algebra, hence, A is a Noetherian module. For the proof of the residual finiteness of L it is sufficient to verify the finiteness of the dimension of its monolithic homomorphic images, i.e. those which contain a unique minimal non-zero ideal, the so-called monolith.

Indeed, if $a \ne 0$ then by Zorn's Lemma there exists a largest ideal T_a in L which does not contain a. In this case the quotient algebra of L by T_a is monolithic because any ideal I in L satisfying $I \supset T_a$, $I \ne T_a$ contains a. If all monolithic homomorphic images of L are of finite dimension then $\dim L/T_a < \infty$. Clearly, $\bigcap_{a \in L} T_a = 0$. It follows that L is embeddable in the Cartesian product of finite-dimensional superalgebras of the form L/T_a.

Without loss of generality we can assume that L itself is monolithic with monolith M. First let us prove that M is a finite-dimensional space.

If char $K \ne 0$ this follows immediately from Lemma 5.2.4, since M is an irreducible module over the finite-dimensional algebra $H = L/A$. Now suppose char $K = 0$. Consider M as an $U(H_+)$-module. By the Poincaré-Birkhoff-Witt Theorem $U(H)$ is a module of finite type over $U(H_+)$, hence, M is a finitely generated H_+-module.

Consider a maximal proper H_+-submodule Q in M. By Lemma 5.2.5 $\dim M/Q < \infty$ since $H_+^2 = 0$. Denote by $f_1, \ldots, f_m$ a basis of the space H_-. As in the proof of Lemma 3.4 it is possible to choose a subspace V of finite codimension in Q such that the elements

$$f_{i_1} \ldots f_{i_s} v, \quad m \ge i_1 > \cdots > i_s \ge 1, \quad v \in V$$

lie in Q. We denote by W the sum of V and the linear span of all these elements. Arguing as in the proof of Lemma 3.4 we find that $f_j f_{i_1} \ldots f_{i_s} v$ is in $U(H_+)W$ for any j. Hence $U(H_+)W$ is an H-submodule in Q. The codimension of $U(H_+)W$ in M is finite and it depends only on the codimension of Q in M. It follows that if $\dim Q = \infty$ then $U(H_+)W \ne 0$, i.e. M contains some proper H-submodule. This proves the finiteness of the dimension of M.

As a result, we have a chain $L \supset A \supset M$ where $\dim L/A < \infty$, $\dim M < \infty$, $A^2 = 0$. It is sufficient to prove the finiteness of the dimension of A provided that A is a module of finite type over the finite-dimensional superalgebra $H = L/A$, containing the minimal submodule M.

We recall that $U(H_+)$ contains a central Noetherian G-homogeneous subalgebra Z such that $U(H)$ is a module of finite type over Z. If $\operatorname{char} K = 0$ then Z is generated by the elements $e_1^n, \ldots, e_s^n$ where $e_i = h_i + A$, $i = 1, \ldots, s$, is a basis of H_+ and $n = |G|$. If $\operatorname{char} K$ is a prime p, then Z is an algebra with a more complicated strucure. To construct Z we consider some homogeneous $x \in H_+$, $d(x) = g$. If y is another homogeneous element in H, $d(y) = h$ then

$$(\operatorname{ad} x)^p(y) = (l_x - \varepsilon(g,r)r_x)^p(y) = (l_{x^p} - \varepsilon(pg,h)r_{x^p})(y)$$

$$= (\operatorname{ad} x^p)(y)$$

where l_x, r_x are the left and right multiplications by x in $U(H)$ respectively. Here, under the consecutive actions of $\operatorname{ad} x$, the value of the form ε does not change since in the $(k+1)$-st step $\operatorname{ad} x$ acts on $(\operatorname{ad} x)^k(y) \in L_{g+h}$ and

$$\varepsilon(g, kg + h) = \varepsilon(g,g)^k \varepsilon(g,h) = \varepsilon(g,h).$$

We assume that f is a p-polynomial annihilating $\operatorname{ad} x$. Then $\operatorname{ad} f(x) = f(\operatorname{ad} x) = 0$. It is easy to see that if $f(x) = u_1 + \cdots + u_k$, $d(u_i) = g_i \in G$, and $g_i \neq g_j$ for $i \neq j$ then $\operatorname{ad} u_1, \ldots, \operatorname{ad} u_k$ are zero in $U(H)$.

Now we construct the subalgebra Z. Let $\{e_1, \ldots, e_s\}$ be a homogeneous basis of H_+. Since H is of finite dimension, for any $x \in \{e_1, \ldots, e_s\}$ there exists a p-polynomial f annihilating x. Then $f(x)$ is a central element in $U(H)$. If $f = u_1 + \cdots + u_k$, $d(u_i) = g_i$, then $u_1, \ldots, u_k$ have the same property. Now the elements $u_1^n, \ldots, u_k^n$ with $n = |G|$ belong to the usual centre of $U(L)$. The subalgebra Z is generated by all $u_1^n, \ldots, u_k^n$ with x running through the basis $\{e_1, \ldots, e_s\}$ of H_+.

Let z be an arbitrary G-homogeneous element in $Z \cap \operatorname{Ann} M$. Then the chain of subspaces

$$zA \supset z^2 A \supset \cdots \supset z^k A \supset \cdots \qquad (10)$$

is a chain of $U(H)$-submodules. Assume that none of the subspaces in the chain (10) is zero. In this case $z^k A \cap M \neq 0$ for any k because M is the monolith in L. We consider also the chain

$$M \subset \operatorname{Ann}_A z \subset \operatorname{Ann}_A z^2 \subset \cdots \subset \operatorname{Ann}_A z^u \subset \cdots.$$

§ 3. Sufficient conditions for residual finiteness

Since A is a Noetherian $U(H)$-module, we have, for a suitable u,

$$\text{Ann}_A z^u = \text{Ann}_A z^{u+1}. \tag{11}$$

Let $0 \neq x \in M \cap z^u A$. Then, for some $a \in A$, we have $x = z^u a$.

Since $z \in \text{Ann } M$, it follows that $zx = 0$. On the other hand, $zx = z(z^u a) = z^{u+1}a$, i.e. $a \in \text{Ann}_A z^{u+1}$. It follows from (11) that $z^u a = x = 0$, a contradiction. Thus in the chain (10) not all the spaces are nonzero, i.e. the action of z on A is nilpotent.

Now if $z_1, \ldots, z_l$ are the generators of Z then there exist polynomials f_i such that $f_i(z_i) = w_i \in Z \cap \text{Ann } M$. We denote by $b_1, \ldots, b_k$ the Z-generators of $U(H)$. Hence any element in $U(H)$ can be represented as a linear combination of monomials of the form

$$b_i z_1^{m_1} \ldots z_l^{m_l} w_1^{t_1} \ldots w_l^{t_l}; \tag{12}$$
$$0 \leq m_i < \deg f_i; \quad t_i \geq 0, \quad i = 1, \ldots, l.$$

If n_i is the nilpotent index for the action of u_i on A then only a finite number of elements of the form (12) act non-trivially on A. Since A is a finitely generated H-module, we deduce that it is finite-dimensional. It follows from $\dim L/A < \infty$ that also $\dim L < \infty$. □

3.6. Residual finiteness and cardinality of the grading group. The condition $|G| < \infty$ is necessary in the previous theorem. This can be illustrated by the following example.

Consider a metabelian Lie algebra $L = P + M$ where P is abelian with basis $\{x, y\}$ and M is an abelian ideal in L spanned by $\{z_i | i \in \mathbb{Z}\}$. The multiplication in L is given by $[x, z_i] = z_{i-1}$, $[y, z_i] = z_{i+1}$, $i \in \mathbb{Z}$. Thus L is a $\mathbb{Z}$-graded algebra where $L_i = \langle z_i \rangle$ for $i \neq \pm 1$, $L_1 = \langle y, z_1 \rangle$, $L_{-1} = \langle x, z_{-1} \rangle$. If we define $\varepsilon(m, n) = 1$ for all $m, n \in \mathbb{Z}$ then L becomes a colour Lie superalgebra with nonzero minimal homogeneous ideal M. Therefore L is a 3-generated monolithic metabelian algebra of infinite dimension. Hence it is not residually finite.

3.7. Lemma. *Let L and M be as in Lemma 3.3 with finite grading group G and suppose L_+^m acts on M as a nilpotent space of transformations for some $m \geq 1$, where $m = 2$ if $\text{char } k = 0$. If M contains a unique minimal L-submodule T, then $\dim M < \infty$.*

Proof. Suppose that $L_+^m M = 0$. Consider the semidirect product $H \curlywedge M$ where $H = L/\text{Ann } M$ is a finite-dimensional colour Lie superalgebra and M is an ideal with zero multiplication. If $\text{char } k = 0$ then $H_+^2 = 0$ and

$(H \curlywedge M)_+^2 \subset M$. Hence $H \curlywedge M$ satisfies the conditions of Theorem 3.5. As T is the minimal L-submodule in M, $T \neq 0$, it follows that T is the monolith in $H \curlywedge M$. By Theorem 3.5 $\dim(H \curlywedge M) < \infty$ and the dimension of M is finite as claimed.

Now let $L_+^m M \neq 0$. We set $M_0 = M$, $M_j = L_+^m M_{j-1}$ for $j \geq 1$. By the hypotheses of the lemma there exists k such that $M_k = 0$. We use induction on k.

First assume that $L = L_+$, i.e. $L_- = 0$. By Theorem 2.5 M is a Noetherian L-module. Consequently, $L^m M$ is a finitely generated L-module, and the pair $(L, L^m M)$ satisfies the hypotheses of the lemma. Using induction enables us to assume that $L^m M$ is of finite dimension. Consider the chain of submodules $P_1 \subset P_2 \subset \cdots$ in M where

$$P_j = \{x \in M \mid b_1 \ldots b_j x = 0 \; \forall b_1, \ldots, b_j \in L^m\}.$$

Our next step is to prove that $\dim M/P_j < \infty$ for any j.

By the hypotheses of the lemma $P_k = M$, hence, $\dim M/P_k < \infty$. Suppose that the finiteness of the codimension of P_{j+1} in M has been already proved. We denote by $e_1, \ldots, e_q$ a homogeneous basis of L modulo L^m and by $a_1, \ldots, a_r$ a finite set of homogeneous generators of L^m as an ideal in L. Let $A_1, \ldots, A_N$ be all the operators in End M of the form $b_1 \ldots b_j$ where any factor is of the form

$$[e_1^{(t_1)}, \ldots, e_q^{(t_q)}, a_i], \qquad 1 \leq i \leq r, \qquad 0 \leq t_1, \ldots, t_q \leq n-1. \tag{13}$$

We set $Q_i = \operatorname{Ker} A_i$ and verify the relation

$$P_j \supset P_{j+1} \cap Q_1 \cap \cdots \cap Q_N = T. \tag{14}$$

Let w be an arbitrary element in T. We want to show that $f_1 \ldots f_j w = 0$ for any $f_1, \ldots, f_j \in L^m$.

Applying Lemma 2.6 to L and to the L-module generated by w in M we see that the vector $f_1 \ldots f_j w$ is contained in the L-module generated by all $h_1 \ldots h_k v$ and $b_1 \ldots b_k v$ where $v \in P_{j+1}$, $h_1, \ldots, h_k \in L^m$, $k \geq j+1$ and $b_1, \ldots, b_j$ are of the form (13). It follows that $f_1 \ldots f_j w = 0$ which proves (14).

It was mentioned above that $\dim L^m M < \infty$. Hence every A_i maps M into the finite-dimensional vector space $L^m M$, and the codimension of $Q_i = \operatorname{Ker} A_i$ in M is finite. Finally, (14) and $\dim M/P_{j+1} < \infty$ imply the finiteness of the codimension of P_j in M.

For $j = 1$ we have obtained the submodule P_1 in M of finite codimension. Furthermore, $L^m P_1 = 0$ and P_1 is generated by a finite set of elements, since M is Noetherian. It follows that the semidirect product $H \curlywedge P_1$ is a finitely generated colour Lie superalgebra, where $H = L/\operatorname{Ann} P_1$. Since $L^m \subset$

Ann P_1, by Theorem 3.5 $H \wedge P_1$ is residually finite. The unique minimal L-submodule T of M lies in P_1 and is the monolith of $H \wedge P_1$. Therefore $\dim P_1$ is finite. Since $\dim M/P_1 < \infty$, the proof is complete in the case $L = L_+$.

Now let $L_- \neq 0$. Consider the chain of subalgebras $H_1 \supset H_2 \supset \cdots$ in L with $H_1 = L_+ \oplus [L_+^m, L_-]$ and $H_{i+1} = L_+ \oplus [L_+^m, (H_i)_-]$ if $i \geq 1$. Obviously, L_+^m acts on M as a nilpotent space of transformations, hence the odd component $(H_c)_-$, for some c, acts trivially on M. This is true, for example, for $c = 2k + 1$, since

$$\underbrace{[L_+^m, \ldots, L_+^m}_{2k+1}, L_-]M = 0.$$

We prove the statement by induction on c. By Lemma 3.3 M is a finitely generated H_i-module for any $i \geq 1$. If $c = 1$ then M is a finitely generated module over the algebra $P = H_1/\mathrm{Ann}_{H_1} M$ with $P_- = 0$ because $\mathrm{Ann}_{H_1} M \supset (H_1)_-$. By the hypotheses of the lemma M contains the minimal nonzero submodule T. Let a be a nonzero element in T. We denote by Q a maximal P-submodule in M not containing a. Now M/Q has a unique minimal P-submodule, hence, as proved above, the dimension of M/Q is finite. Any P-submodule in M is also an H_1-submodule, and, if $\dim M = \infty$, by Lemma 3.4 Q possesses an L-submodule N of finite codimension. Since $N \subset Q$, we get $a \notin N$. On the other hand, $T \subset N$, because T is the unique minimal L-submodule, and a is in T. This contradiction shows that $\dim M < \infty$.

Now suppose $c > 1$. As before, we choose a nonzero element a in the minimal L-submodule T and denote by Q a largest H_1-submodule in M which does not contain a. Clearly, $\bar{a} = a + Q$ lies in every nonzero H_1-submodule of M/Q. Hence, by the inductive hypothesis, M/Q is a finite-dimensional vector space. By Lemma 3.4 M has an L-submodule of finite codimension which does not contain a. Since T is the monolith of M and a is in T, the unique L-submodule which does not contain a must be zero. It follows that $\dim M < \infty$, proving the lemma. □

We translate the definition of residual finiteness from the language of superalgebras into the language of its modules. A module M over a colour Lie superalgebra L is called *residually finite* if for any $0 \neq v \in M$ there exists an L-submodule V such that $\dim M/V < \infty$ and $v \notin V$.

3.8. Theorem. *Let L be a finitely generated soluble colour Lie superalgebra over a field K, $\mathrm{char}\, K \neq 2$, graded by a finite group G. Let M be a finitely generated L-module and assume that the representation of L on M, for any g,*

$h \in G$, $r \in G_+$, satisfies identities of the form

$$[x, y^{(n)}, z] = \sum_{j=1}^{n} \alpha_j [y^{(j)}, x, y^{(n-j)}, z], \quad d(x) = g, \quad d(y) = r, \quad d(z) = h \quad (15)$$

and

$$[y^{(n)}, z]v = \sum_{j=1}^{n} \lambda_j y^j [y^{(n-j)}, z]v, \quad v \in M_g, \quad d(y) = r, \quad d(z) = h \quad (16)$$

with $\alpha_j, \lambda_j \in K, j = 1, \ldots, n$. If L_+^m acts on M as a nilpotent space of transformations and $m = 2$ in case char $K = 0$, then M is a residually finite L-module.

Proof. The theorem follows from Lemma 3.7 since for any nonzero v in M a largest L-submodule not containing v has finite codimension. □

Now we consider a colour Lie superalgebra L with the set of identities (15). It was shown earlier (see Subsection 2.8) that the adjoint representation of L satisfies the identities of the form (16). Therefore the following result is an immediate consequence of Theorem 3.8.

3.9. Theorem. *Let L be a finitely generated soluble colour Lie superalgebra over a field K, char $K \neq 2$, with identities of the type (15) for any $g, h \in G$, $r \in G_+$. If L_+^m acts on L as a nilpotent space of transformations, where $m = 2$ if char $K = 0$, and if the grading group G is finite, then L is residually finite.* □

§4. Representability of Lie superalgebras by matrices

An algebra A over a field K is called *representable* if there exist an extension $\tilde{K}$ of K and a finite-dimensional $\tilde{K}$-algebra B such that A is embeddable in B considered as a K-algebra. If G is a commutative semigroup and A is a G-graded algebra then the embedding $A \to B$ must be a homomorphism of G-graded algebras.

The main purpose of this section is to prove the representability of a finitely generated soluble colour Lie superalgebra $L = L_+ \oplus L_-$ with a family of identities (3.15) provided that L_+^m acts on L as a nilpotent space of transformations and K is an infinite field. By Theorem 3.9 L is then residually finite. This connection between residual finiteness and representability of finitely generated algebras over an infinite field is not a mere coincidence as will be shown later. First we prove a result for arbitrary graded algebras over a field K.

§4. Representability of Lie superalgebras by matrices

4.1. Theorem. *Let K be an arbitrary field, G a commutative semigroup and let R be an associative commutative Noetherian algebra with unity over K. Consider a G-graded algebra $A = \bigoplus_{g \in G} A_g$ over R. If A is an R-module generated by finitely many elements then A is a representable K-algebra.*

Proof. We use the notation and some results from the theory of primary decomposition of commutative rings (see [Bourbaki, 1961], Chapter IV). Let $\mathrm{Ass}_R(A)$ be the set of all prime ideals in R associated with A. Then $\mathrm{Ass}_R(A)$ is a finite set ([Bourbaki, 1961], Chapter IV, §1, Sect. 4, Theorem 2). Denote by $P_1, \ldots, P_n$ all maximal elements of this set and define

$$S = (R \setminus P_1) \cap \ldots \cap (R \setminus P_n).$$

Consider the ring $Q = S^{-1}R$ and the R-algebra $B = S^{-1}A$ with the canonical homomorphism of R-algebras $f: A \to B$. Then f is a monomorphism ([Bourbaki, 1961], Chapter II, §2, Sect. 2, Proposition 4, Chapter IV, §1, Sect. 1, Corollary 2). One can transfer the grading from A to B. Hence, we may assume that $B = \oplus B_g$ is a graded Q-algebra. Furthermore, Q is a semilocal ring, its set of maximal ideals being $\{S^{-1}P_i | 1 \leq i \leq n\}$ ([Bourbaki, 1961], Chapter II, §3, Sect. 5, Proposition 17) and

$$\mathrm{Ass}_Q(B) = \{S^{-1}P | P \in \mathrm{Ass}_R(A)\}$$

([Bourbaki, 1961], Chapter IV, §1, Sect. 2, Proposition 5). To prove the theorem it is sufficient to show that B is a representable K-algebra. Therefore, without loss of generality we assume that R is semilocal and that all maximal ideals of R belong to $\mathrm{Ass}_R(A)$.

Let $l_R(A)$ denote the maximal length of descending chains of prime ideals of $\mathrm{Ass}_R(A)$. We shall prove the theorem by induction on $l_R(A)$.

First let $l_R(A) = 1$. In this case A is an R-module of finite length ([Bourbaki, 1961], Chapter IV, §2, Sect. 5, Proposition 7). Since a commutative ring with a faithful module of finite length is Artinian, $\bar{R} = R/\mathrm{Ass}_R(A)$ is a semilocal Artinian ring. It follows that $\bar{R} = \bigoplus_{i=1}^n R_i$ where every R_i is a local Artinian ring ([Bourbaki, 1961], Chapter IV, §2, Sect. 5, Corollary 1 of Proposition 9). Clearly, $A = \bigoplus_{i=1}^n R_i A$ and every summand $R_i A$ is a G-graded algebra over R_i. Hence, it is sufficient to consider R as a local Artinian ring. Let P denote the maximal ideal in R. Then R/P is a field containing K. There exists a subfield $\tilde{K}$ in R such that $R = \tilde{K} + P$ and $\tilde{K} \cong R/P$ ([Zariski, Samuel, 1960], Chapter VIII, §12, Theorem 27). It was proved earlier that A is an R-module of finite length. Therefore, A possesses a composition series every factor of which is a one-dimensional vector space over R/P. Since $\tilde{K} \subset R$, $A = \bigoplus_{g \in G} A_g$ is a G-graded finite-dimensional $\tilde{K}$-algebra and for $l_R(A) = 1$ the proof is complete.

Now let $l_R(A) > 1$. Then $H = \text{Ass}_R(A) \setminus \{P_1, \ldots, P_n\}$ is nonempty. We denote by $Q_1, \ldots, Q_m$ all maximal elements of H and set

$$T = (R \setminus Q_1) \cap \ldots \cap (R \setminus Q_m).$$

Consider $D = T^{-1}R$, $C = T^{-1}A$ as R-algebras and the canonical homomorphism $f: A \to C$ as a homomorphism of R-algebras. By Proposition 1.11 from [Atiyah, Macdonald, 1969] $P_i \cap T \neq \emptyset$ for every $i = 1, \ldots, n$. Then $\text{Ass}_D(C) = \{T^{-1}Q | Q \in H\}$ ([Bourbaki, 1961], Chapter IV, § 1, Sect. 2, Proposition 5) and $l_D(C) < l_R(A)$. By the induction hypothesis the K-algebra C is representable.

Now let $U = \text{Ker} f$. Since f is a homomorphism of graded R-algebras, U is a graded R-algebra and $\text{Ass}_R(U) = \{P_1, \ldots, P_n\}$ ([Bourbaki, 1961], Chapter IV, § 1, Sect. 2, Proposition 6). Moreover, as an R-module, U is of finite length ([Bourbaki, 1961], Chapter IV, §2, Sect. 5, Proposition 7). Suppose that $I = P_1 \cap \ldots \cap P_n$. Then I is the Jacobson radical of R, hence, $I^r U = 0$ for some $r > 0$. By the Artin—Rees Lemma ([Bourbaki, 1961], Chapter III, § 3, Sect. 1, Corollary 1) it follows that $I^t A \cap U = 0$ for some $t \geq r$. Obviously $\bar{A} = A/I^t A$ is a module of finite length over R. It has been proved above that $\bar{A}$ is a representable K-algebra. Since $I^t A \cap U = 0$, A can be embedded in the direct product of two representable algebras $\bar{A}$ and A/U. Let $\bar{A}$ be embeddable into a finite-dimensional algebra over a field K_1 and let A/U be embeddable into a finite-dimensional algebra over a field K_2 with $K_i \supset K$, $i = 1, 2$. Then there exists a field $\tilde{K}$ containing both K_1 and K_2. Hence, A is embeddable in some $\tilde{K}$-algebra of finite dimension, proving the theorem. □

The following result is a principal tool in establishing representability. Recall that an algebra A is residually an algebra in the family $\{A_\alpha\}$, $\alpha \in I$, if for any $x \neq a$ in A there exists a homomorphism φ of A such that $\varphi(A) \in \{A_i\}$ and $\varphi(x) \neq 0$.

4.2. Theorem. *Let K be an infinite field, and let $A = \bigoplus_{g \in G} A_g$ be a G-graded K-algebra generated by a finite set of elements, where G is a finite commutative semigroup. Then A is a representable algebra if, and only if, A is residually a finite-dimensional K-algebra of bounded dimension.*

Proof. First suppose A can be embedded in a finite-dimensional graded $\tilde{K}$-algebra B for some extension $\tilde{K}$ of K. Since A is finitely generated one may assume that a finite set of elements generates $\tilde{K}$ over K. It follows that there exists a subfield P such that $P = K(x_1, \ldots, x_m)$ is a purely transcendental extension of K by $x_1, \ldots, x_m$ and $\tilde{K}$ is an algebraic extension of P. Let $\lambda_1, \ldots, \lambda_k$ be a basis of $\tilde{K}$ over P and let $e_1, \ldots, e_n$ be a homogeneous basis of B as $\tilde{K}$-algebra. Without loss of generality assume that the elements $e_1, \ldots, e_n$

generate the K-algebra A. Clearly, B is the P-linear span of the finite set $T = \{\lambda_i e_j | i = 1, \ldots, k; j = 1, \ldots, n\}$. Denote by C the K-algebra generated by T. Any nonzero element $a \in C$ can be written in the form

$$a = \sum_i \frac{f_i(x_1, \ldots, x_m)}{h_i(x_1, \ldots, x_m)} t_i \tag{1}$$

where $t_i \in T$, $f_i = f_i(x_1, \ldots, x_m)$ and $h_i = h_i(x_1, \ldots, x_m)$ are some polynomials in variables the $x_1, \ldots, x_m$. Since C is a finitely generated K-algebra, every h_i is a product of polynomials from a finite set H, and H does not depend on a. Suppose a substitution $\varphi: x_j \mapsto \xi_j \in K$, $j = 1, \ldots, m$, satisfies the condition $h(\xi_1, \ldots, \xi_m) \neq 0$ for all $h \in H$. Then φ gives rise to a homomorphism of C onto some G-graded K-algebra of dimension kn. Moreover, $\varphi(a) \neq 0$ for an element a of the form (1) if $f_i(\xi_1, \ldots, \xi_m) \neq 0$ for some i. Since K is an infinite field, for every $0 \neq a \in C$ there exists a homogeneous ideal of codimension kn which does not contain a. Since $A \subset C$, we find that A is residually a K-algebra of dimension at most kn.

Now we assume that A is residually an algebra of bounded dimension and prove the representability of A.

If $C = \bigoplus_{g \in G} C_g$ is a finite-dimensional G-graded algebra, $\dim C \leq n$, then C can be embedded in an N-dimensional algebra $C' = \bigoplus_{g \in G} C'_g$ with $\dim C'_g = n$ for any $g \in G$ where $N = mn$ and $m = |G|$. Indeed, it is sufficient to set $C' = C \oplus T$ where $T = \bigoplus_{g \in G} T_g$, $\dim T_g = n - \dim C_g$ and $tx = 0$ for any $t \in T, x \in C'$.

It follows that A is embeddable in the Cartesian product $\prod_{i \in I} A_i$ of N-dimensional algebras A_i, and for any $i \in I$, $g \in G$ we have $\dim(A_i)_g = n$. We enumerate all elements of the semigroup G and define a grading on the set $\{1, \ldots, N\}$ as follows

$$d(1) = \cdots = d(n) = g_1, \ldots, d(mn - n + 1) = \cdots = d(mn) = g_m.$$

Then for every A_i one can choose a homogeneous basis $e_1^i, \ldots, e_N^i$ such that $d(e_j^i) = d(j)$. Let γ_{jk}^{ir} be the structure constants of A_i, i.e.

$$e_j^i e_k^i = \sum_{r=1}^N \gamma_{jk}^{ir} e_r^i.$$

Since A is a G-graded algebra, $\gamma_{jk}^{ir} = 0$ for $d(r) \neq d(j) + d(k)$.

Consider the Cartesian power $\tilde{R} = K^I$ and the free $\tilde{R}$-module

$$\tilde{A} = \tilde{R} v_1 \oplus \cdots \oplus \tilde{R} v_N$$

with basis $v_1, \ldots, v_N$.

Transform A into a G-graded $\tilde{R}$-algebra by setting $d(v_j) = d(j)$, $j = 1, \ldots, N$ and

$$v_j v_k = \sum_{r=1}^{N} \Gamma_{jk}^r v_r,$$

where $\Gamma_{jk}^r \in \tilde{R}$, $\Gamma_{jk}^r(i) = \gamma_{jk}^{ir}$. Since $\Gamma_{jk}^r = 0$ for $d(j) + d(k) \neq d(r)$, this defines a G-graded algebra.

Define an embedding of the K-algebra $C = \prod_{i \in I} A_i$ into $\tilde{A}$ in the following way. For $f \in C$, $i \in I$ put

$$f(i) = b_1^i e_1^i + \cdots + b_N^i e_N^i$$

where $b_j^i \in K$, $j = 1, \ldots, N$. Define $\varphi(f) = B_1 v_1 + \cdots + B_N v_N$ with $B_j(i) = b_j^i$, $j = 1, \ldots, N$, $i \in I$. A direct verification shows that φ is a homomorphism of K-algebras. If A is in C then $\varphi(A)$ is a finitely generated subalgebra of $\tilde{A}$.

Now let a be an arbitrary element in A. Write

$$\varphi(a) = r_1 v_1 + \cdots + r_N v_N$$

where $r_j \in \tilde{R}$, $j = 1, \ldots, N$. We denote by R the subalgebra of $\tilde{R}$ generated by all $r_1, \ldots, r_N$ with a running through the finite set of generators of A. Obviously, $\varphi(A) \subset Q$ where

$$Q = R v_1 \oplus \ldots \oplus R v_N = \bigoplus_{g \in G} Q_g$$

is a G-graded algebra over R, R being a finitely generated commutative K-algebra. By Theorem 4.1 the K-algebra Q is representable. It follows that A is also representable. □

4.3. Corollary. *Suppose that R and A satisfy the hypotheses of Theorem 4.1. If, in addition, K is an infinite field and G a finite semigroup then A is a residually finite-dimensional R-module of bounded dimension over K.*

As in the case of algebras we say that an R-module A is residually a module in the family $\{A_i\}$, $i \in I$, if for any $0 \neq x \in A$ there is some homomorphism φ of A such that $\varphi(A) \in \{A_i\}$ and $\varphi(x) \neq 0$.

Proof. Consider the R-module $B = \bigoplus_{g \in G} B_g$ with $B_g = A_g$ if $g \neq 0$ in G and $B_0 = A_0 \oplus R$. Then A is a left R-module, and the action of $r \in R$ on A transforms a into ra. Defining the product ar in B as ra, we impose the structure of a G-graded R-algebra on B. By Theorem 4.1 the K-algebra B is representable. Since R is finitely generated, B is also finitely generated. By Theorem 4.2 there exists n such that for any nonzero b in B one can choose a homogeneous ideal Q in B of codimension less than n which does not contain b. Therefore $Q \cap A$ is an R-module in A and $\dim A/Q \cap A \leq n$. □

4.4. Representable modules. Extension of the ground field. Consider a colour Lie superalgebra L over K and a module M over L. We call M *representable* if M can be embedded in an L-module $\tilde{M}$ satisfying the following conditions:
1) there exists an extension $\tilde{K}$ of K such that $\tilde{M}$ is a finite-dimensional vector space over $\tilde{K}$;
2) the action of L on $\tilde{M}$ is $\tilde{K}$-linear, i.e. $a\lambda v = \lambda a v$ for any $a \in L, \lambda \in \tilde{K}, v \in \tilde{M}$.

It is easy to see that M is representable if, and only if, $L/\operatorname{Ann} M \curlywedge M$ is a representable algebra. It follows, for a module M of finite type over a finitely generated algebra L, that the representability of M is equivalent to M being a residually finite-dimensional L-module of bounded dimension, provided that K is an infinite field.

Indeed, if M is a representable L-module then $H = L/\operatorname{Ann} M \curlywedge M$ is a finitely generated representable colour Lie superalgebra. By Theorem 4.2 H is a residually n-dimensional algebra in the family $\{H_i = H/Q_i\}, i \in I$. Hence, $M_i = M/M \cap Q_i, i \in I$, is a family of finite-dimensional L-modules of restricted dimension, and M can be embedded in $\prod_{i \in I} M_i$.

Now suppose that $M_i, i \in I$, is a family of finite-dimensional L-modules, $\dim M_i \leq n$, and M is a submodule in $\prod_{i \in I} M_i$. We may assume that every M_i is a quotient module of M. In this case

$$\operatorname{Ann} M = \bigcap_{i \in I} \operatorname{Ann} M_i.$$

Let $Q_i = \operatorname{Ann} M_i$. Then the semidirect product $H_i = L/Q_i \curlywedge M_i$ is a homomorphic image of $H = L/\operatorname{Ann} M \curlywedge M$. Moreover, $\dim H_i \leq n^2 + n$ and H can be embedded in $\prod_{i \in I} H_i$. Being finitely generated, H is representable by Theorem 4.2. Hence, M is a representable module.

Therefore Corollary 4.3 shows that A (the algebra in Theorem 4.1) is representable both as a K-algebra and as an R-module, if some additional restrictions are imposed on K, R and G.

As in the case of residual finiteness it is appropriate to prove some representability results in terms of representable modules.

We need an observation concerning representability of superalgebras over a field K. Suppose that L can be embedded as a K-algebra in a finite-dimensional $\tilde{K}$-algebra $\tilde{L}$ for some extension $\tilde{K}$ of K. Consider the tensor product $H = L \otimes_K \tilde{K}$ which is a $\tilde{K}$-algebra. If φ is an embedding of L in $\tilde{L}$ as K-algebra then φ extends to a homomorphism of H into $\tilde{L}$ as $\tilde{K}$-algebra by setting $\varphi(a \otimes \lambda) = \lambda \varphi(a), a \in L, \lambda \in \tilde{K}$. The kernel Q of this homomorphism is an ideal in H and dim $H/Q < \infty$ over $\tilde{K}$. The representability of L is equivalent to the existence of an ideal Q in the $\tilde{K}$-algebra $L \otimes \tilde{K}$ such that $L \cap Q = 0$ and the codimension of Q in $L \otimes \tilde{K}$ is finite over $\tilde{K}$.

Similarly, the representability of the L-module M is equivalent to the existence of a submodule T of finite $\tilde{K}$-codimension in the $L \otimes \tilde{K}$-module $M \otimes \tilde{K}$ which has trivial intersection with $M = M \otimes 1$.

All further results of this section hold for the case of an infinite field K. First we prove a statement which generalizes Theorem 3.5, if we take Theorem 4.2 into account.

4.5. Theorem. *Let $L = \bigoplus_{g \in G} L_g$ be a finitely generated colour Lie superalgebra over an infinite field K, char $K \neq 2$, and let G be a finite group. Suppose that L contains an abelian ideal A of finite codimension and, in addition, L_+^2 lies in A if char $K = 0$. Then A is a representable algebra.*

Proof. Let $H = L/A$. The natural action of L on A makes A into an H-module. It was remarked in the proof of Theorem 3.5 that A is a finitely generated U-module where $U = U(H_+)$ is the universal enveloping algebra for the even component of H. Let $e_1, \ldots, e_m$ be a homogeneous basis of H_+. We claim that, for any $i = 1, \ldots, m$, there exists a polynomial $f_i = f_i(e_i)$ which is in the center of U and $d(f_i) = 0$. First consider the case that char $K = 0$. Then $f_i = e_i^n$ where $n = |G|$. Indeed, if $d(e_i) = g_1$, $d(e_j) = g_2$ then

$$f_i e_g = e_i^n e_j = \varepsilon(g_1, g_2) e_i^{n-1} e_j e_i = \cdots = \varepsilon(ng_1, g_2) e_j e_i^n = e_j e_i^n = e_j f_i$$

and $d(f_i) = nd(e_i) = 0$ in G. If char $K = p > 2$ then we consider two homogeneous elements x, y in H_+, $d(x) = g_1$, $d(y) = g_2$. Denote by r_x and l_x the right and the left multiplication by x in $U(H_+)$. Then ad $x(y) = (l_x - \lambda r_x)(y)$ where $\lambda = \varepsilon(g_1, g_2)$. If $z = (\text{ad } x)^k(y)$ then ad $x(z) = (l_x - \mu r_x)(z)$ where

$$\mu = \varepsilon(g_1, kg_1 + g_2) = (\varepsilon(g_1, g_1))^k \varepsilon(g_1, g_2) = \varepsilon(g_1, g_2) = \lambda.$$

It follows that

$$(\text{ad } x)^p(y) = (l_x - \lambda r_x)^p(y) = (l_{x^p} - \lambda^p r_{x^p})(y) = (\text{ad } x^p)(y)$$

since $\lambda^p = \varepsilon(pg_1, g_2)$ and $d(x^p) = pg_1$. Therefore for any p-polynomial h we have $[h(x), y] = h(\text{ad } x)(y)$.

Since G is finite, the sequence $e_i, e_i^p, e_i^{p^2}, \ldots$ contains infinitely many elements of the same degree in G. Since H is of finite dimension, there exists a p-polynomial h_i such that $h_i(e_i)$ is a homogeneous element in $U(H_+)$ and $[h_i(e_i), e_j] = 0$ for all $j = 1, \ldots, m$. Now we set $f_i = (h_i(e_i))^n$. Clearly, $d(f_i) = 0$ and $f_i e_j = e_j f_i$, as claimed.

§4. Representability of Lie superalgebras by matrices

We denote by Z the subalgebra of $U(H_+)$ generated by $f_1, \ldots, f_m$. By the Poincaré-Birkhoff-Witt Theorem $U(H_+)$ is a module of finite type over Z. By Corollary 4.3, A contains a family of Z-submodules A_α such that dim $A/A_\alpha \leq N$ and $\bigcap_\alpha A_\alpha = 0$.

Let B be one of these submodules A_α. We shall prove that B contains some U-submodule C of codimension at most $\varphi(N)$ where φ is a function not depending on B.

Since dim $A/B \leq N$, one can construct, for any f_i, a polynomial $t_i = t_i(f_i)$ such that $t_i A \subset B$ and $\deg t_i$ depends only on N. Set $C = t_1 A + \cdots + t_m A$. Then C is a U-submodule in B. If $a_1, \ldots, a_n$ generate the U-module A, then A, modulo C, coincides with the linear span of all the elements

$$e_1^{j_1} \ldots e_m^{j_m} a_i$$

with $j_k \leq \deg f_k \cdot \deg t_k$. Thus C has the required properties, and we can assume that A contains a family of H_+-submodules C_α such that for some N the codimension of each C_α in A is at most N and $\bigcap_\alpha C_\alpha = 0$. It was shown in the proof of Theorem 3.5 that the finitely generated H-module A with H_+-submodule Q of codimension N contains an H-submodule T in Q such that the codimension of T in A is finite and does not depend on Q. Therefore A, as an H-module, is residually a module of the form $\{A/T_\alpha\}$ where the dimensions are bounded. Any T_α is an ideal in L and $\bigcap_\alpha T_\alpha = 0$. By Theorem 4.2 L is representable, proving Theorem 4.5. □

4.6. Identities of superalgebras and extension of the ground field. Suppose that K is an infinite field and L is a colour Lie superalgebra over K. Consider an extension $\tilde{K}$ of K. Using the decomposition $L = \bigoplus_{g \in G} L_g$ one can construct a $\tilde{K}$-algebra $\tilde{L}$ by setting

$$\tilde{L} = L \otimes_K \tilde{K} = \bigoplus_{g \in G} (L_g \oplus \tilde{K}).$$

The multiplication in $\tilde{L}$ is defined in a natural way: $[x \otimes \alpha, y \otimes \beta] = [x, y] \otimes \alpha\beta$. The verification of the Jacobi and anticommutative identities is straightforward.

Now suppose that a multihomogeneous identity $f(x_1, \ldots, x_n) = 0$ holds in L. It is obvious that $f(a_1 \otimes \alpha_1, \ldots, a_n \otimes \alpha_n) = 0$ in $\tilde{L}$ for any $a_1, \ldots, a_n \in L$, $\alpha_1, \ldots, \alpha_n \in \tilde{K}$. Moreover, f is equal to zero for any substitution of elements in $\tilde{L}$ for $x_1, \ldots, x_n$.

To prove this claim we replace f by a system of identities. Let the degree of f in x_1 be equal to $m > 1$. Denote by $f_i = f_i(x_1, y_1, x_2, \ldots, x_n)$ the sum of monomials of degree i in y_1 in the polynomial $f(x_1 + y_1, x_2, \ldots, x_n)$ where $d(y_1) = d(x_1)$, $i = 1, \ldots, m-1$. Since K is an infinite field, all f_i vanish identically on L, $i = 1, \ldots, m - 1$ (see Chapter 1, Subsection 1.2.3). The identities f_i are called partial linearizations of f relative to the variable x_1. Clearly, the system of identities $f = 0, f_1 = 0, \ldots, f_{m-1} = 0$ is equivalent to $f = 0$. We continue the process of partial linearization of the identities $f_1 = 0, \ldots, f_{m-1} = 0$ relative to all variables. As a result we obtain a family of identities $\{f_\alpha = 0\}$ which is equivalent to $f = 0$. Note that $f = 0$ is one of the members of this family.

After that we apply this procedure to the variable x_2 in f if the degree of f in x_2 is greater than 1, and so on. Finally, we get a family of identities $\{f_\lambda = 0\}$ which is equivalent to f. This family has an additional property. If H is an algebra with basis B then in order to verify the family of identities $\{f_\lambda = 0\}$ on H it is sufficient to check $f_\lambda(a_1, \ldots, a_k) = 0$ for any λ only for elements $a_1, \ldots, a_k$ in B. As usual, if $f_\lambda = f_\lambda(z_1, \ldots, z_k)$ then $d(a_i) = d(z_i)$, $i = 1, \ldots, k$. There is no need to substitute $a_i + a'_i$, $a_i, a'_i \in B$, for z_i in f_λ. It follows that the family of identities $\{f_\lambda = 0\}$ holds in L if, and only if, it holds in $\tilde{L}$. This proves the following result.

Proposition. *Let L be a colour Lie superalgebra over an infinite field K, and $\tilde{K}$ an extension of K. If $f = f(x_1, \ldots, x_n) = 0$ holds on L identically then it holds identically on the $\tilde{K}$-algebra $\tilde{L} = L \otimes \tilde{K}$.* □

4.7. Lemma. *Suppose that L is a finitely generated soluble colour Lie superalgebra over an infinite field K graded by a finite group G. Let the family of identities of the form*

$$[x, y^{(n)}, z] = \sum_{j=1}^{n} \alpha_j [y^{(j)}, x, y^{(n-j)}, z],$$

$$d(x) = g_1, \qquad d(y) = g_2, \qquad d(z) = g_3,$$

hold on L where g_1, g_2, g_3 are arbitrary elements in G. Let M be a module of finite type over L. Suppose that on every homogeneous component M_g an identity of the form

$$[y^{(n)}, z]v = \sum_{j=1}^{n} \lambda_j y^j [y^{(n-j)}, z]v,$$

$$v \in M_g, \qquad d(y) = g_1, \qquad d(z) = g_2,$$

holds for any g_1, g_2 in G. Suppose further that L^m acts on M as a nilpotent space of transformations and $m = 2$ if char $K = 0$. Then M is a residually N-dimensional module for some N.

Proof. First suppose that $L^m M = 0$. Then $L/L^m \leftthreetimes M$ satisfies all hypotheses of Theorem 4.5. Hence, this algebra is representable. By Theorem 4.2 M is residually an L-module of bounded dimension.

Now let $L^m M \neq 0$. By the hypotheses of the lemma there exists r such that

$$\underbrace{L^m \ldots L^m}_{r} M = T \neq 0, \qquad L^m T = 0.$$

We denote by $I_j(M)$ the intersection of all L-submodules in M with codimension not larger than j. Using induction on r we conclude that $I_j(M) \subset T$ for some j. By Theorem 2.5 T is a finitely generated L-module. As before, T can be embedded in a finite-dimensional module over some extension $\tilde{K}$ of the field K. It has been shown in Subsection 4.4 that the $L \otimes \tilde{K}$-module $\tilde{T} = T \otimes K$ contains a submodule Q over $\tilde{K}$ of finite codimension in $\tilde{T}$. Moreover $Q \cap T = 0$.

Set $\tilde{M} = M \otimes \tilde{K}$ and consider the quotient module $P = \tilde{M}/Q$ over the superalgebra $\tilde{L} = L \otimes \tilde{K}$. Since $M \cap \tilde{T} = T$ it follows that $Q \cap M = 0$. Thus M can be embedded in P. By Proposition 4.6 all identities of the algebra L hold in $\tilde{L}$. Similarly, in the representation of $\tilde{L}$ on $\tilde{M}$ we find the same identities as in the representation of L on M. Hence, $\tilde{L}$ and its module P satisfy all hypotheses of the lemma. We can apply the inductive hypothesis to the quotient module of P by $\tilde{T}/Q$ since

$$\underbrace{\tilde{L} \ldots \tilde{L}}_{r} \tilde{M} \subset \tilde{T}.$$

It follows that $I_k(P) \subset \tilde{T}/Q$ for some k. We consider the descending chain

$$I_k(P) \supset I_{k+1}(P) \supset \cdots . \qquad (2)$$

By Theorem 3.8 the intersection of all terms in (2) is zero. On the other hand, $\dim I_k(P) < \infty$ over $\tilde{K}$. Consequently, the chain (2) breaks off after finitely many steps, i.e. $I_{k+t}(P) = 0$ for some t. This yields the representability of the $\tilde{L}$-module P. Since $L \subset \tilde{L}$ and $M \subset P$, M is also a representable L-module. Applying Theorem 4.2 to $L/\mathrm{Ann}\, M \leftthreetimes M$ we see that M is a residually N-dimensional L-module, and Lemma 4.7 is proved. □

4.8. Theorem. *If a colour Lie superalgebra L and its module M are as in Theorem 3.8 and the ground field K is infinite then M is a representable L-module.*

Proof. By Lemma 2.7 L_+ is a finitely generated algebra and L_- is an L_+-module of finite type. Without loss of generality one can assume that M is a faithful L-module. By the hypotheses of the theorem L_+^m acts on M as a nilpotent space of transformations. Since M is a faithful module, it follows that

$$[\underbrace{L_+^m, \ldots, L_+^m}_{k}, L_-] = 0$$

for some k. By Lemma 3.3 M is a module of finite type over L_+ which is finitely generated. Then, by Lemma 4.7, the intersection of all L_+-submodules in M of codimension less than N is zero for some integer N. Now Lemma 3.4 shows that M is a residually finite-dimensional L-module of bounded dimension. In Subsection 4.4 it was shown that in this case M is a representable L-module. □

The theorem just proved immediately implies the following.

4.9. Theorem. *Let L be a finitely generated soluble colour Lie superalgebra over an infinite field K, char $K \neq 2$, graded by a finite group G. If for any g, $h \in G$, $r \in G_+$, L satisfies an identity of the form*

$$[x, y^{(n)}, z] = \sum_{j=1}^{n} \alpha_j [y^{(j)}, x, y^{(n-j)}, z],$$

$$d(x) = g, \qquad d(y) = r, \qquad d(z) = h,$$

and if L_+^m acts on L as a nilpotent space of transformations with $m = 2$ if char $K = 0$, then L is representable. □

All results to follow in this section will concern ordinary $\mathbb{Z}_2$-graded Lie superalgebras. However, before we proceed we give a corollary of the previous results to associative algebras.

4.10. Representability of associative algebras and weak identities. In the second section (see Subsection 2.1) the definition of the identity of the representation of a colour Lie superalgebra on its module M was given. For ordinary Lie algebras we consider a more general concept.

Suppose that L is a Lie algebra over a field K and A is an associative enveloping algebra for L, i.e. L is a subalgebra of the Lie algebra $[A]$ and generates A in the associative sense. A relation $f(x_1, \ldots, x_n) = 0$ is said to be a *weak identity* on A or an identity of the pair (A, L) if the associative polynomial $f(x_1, \ldots, x_n)$ vanishes after substituting arbitrary elements $a_1, \ldots, a_n \in L$ for $x_1, \ldots, x_n$. Comparing these two definitions we see that an identity of the representation $\rho\colon L \to \operatorname{End} M$ is a weak identity of the associative subalgebra generated in $\operatorname{End} M$ by all $\rho(x)$, $x \in L$.

Sometimes we shall be speaking about the weak identities of an associative algebra A without mentioning the Lie subalgebra L in $[A]$. However, if we speak about several weak identities holding in A simultaneously then the Lie algebra L for all these identities must be the same. Furthermore, we will assume that a finitely generated algebra with a weak identity is an associative envelope with a weak identity for some finitely generated Lie algebra.

Theorem. *Suppose that A is a finitely generated associative algebra over an infinite field K, char $K \neq 2$, satisfying a weak identity of the form*

$$xy^n = \sum_{i=1}^{n} \beta_i y^i x y^{n-i}. \tag{3}$$

Moreover let A satisfy one of the weak identities

$$[x_1, y_1] \ldots [x_m, y_m] = 0 \tag{4}$$

if char $K = 0$, *or*

$$[x, y^{(n)}, z] = \sum_{j=1}^{n} \alpha_j [y^{(j)}, x, y^{(n-j)}, z] \tag{5}$$

if char $K > 2$. *Then A is representable.*

Proof. First we observe that (3) is another form of the relation

$$[y^{(n)}, x] = \sum_{j=1}^{n} \lambda_j y^j [y^{(n-j)}, x]. \tag{6}$$

Indeed, by induction on k it is easy to prove that the difference $xy^k - (-1)^k [y^{(k)}, x]$ is a linear combination of $y^i [y^{(k-i)}, x]$ with $i \geq 1$. Using this fact it is easy to pass from (3) to (6). To obtain (3) from (6) it sufficient to open all commutator brackets.

Now assume $A = \operatorname{alg}\{a_1, \ldots, a_t\}$ and let L be the Lie algebra generated by $a_1, \ldots, a_t$ such that (3) together with (4) or (5) holds for all elements in L. Denote by M a free left A-module of rank 1. If char $K \neq 0$ then the identities (5) and (6) hold for the representation of L on M. By Lemma 2.4 there exists q such that L^q acts on M as a nilpotent space of transformations. If char $K = 0$ then by (4) L^2 acts nilpotently on M. By Theorem 4.8 M is a faithful representable L-module. In other words, there exists an extension $\tilde{K}$ of K and a finite-dimensional vector space V over $\tilde{K}$ which contains M and which is a $\tilde{K}$-linear span of M. Furthermore, there exists a faithful $\tilde{K}$-linear representation of L on V, $\rho: L \to \operatorname{End} V$, and the restriction of $\rho(x)$ on M for any x in L coincides with the action of x on the free A-module M.

We extend ρ to A by setting

$$\rho(f(a_1,\ldots,a_t)) = f(\rho(a_1),\ldots,\rho(a_t))$$

and prove that this is well-defined. If $f(a_1,\ldots,a_t) = 0$ and $b = f(\rho(a_1),\ldots,\rho(a_t))$ then it is sufficient to check that b is the zero transformation on V. Let v be an element in M. Then $\rho(x)v = xv$ for any x in L. Hence,

$$bv = f(\rho(a_1),\ldots,\rho(a_t))v = f(a_1,\ldots,a_t)v = 0.$$

Since all $\rho(a_i)$ commute with the scalars in $\tilde{K}$ and V is a $\tilde{K}$-linear span of M, b is equal to zero in $\operatorname{End} V$. Hence ρ is a faithful representation of A as $\tilde{K}$-linear transformations on the finite-dimensional vector space V over $\tilde{K}$, and the proof of the theorem is complete. $\square$

Remark. In Theorem 4.10 the characteristic of the ground field K is different from two. This requirement appears because we use some previous results where $\operatorname{char} K \neq 2$. In fact this restriction in Theorem 4.10 is not essential since all necessary auxiliary results may be proved by the same methods for $\operatorname{char} K = 2$ provided that L is an ordinary Lie algebra.

Concerning Theorems 3.9 and 4.9 a natural question arises: does there exist another family of identities on a finitely generated colour Lie superalgebra which implies representability or residual finiteness? For ordinary $\mathbb{Z}_2$-graded Lie superalgebras over an infinite field with solubility conditions a negative answer to this question is given in the following. Recall (see Subsection 2.13) that a variety $\mathfrak{V}$ is called *locally representable* (*locally residually finite*) if any finitely generated algebra in $\mathfrak{V}$ is representable (residually finite).

4.11. Theorem. *Let $\mathfrak{V}$ be a locally soluble variety of Lie superalgebras over an infinite field K, $\operatorname{char} K > 2$. Then the following conditions are equivalent:*
a) *$\mathfrak{V}$ is locally representable;*
b) *$\mathfrak{V}$ is locally residually finite;*
c) *$\mathfrak{V}$ satisfies, for some n, the three identities of the form*

$$[x, y^{(n)}, z] = \sum_{j=1}^{n} \alpha_j^i [y^{(j)}, x, y^{(n-j)}, z] \qquad (7)$$

with $i = 1, 2, 3$ where $\alpha_j^i \in K$, $d(y) = 0$ and $d(x) = d(y) = 0$ if $i = 1$, $d(x) = 0$, $d(z) = 1$ if $i = 2$, $d(x) = d(z) = 1$ if $i = 3$.

Proof. By Theorem 4.2 a) implies b). Now suppose $\mathfrak{B}$ is a locally residually finite variety of Lie superalgebras. In Section 1, for any group G we have constructed examples of colour Lie superalgebras $B(g, h)$ and $Q(g, h)$ (see Subsection 1.3). For every g, h in the grading group G, $B(g, h)$ is not residually finite because it contains the monolith and $\dim B(g, h) = \infty$. Now $Q(g, h)$ contains a central subalgebra with basis $\{c_0, c_1, \ldots\}$. Let C be the linear span of the c_i with $i \geq 1$. Then, obviously, C is an ideal in $Q(g, h)$ and the quotient algebra $Q(g, h)/C$ contains a subalgebra isomorphic to $B(g, h)$. It follows that $Q(g, h)/C$ is not residually finite and hence not in $\mathfrak{B}$. Therefore $\mathfrak{B}$ cannot contain $Q(g, h)$ either. By Proposition 2.9 identities of the form (7) hold in the variety $\mathfrak{B}$, i.e. b) implies c).

Finally, we prove that c) implies a). In the proof of Theorem 2.14 we already remarked that the identities (7) imply a similar identity with $d(x) = 1$, $d(z) = 0$. Hence, we can apply Lemma 2.7 to any finitely generated Lie superalgebra $L = L_0 \oplus L_1$ in $\mathfrak{B}$. It follows that L_0 is a soluble Lie algebra with a finite set of generators. Moreover, L is an L_0-module of finite type. By Lemma 2.4 L_0^q acts on L_0 and on L_1 as a nilpotent space of transformations for some q. To complete the proof it is now sufficient to apply Theorem 4.9. □

Note that for an infinite field K of positive characteristic the family of identities (7) is equivalent to each of the (equivalent) conditions of Theorem 2.14 in Section 2.

If the ground field has characteristic zero, then we obtain a complete description of locally residually finite and locally representable varieties of Lie superalgebras. It will be shown that solubility is a necessary condition for the local residual finiteness of a variety of Lie superalgebras. First we prove some auxiliary results.

4.12. Lemma. *Let $L = L_0 \oplus L_1$ be a Lie superalgebra over a field of characteristic zero with identities of the form (7). Suppose that for any q there exists $N \in \mathbb{N}$ such that*

$$(\mathrm{ad}([a_1, b_1]) + \cdots + [a_q, b_q]))^N = 0$$

on L for all $a_1, \ldots, a_q, b_1, \ldots, b_q$ in L_0. Then L is locally soluble.

Proof. We construct a subalgebra in L of the form

$$A = A_0 \oplus A_1 = L_0 \oplus [L_0, \ldots, L_0, L_1]$$

such that for some k we have $(\operatorname{ad} b^2)^{k+1} = 0$ in A for any b in A_1. First consider the case where both for $i = 2$ and for $i = 3$ not all coefficients α_j^i in (7) are zero. As was shown in Subsection 2.2, L satisfies the following identities

$$[x, y^{(k)}, Z] = \sum_{j=1}^{k} \lambda_j [y^{(j)}, x, y^{(k-j)}, Z], \qquad d(x) = d(y) = 0, \qquad (8)$$

$$[x, y^{(k)}, Z] = \sum_{j=1}^{k} \mu_j [y^{(j)}, x, y^{(n-j)}, Z], \qquad d(x) = 1, d(y) = 0, \qquad (9)$$

where $Z = [t^{(r)}, z]$, $d(z) = 1$, $d(t) = 0$. Furthermore, $\lambda_k \neq 0$, $\mu_k \neq 0$.

Denote by M the L_0-submodule in L_1 generated by all the products $[a^{(r)}, b]$, $a \in L_0$, $b \in L_1$. Then in the representation of L_0 on L_1/M we have the weak identity $x^r = 0$. Let us prove that for any Lie algebra H over a field of characteristic zero and an H-module W the weak identity $x^r = 0$ for the representation of H on W implies a weak identity $x_1 \ldots x_s = 0$ for some $s \geq r$. Without loss of generality one can assume that W is a faithful H-module.

By Theorem 4.1 in [Kostrikin, 1990] H is nilpotent since $(\operatorname{ad} x)^{2r+1} = 0$ on H for any x. First we assume that $H^2 = 0$ and consider the equation $(x_1 + \cdots + x_r)^r = 0$ in End W with $x_1, \ldots, x_r$ in H. The left hand side of this equation contains only one monomial depending on all variables $x_1, \ldots, x_r$. It equals $r! \, x_1 \ldots x_r$. Since char $K = 0$, it follows that $x_1 \ldots x_r = 0$. Hence, for an abelian Lie algebra H the statement is proved.

Now assume $H^t = 0$, $t \geq 2$. We use induction on t. Denote by C the center of H. Then, as before, C acts on W as a nilpotent space of transformations. Consider the chain of H-modules

$$W = W_0 \supset W_1 \supset \cdots \supset W_r = 0,$$

where $W_j = CW_{j-1}$ for $j = 1, \ldots, r$. For any $j \geq 0$ the space W_j/W_{j+1} is an H/C-module and, by induction, H/C acts nilpotently on W_j/W_{j+1}. This implies the nilpotence of the action of H on W, as claimed.

Therefore we may assume that

$$[\underbrace{L_0, \ldots, L_0}_{s}, L_1] \subset M \qquad (10)$$

for some s.

We assert that M is the linear span of elements of the form $[a^{(r)}, b]$, with $a \in L_0$, $b \in L_1$. If $a, c \in L_0$, $b \in L_1$ then

$$[c, a^{(r)}, b] - [a^{(r)}, c, b] = \sum_{j=0}^{r-1} [a^{(j)}, [c, a], a^{(r-j-1)}, b]. \qquad (11)$$

Since K is infinite any partial linearization of the element $[t^{(r)}, z]$ lies in M where $t \in L_0$, $z \in L_1$. Denote by $f(y, t, z)$ the partial linearization of $[t^{(r)}, z]$ of degree 1 in y, that is, the sum of monomials of degree 1 in y in the product $[(t + y)^{(r)}, z]$. Then for any $y, t \in L_0$, $z \in L_1$, we have $f(y, t, z) \in M$. On the other hand, the sum on the right hand side of (11) is equal to $f([c, a], a, b)$. Hence M is the linear span of the elements $[a^{(r)}, b]$ with $a \in L_0$, $b \in L_1$. This enables us to rewrite (8), (9) in the form

$$[x, y^{(k)}, v] = \sum_{j=1}^{k} \lambda_j [y^{(j)}, x, y^{(k-j)}, v], \tag{12}$$

$$[z, y^{(k)}, v] = \sum_{j=1}^{k} \mu_j [y^{(j)}, x, y^{(k-j)}, v], \tag{13}$$

respectively, where $x, y \in L_0$, $z \in L_1$, v is an arbitrary element in M and $\lambda_k \neq 0$, $\mu_k \neq 0$.

Consider the subalgebra

$$A = A_0 \oplus A_1 = L_0 \oplus \underbrace{[L_0, \ldots, L_0, L_1]}_{s}$$

in L. Now assume that for all $a \in A_0$, $b \in A_1$ we have $[a, b] = 0$. According to (10), $A_1 \subset M$, hence (12) and (13) hold for $y = a$, $v = b$, $x = c_0 \in A_0$, $z = c_1 \in A_1$. Since λ_k and μ_k are nonzero and $[a, b] = 0$, it follows from (12) and (13) that $[a^{(k)}, c, b] = 0$ for any c in A. Substituting $a = [b, b]$ into this equation we obtain $(\operatorname{ad} b)^{2k+1} = 0$ on A, since $\operatorname{ad}[x, x] = 2(\operatorname{ad} x)^2$ for any odd element x in a Lie superalgebra. Thus we have proved the equality $(\operatorname{ad} b^2)^{k+1} = 0$ holding on A with $b^2 = [b, b]$ and $b \in A_1$.

If all the α_j^i are zero in (7) for $i = 2$ then the weak identity $x^{n+1} = 0$ holds in the representation of L_0 on its module L_1, hence $\underbrace{[L_0, \ldots, L_0, L_1]}_{s} = 0$ for some s, therefore $A_1 = 0$. If the coefficients α_j^2 are not equal to zero simultaneously then we have (8). As was shown above $(\operatorname{ad} b^2)^{k+1}$ is the zero map on A_0 for any $b \in A_1$. If, in addition, all the α_j^3 are zero then (9) does not hold, but for any $x \in L_1$ we have $[x, (x^2)^{(n)}, z] = 0$. Therefore, $(\operatorname{ad} b^2)^{(n+1)} = 0$ on L_1 for any $b \in L_1$.

Now suppose that $a_1, \ldots, a_q$ are elements in A_0 and $(\operatorname{ad} a_j)^T = 0$ on A for all $j = 1, \ldots, q$ and suitable $T \in \mathbb{N}$. We want to prove that there exists a number $P = P(q, T)$ such that $(\operatorname{ad}(a_1 + \cdots + a_q))^P = 0$ on A. Consider the Lie algebra H generated by $a_1, \ldots, a_q$. If $b \in H^2$ then $b = [a_1, b_1] + \cdots + [a_q, b_q]$ with $b_1, \ldots, b_q \in H$. By the hypotheses of the lemma one can find some N such that $[b^{(N)}, u] = 0$ for any $u \in A$ and N does not depend on b. In other words, in the representation of H^2 on A the weak identity $x^N = 0$ holds. It was shown that there exists m depending only on N such that $[c_1, \ldots, c_m, u] = 0$ provided that $c_1, \ldots, c_m \in H^2$ and $u \in A$.

Since $(\operatorname{ad} a_j)^T = 0$, the elements

$$[a_1^{(t_1)}, \ldots, a_q^{(t_q)}, a_i], \qquad t_1, \ldots, t_q \leq T - 1, \qquad \sum_j t_j > 0, \tag{14}$$

generate H^2 as a Lie algebra. Using the Poincaré-Birkhoff-Witt Theorem one can write the element $(\operatorname{ad}(a_1 + \cdots + a_q))^P(v)$ as a linear combination of the products

$$[b_1, \ldots, b_t, a_1^{(j_1)}, \ldots, a_q^{(j_q)}, v], \qquad j_1, \ldots, j_q \leq T - 1 \tag{15}$$

with $b_1, \ldots, b_t$ of type (14). Since the degree of each element of the form (14) (over all variables a_i) does not exceed q^T, we have $t \geq (P/qT) - 1$. It follows that $(\operatorname{ad}(a_1 + \cdots + a_q))^P = 0$ on A if $P \geq (m+1)qT$ since for this value of P we have $t \geq m$ and $[c_1, \ldots, c_m, u] = 0$ for any $c_1, \ldots, c_m \in H^2$, $u \in A$.

We conclude, from what has been proved earlier, that there exists t such that for any a, b, c in A_1 the equality $(\operatorname{ad}(a^2 + b^2 + c^2))^t = 0$ holds on A since $(\operatorname{ad} a^2)^{k+1} = (\operatorname{ad} b^2)^{k+1} = (\operatorname{ad} c^2)^{k+1} = 0$. Therefore, $(\operatorname{ad}[a, b])^t = 0$ for any $a, b \in A_1$, because $[a, b] = (a+b)^2 - a^2 - b^2$. It follows that the sum $[a_1, b_1] + \cdots + [a_q, b_q]$ is a nilpotent transformation on A for any $a_i, b_i \in A_1$, $i = 1, \ldots, q$. In other words, every element in $[A_1, A_1]$ acts on A as a nilpotent transformation.

Now we are able to prove the local solubility of the Lie superalgebra

$$H = H_0 \oplus H_1 = [A_1, A_1] \oplus [A_0, A_1]. \tag{16}$$

Suppose that $B = B_0 \oplus B_1$ is a subalgebra in H generated by $x_1, \ldots, x_q \in H_0$, $y_1, \ldots, y_q \in H_1$. First we prove that B_0 acts on B_1 as a nilpotent space of transformations. If b is an arbitrary element in B_0 then it can be written in the form

$$b = x + a_1 + \cdots + a_q + b_1 + \cdots + b_q$$

where x is a linear combination of $x_1, \ldots, x_q$, $a_i = [x_i, t_i]$, $b_i = [y_i, z_i]$ with some t_i and z_i in B_0 and B_1, respectively, $i = 1, \ldots, q$. Every element in $[A_1, A_1]$ acts nilpotently on A, hence there exists T such that $(\operatorname{ad} x_j)^T = 0$, $(\operatorname{ad} b_j)^T = 0$, $j = 1, \ldots, q$. By the hypotheses of the lemma $(\operatorname{ad} a_j)^t = 0$ for some t and $j = 1, \ldots, q$. Clearly, we can choose $t = T$. Then, as has been shown earlier, there exists $P = P(3q, T)$ such that $(\operatorname{ad} b)^P = 0$. Since P is independent of b, the weak identity $x^P = 0$ for the representation of B_0 on B implies the nilpotence of the action of B_0 on B and in particular on B_1, as required.

The nilpotence of the action of B_0 on B_1 and the inclusion $B^2 \subset B_0 \oplus [B_0, B_1]$ imply $B^{(j)} \subset B_0$ where $B^{(j)} = [B^{(j-1)}, B^{(j-1)}]$ is the j-th term of the derived series. Thus, in order to prove the solubility of B, it is sufficient to verify the solubility of the even component L_0 of the initial superalgebra L.

If all $\alpha_j^1 = 0$ in (7) then L_0 is nilpotent by Theorem 4.1 in [6].

Hence we can apply the method of separating variables (see Subsection 2.2) to the Lie algebra L_0. As a result we obtain a relation of the form

$$[x, y^{(k)}, Z] = \sum_{j=1}^{k} \gamma_j [y^{(j)}, x, y^{(k-j)}, Z] \tag{17}$$

where $Z = [t^{(r)}, z]$, $x, y, z, t \in B_0$ and $\gamma_k \neq 0$. If we denote by Q the ideal in L_0 generated by all $[a^{(r)}, b]$, $a, b \in B_0$, then, as before, Q is the linear span of the products $[a^{(r)}, b]$, $a, b \in L_0$. Furthermore L_0 acts on L_0/Q as a nilpotent space of transformations, i.e. $L_0^p \subset Q$ for some p. Obviously, (17) holds not only for $Z = [t^{(r)}, z]$. It is also true for any $Z = [a_1^{(r)}, b_1] + \cdots + [a_m^{(r)}, b_m]$, i.e. for all $Z \in Q$, $x, y \in L_0$. Thus replacing y by Z in (17) and taking into account that $\gamma_k \neq 0$, we obtain $(\operatorname{ad} Z)^k = 0$ on Q. Now the solubility of L_0 follows since $L_0^p \subset Q$ and Q is nilpotent.

From the local solubility of H we deduce now that $A^{(N)} \subset H$ for some N. Indeed, H is an ideal in A containing $[A_0, A_1]$. Therefore $(A/H)^2$ is an ordinary Lie algebra satisfying the hypotheses of the lemma. As before, $(A/H)^2$ is a soluble Lie algebra. It follows that $A^{(N)} \subset H$ for some N.

By the construction of A we have $L^{(s)} \subset A$. Hence $L^{(q)} \subset H$ for some q where H is locally soluble. Now two more lemmas are needed for proving the local solubility of L.

4.13. Lemma. *Let $L = L_0 \oplus L_1$ be a finitely generated Lie superalgebra as in Lemma 4.12 and suppose L^2 is locally soluble. Then L_0 is a finitely generated Lie algebra.*

Proof. Suppose that even elements $y_1, \ldots, y_m$ and odd elements $x_1, \ldots, x_m$ generate L. Denote by H the subalgebra in L_0 generated by $y_1, \ldots, y_m$ and by M the H-submodule in L generated by $x_1, \ldots, x_m$. We want to check that M is the linear span of the elements

$$[y_1^{(j_1)}, \ldots, y_m^{(j_m)}, a] \tag{18}$$

with

$$a = [h_1, \ldots, h_k, x_i], \quad k < k_0, \tag{19}$$

where every $h_1, \ldots, h_k$ is a commutator of the form

$$[y_1^{(t_1)}, \ldots, y_m^{(t_m)}, y_{i_1}, y_{i_2}], \qquad t_1, \ldots, t_m \leq n - 1. \tag{20}$$

Every element in H^2 is a sum of commutators with at most $m + 1$ summands. By the hypotheses of Lemma 4.12 there exists N such that $(\operatorname{ad} x)^N = 0$ on L for any x in H. Hence, for some k_0, we have the equality

$$[\underbrace{H^2, \ldots, H^2}_{k_0}, L] = 0. \tag{21}$$

We consider the descending chain of H-submodules

$$M = M_0 \supset M_1 \supset \cdots \supset M_{k_0} = 0 \tag{22}$$

where $M_j = [H^2, M_{j-1}]$ if $j \geq 1$. Since H modulo H^2 coincides with the linear span of $y_1, \ldots, y_m$ and H^2 is generated as an ideal in H by all the products $[y_{i_1}, y_{i_2}]$, one can apply Lemma 2.6 to H and the module M with descending chain (22). It follows that M is the linear span of elements of type (18) satisfying (19) and (20). Denote by B a finite set of elements of the form (18) with the additional restriction $j_1, \ldots, j_m \leq n - 1$. If $M^1 = M$ and $M^r = [M, M^{r-1}]$ for $r \geq 2$ then M^r is the linear span of elements $[b_1, \ldots, b_r]$ with $b_1, \ldots, b_r$ of the form (18)–(20). We shall prove that M^r is an H-module generated by the products $[b_1, \ldots, b_r]$ with $b_1, \ldots, b_r \in B$.

For $r = 1$ this is clear. If $r > 1$ then by induction hypothesis M^r is the linear span of elements

$$[v, y_{i_1}, \ldots, y_{i_s}, b_2, \ldots, b_r]$$

with $v \in M$, $b_2, \ldots, b_r \in B$. It follows by the Jacobi identity for Lie superalgebras that M lies in the H-module generated by the products $[v, b_2, \ldots, b_r]$ where $b_2, \ldots, b_r \in B$ and v is a product of the form (18)–(20). Define a function v on the elements of type (18) by setting $v(v) = v(a) = k$ if

$$v = [y_1^{(j_1)}, \ldots, y_m^{(j_m)}, a]$$

with a of the form (19). Denote by V_k the linear span of all elements v of the form (18)–(20) with $v(v) \geq k$. Obviously V_k is an H-module and $V_{k_0} = 0$. (For $k \geq k_0$ the commutator (19) equals zero by (21).)

If $u = [v, b_2, \ldots, b_r]$ where $b_2, \ldots, b_r$ are in B and v is of the form (18)–(20) with $j_1, \ldots, j_m \leq n - 1$ then $v \in B$ by construction and u has the required form. If one of $j_1, \ldots, j_m$ is larger than $n - 1$ and v is in V_{k_0-1} then we may assume that all the $y_1, \ldots, y_m$ in the expression for v commute. Using (7) we conclude that u is in the H-module generated by $[b_1, \ldots, b_r]$, $b_i \in B$, $i = 1, \ldots, r$.

§4. Representability of Lie superalgebras by matrices

Finally, let v be in V_k with $k < k_0 - 1$. If $j_m \geq n$ in v then

$$v \equiv [y_m^{(j_m)}, y_1^{(j_1)}, \ldots, y_{m-1}^{(j_{m-1})}, a] \pmod{V_{k+1}}. \tag{23}$$

Using (7) and (23) one can lower the degree of v in y_m. Similarly, we can lower the degree of v in any variable y_t if $j_t \geq n$ for an arbitrary value of $v(v)$. Thus we have proved that each M^r as an H-module is generated by the commutators $[b_1, \ldots, b_r]$ with $b_1, \ldots, b_r \in B$.

Suppose that $C = C_0 \oplus C_1$ is the subalgebra in L generated by the finite set B. Then C_0 together with $y_1, \ldots, y_m$ generate L_0 as a Lie algebra. Therefore, to complete the proof of the lemma it is sufficient to verify that C_0 is finitely generated. An important difference between the Lie superalgebras C and L is that C is generated by a finite set of odd elements.

If $B = \{b_1, \ldots, b_t\}$ then C^2, as a Lie superalgebra, is generated by the products

$$[b_1^{(i_1)}, \ldots, b_t^{(i_t)}, b_j], \quad i_1 + \cdots + i_t \geq 1. \tag{24}$$

If one of the $i_1, \ldots, i_{t-1}$ is greater than 1 then (24) is an element in $(C^2)^2$ since any b_i, $i = 1, \ldots, t-1$, is in L_1 and $(\text{ad } b_i)^2 = \frac{1}{2} \text{ad } b_i^2$ with $b_i^2 \in C^2$. If $i_t > 2$ then (24) is also in $(C^2)^2$. It follows that C^2 is generated by the elements of the form (24) with $i_1, \ldots, i_t \leq 2$. Hence C^2 is a finitely generated Lie superalgebra. Furthermore, C^2 is a subalgebra in L^2 which is locally soluble by the hypothesis of the lemma. Consequently, C is a soluble Lie superalgebra satisfying the identities (7). By Lemma 2.7 C_0 is a finitely generated Lie algebra, and the proof of Lemma 4.13 is complete. □

We continue with the proof of Lemma 4.12. So far we have shown the local solubility of $L^{(q)}$ for some q. Now it is sufficient to verify that the local solubility of $L^{(q)}$ implies the same property for $L^{(q-1)}$, $q \geq 1$. Therefore, application of the next lemma completes the proof of Lemma 4.12.

4.14. Lemma. *Suppose that L is a Lie superalgebra as in Lemma 4.12. If L^2 is locally soluble then L is also locally soluble.*

Proof. Suppose that a finite set $\{y_1, \ldots, y_m, x_1, \ldots, x_m\}$ generates L, where $y_1, \ldots, y_m$ are even and $x_1, \ldots, x_m$ are odd elements. By Lemma 4.13 L_0 is a finitely generated Lie algebra. If L_0 is generated by k elements then any element in L_0 is a sum of at most k commutators. It follows by the hypotheses of Lemma 4.12 that one can choose a number N such that $(\text{ad } x)^N = 0$ on L for any x in L_0^2.

In the first part of the proof of Lemma 4.12, it was shown that $L^{(q)} \subset H$ with a subalgebra $H = [A_1, A_1] \oplus [A_0, A_1]$ defined in (16), $A_0 = L_0$ and

$A_1 = [L_0, \ldots, L_0, L_1]$. Moreover it was proved for any $a, b \in A_1$ that $(\operatorname{ad}[a,b])^t = 0$ on A and that t does not depend on a, b. Furthermore, if $a_1, \ldots, a_q \in L_0$, $(\operatorname{ad} a_1)^T = \cdots = (\operatorname{ad} a_q)^T = 0$ then there exists $P = P(q, T)$ such that $(\operatorname{ad}(a_1 + \cdots + a_q))^P = 0$.

Now, since $H_0 = [A_1, A_1]$, any x in H_0 is a sum $x = [x_1, c_1] + \cdots + [x_m, c_m] + b$ with $b \in L_0^2$ and $c_1, \ldots, c_m \in A_1$. Therefore, there exists r such that $(\operatorname{ad} x)^r = 0$ on H for arbitrary $x \in H_0$. Now the inclusion $H^{(r)} \subset H_0$ follows immediately and thus, by Theorem 4.1 in [Kostrikin, 1990], H_0 itself is nilpotent. Consequently, H is a soluble Lie superalgebra. Since $L^{(q)} \subset H$, L is also soluble, completing the proof of Lemma 4.14. □

Lemma 4.12 shows that, for Lie superalgebras over a field of characteristic zero, one can drop the solubility condition in Theorem 4.9.

We now investigate some properties of locally representable and locally residually finite varieties of Lie superalgebras. To start with we consider the case of ordinary Lie algebras. We denote by var L the variety generated by L.

4.15. Lemma. *Let $\mathfrak{V}$ be a locally residually finite variety of Lie algebras over a field K of characteristic zero. If L is a finite-dimensional algebra in $\mathfrak{V}$ then L is soluble.*

Proof. First verify that $\mathfrak{V}$ does not contain the matrix algebra $\mathfrak{sl}(2, K)$. It is known (see [Bahturin, 1987a], Theorem 5.6.5) that a basis of identities holding in $\mathfrak{sl}(2, K)$ can be taken to consist of the two identities

$$\sum_{\sigma \in \operatorname{Sym}(4)} (-1)^\sigma [x_{\sigma(1)}, \ldots, x_{\sigma(4)}, x_0] = 0, \tag{25}$$

$$[z^{(3)}, x, y] = [[z^{(3)}, x], y] + [x, z^{(3)}, y]. \tag{26}$$

Hence any Lie algebra satisfying (25) and (26) lies in var($\mathfrak{sl}(2, K)$).

Suppose that L is a center-by-metabelian Lie algebra, that is the identity

$$[x_1, [x_2, x_3], x_4, x_5] = 0 \tag{27}$$

holds in L. It is easy to see that (27) implies (25). Now (26) is equivalent to $[z, [z, x], z, y] = 0$ and the latter is a consequence of (27). Thus we conclude that any center-by-metabelian Lie algebra belongs to var($\mathfrak{sl}(2, K)$).

Suppose that $\mathfrak{V}$ contains $\mathfrak{sl}(2, K)$. Then every center-by-metabelian Lie algebra lies in $\mathfrak{V}$. However, a locally residually finite variety cannot contain the algebra $B(0,0)$ from Subsection 1.3 since $B(0,0)$ is not residually finite. It follows that $\mathfrak{V} \not\ni \mathfrak{sl}(2, K)$.

Now we assume $L \in \mathfrak{B}$ and $\dim L < \infty$. If L is not soluble then the quotient algebra of L by its radical is a semisimple Lie algebra. Hence, without loss of generality, we may assume that L is semisimple. Denote by $\overline{K}$ the algebraic closure of the field K. It has been shown earlier in Subsection 4.4 that all identities of L also hold in $\overline{L} = L \otimes \overline{K}$. It follows that $\overline{L} \in \mathfrak{B}$. On the other hand, $\overline{L}$ is a semisimple Lie algebra over the algebraically closed field $\overline{K}$, therefore $\overline{L}$ contains $\mathfrak{sl}(2, \overline{K})$ as a subalgebra. Since $\mathfrak{sl}(2, K) \subset \mathfrak{sl}(2, \overline{K})$, we conclude $\mathfrak{sl}(2, K) \in \mathfrak{B}$, a contradiction. It follows that every finite-dimensional Lie algebra in $\mathfrak{B}$ is soluble. $\square$

4.16. Lemma. *Let $L = L_0 \oplus L_1$ be a free Lie superalgebra of a locally residually finite variety $\mathfrak{B}$ over a field of characteristic zero with free generators x_1, $x_2, \ldots, y_1, y_2, \ldots$ where $d(x_i) = 0$, $d(y_i) = 1$, $i = 1, 2, \ldots$ Suppose that $f = f(x_1, \ldots, x_k)$ is a homogeneous element in L_0 of total degree $t > 1$ in $x_1, \ldots, x_k$. Then $(\operatorname{ad} f)^N = 0$ for some N.*

Proof. Let z be one of the free generators of L different from $x_1, \ldots, x_k$ and let x be an even free generator of L different from $x_1, \ldots, x_k, z$. Denote by H the ideal in L generated by $[x, z] + [f, x, z]$. We assert that $[x, z] \in H$.

Assume $[x, z] \notin H$. Since L/H is a residually finite Lie superalgebra, it must contain an ideal T of finite codimension such that $T \supset H$ and $[x, z] \notin T$. The quotient algebra $L/T = A = A_0 \oplus A_1$ is of finite dimension. Hence A_0 is a finite-dimensional Lie algebra in $\mathfrak{B}$. Clearly the variety of Lie algebras var A_0 is locally residually finite. By Lemma 4.15 A_0 is a soluble Lie algebra. Since $\operatorname{char} K = 0$, A_0^2 acts on both A_0 and A_1 as a nilpotent space of transformations. By the hypothesis of the lemma $\deg f > 1$, hence $f + T \in A_0^2$. It follows that $[f^{(j)}, x, z] \in T$ for some j. On the other hand, $T \supset H \ni [x, z] + [f, x, z]$. Therefore all elements $[f^{(j-1)}, x, z], \ldots, [f, x, z], [x, z]$ lie in T, a contradiction, which proves that $[x, z]$ is in H. It follows that the relation

$$[x, z] + h_0([x, z] + [f, x, z]) + \cdots + h_r([x, z] + [f, x, z]) = 0 \quad (28)$$

holds in L with each h_i being a homogeneous polynomial in $\operatorname{ad} x_1, \ldots, \operatorname{ad} x_k$ of degree i. Since $\operatorname{char} K = 0$ any multihomogeneous component in $x_1, \ldots, x_k$ on the left hand side of (28) is zero. In particular, $h_0([x, z]) + [x, z] = 0$, that is, $h_0 = -1$.

The total degree of f is equal to t. Therefore $h_i([x, z]) = 0$ for all $i = 1, \ldots, t - 1$ and

$$h_t([x, z]) = [f, x, z]. \quad (29)$$

Let φ be an endomorphism of L which maps z to $[x, z]$, x to f and leaves x_i fixed, $i = 1, \ldots, k$. We apply φ to the equation (29) and obtain $h_t([f, x, z]) = [f^{(2)}, x, z]$. Setting the homogeneous component of degree $2t + 2$ on the left hand side of (28) equal to zero we obtain the equality

$$h_{2t}([x,z]) = -h_t([f,x,z]) = -[f^{(2)},x,z].$$

Now we assume that the equality $h_{jt}([x,z]) = \pm[f^{(j)},x,z]$ with $j \geq 2$ holds. Applying φ to this equality we get the relation $h_{jt}([f,x,z]) = \pm[f^{(j+1)},x,z]$. Since the homogeneous component of degree $2 + t(j+1)$ on the left hand side of (28) is zero we get

$$h_{(j+1)t}([x,z]) = -h_{jt}([f,x,z]) = \pm[f^{(j+1)},x,z].$$

The number of summands on the left hand side of (28) is finite. Therefore, for some N we obtain

$$h_{(N-2)t}([x,z]) = \pm[f^{(N-2)},x,z], \quad h_{(N-2)t}([f,x,z]) = 0.$$

It follows that

$$[f^{(N-1)},x,z] = \varphi([f^{(N-2)},x,z] = \pm\varphi(h_{(N-2)t}([x,z]))$$
$$= \pm h_{(N-2)t}([f,x,z]) = 0.$$

Replacing the free generator x by f we obtain $(\operatorname{ad} f)^N = 0$ as required, and the proof is complete. $\square$

Now we are prepared to give a complete description of locally residually finite and locally representable varieties of Lie superalgebras over a field of characteristic zero.

4.17. Theorem. *For the variety $\mathfrak{B}$ of Lie superalgebras over a field K of characteristic zero the following conditions are equivalent.*
a) *$\mathfrak{B}$ is locally representable;*
b) *$\mathfrak{B}$ is locally residually finite;*
c) *$\mathfrak{B}$ satisfies, for some n, a family of identities*

$$[x, y^{(n)}, z] = \sum_{j=1}^{n} \alpha_j^i [y^{(j)}, x, y^{(n-j)}, z] \tag{30}$$

where $i = 1, 2, 3$, $\alpha_j^i \in K$. For $i = 1$ all variables x, y, z are even, for $i = 2$ $d(x) = d(y) = 0$, $d(z) = 1$, and $d(x) = d(z) = 1$, $d(y) = 0$, if $i = 3$. Furthermore, a finitely generated Lie superalgebra in $\mathfrak{B}$ satisfies the following identities:

$$[[x_1, y_1], \ldots, [x_m, y_m], z] = 0,$$
$$d(x_i) = d(y_i) = 0, \quad i = 1, \ldots, m, \tag{31}$$

for some m.

Proof. By Theorem 4.2 a locally representable variety of Lie superalgebras over K is also locally residually finite. Hence a) implies b).

Now we assume that $\mathfrak{B}$ is a locally residually finite variety of Lie superalgebras. As in the case where the ground field has positive characteristic (see proof of Theorem 4.11), $\mathfrak{B}$ cannot contain the superalgebras $B(g,h)$, $Q(g,h)$ constructed in Subsection 1.3 for any g, h in the grading group $\mathbb{Z}_2$. Therefore, by Proposition 2.9, $\mathfrak{B}$ satisfies (30). Now suppose that $L = L(\mathfrak{B})$ is a free superalgebra of the variety $\mathfrak{B}$ with the set $\{x_1, x_2, \ldots, y_1, y_2, \ldots\}$ of free generators where $d(x_i) = 0$, $d(y_i) = 1$, $i = 1, 2, \ldots$. Consider the polynomial $f = [x_1, x_2] + \cdots + [x_{2q-1}, x_{2q}]$. By Lemma 4.16 one can find N such that $(\mathrm{ad}\, f)^N = 0$ on L. Since $x_1, \ldots, x_{2q}$ are free generators, for any $a_1, \ldots, a_q$, $b_1, \ldots, b_q$ the equality $(\mathrm{ad}([a_1, b_1] + \cdots + [a_q, b_q]))^N = 0$ holds in L_0. By Lemma 4.12 L is locally soluble. It follows that every finitely generated Lie superalgebra H in $\mathfrak{B}$ is soluble. We want to prove that (31) holds on H.

If $H = H_0 \oplus H_1$ then by Lemma 2.7 H_0 is a finitely generated Lie superalgebra. We denote by $a_1, \ldots, a_q$ generators of H_0. Then any element h in H_0^2 can be written as a sum $h = [a_1, b_1] + \cdots + [a_q, b_q]$ with some $b_1, \ldots, b_q \in H_0$. As before, $(\mathrm{ad}\, h)^N = 0$ on H. In other words, the weak identity $z^N = 0$ holds for the representation of H_0^2 on H. It has been remarked repeatedly that this implies a weak identity $z_1 \ldots z_m = 0$ for the representation of H_0^2 on H, i.e. the identity (31) holds in the Lie superalgebra H as claimed. Thus, c) follows from b).

Finally, suppose that $\mathfrak{B}$ is a variety satisfying c) and let L be a finitely generated Lie superalgebra in $\mathfrak{B}$. By Lemma 4.12 L is soluble. Hence L is representable by Theorem 4.9, i.e. $\mathfrak{B}$ is locally representable. $\square$

4.18. Examples of locally representable varieties of Lie superalgebras. We give several examples of locally representable varieties of Lie superalgebras. If $\mathfrak{B}_1$ and $\mathfrak{B}_2$ are two varieties then their *product* $\mathfrak{B}_1 \mathfrak{B}_2$ consists of all Lie superalgebras L such that some ideal H of L lies in $\mathfrak{B}_1$ and the quotient algebra L/H lies in $\mathfrak{B}_2$. Denote by $\mathfrak{N}_k$ the variety which consists of all nilpotent Lie superalgebras over K with nilpotency index at most k. One can define $\mathfrak{N}_k$ by the single non-graded identity

$$[x_1, \ldots, x_{k+1}] = 0.$$

If we denote by $\mathfrak{A}$ the variety of all abelian K-superalgebras then $\mathfrak{N}_k \mathfrak{A}$ is defined by the non-graded identity

$$[[x_1, y_1], \ldots, [x_{k+1}, y_{k+1}]] = 0, \tag{32}$$

while $\mathfrak{A}\mathfrak{N}_m$ can be defined by

$$[[x_1, \ldots, x_{m+1}], [y_1, \ldots, y_{m+1}]] = 0. \tag{33}$$

The intersection $\mathfrak{B}_1 \cap \mathfrak{B}_2$ of two varieties consists of all algebras which belong both to $\mathfrak{B}_1$ and $\mathfrak{B}_2$. Hence the variety of Lie superalgebras $\mathfrak{N}_k \mathfrak{A} \cap \mathfrak{A}\mathfrak{N}_m$ consists of all algebras satisfying the identities (32) and (33).

We show that $\mathfrak{B} = \mathfrak{N}_k \mathfrak{A} \cap \mathfrak{A}\mathfrak{N}_m$ is a locally representable variety of Lie superalgebras over any infinite field K.

Obviously, by (32) $\mathfrak{B}$ satisfies (31). Now (33) yields

$$[[y^{(m)}, x], [y^{(m)}, z]] = 0 \qquad (34)$$

for any x, y, z. Reformulating (34) by means of the Jacobi identity one obtains an identity of form (30). Restricting the parities of x, y, z as in Theorem 4.17, one may apply this result. Thus $\mathfrak{B} = \mathfrak{N}_k \mathfrak{A} \cap \mathfrak{A}\mathfrak{N}_m$ is locally representable provided that char $K = 0$. If char $K > 2$, this variety is locally representable by Theorem 4.11, since it is locally soluble.

In Theorem 4.17 we require the validity of an identity of the form (31) only for finitely generated Lie superalgebras. A natural question is: does an identity of the form (31) hold in every Lie superalgebra in a locally representable variety $\mathfrak{B}$ if char $K = 0$? The following example shows that the answer is negative.

Let $X = \{x_1, x_2, \ldots\}$ be an infinite set of variables and let $\Lambda = \Lambda(X)$ be the Grassmann algebra generated by X. Recall that the multiplication in the associative algebra Λ is subject to the relations: $x_i x_j = -x_j x_i$, $i, j \geq 1$. Hence Λ is the linear span of all products

$$x_{j_1} \ldots x_{j_k}, \qquad 1 \leq j_1 < \ldots < j_k. \qquad (35)$$

Now Λ may be graded by the group $\mathbb{Z}_2$, $\Lambda = \Lambda_0 \oplus \Lambda_1$, where Λ_0 is the linear span of all elements (35) with even k and Λ_1 is the same with k odd. One can make Λ into a Lie algebra $[\Lambda]$ by setting $[x, y] = xy - yx$. First prove that any product $h_1 \ldots h_m$ is zero in Λ if $h_1, \ldots, h_m$ lie in $[A, A]$ where A is some m-generated Lie subalgebra in $[\Lambda]$.

Let A be generated as a Lie algebra by some elements $y_1, \ldots, y_m$. Then $y_i = a_i + b_i$ where $a_i \in \Lambda_0$, $b_i \in \Lambda_1$, $i = 1, \ldots, m$. Clearly, $[a_i, b_j] = [b_i, b_j] = 0$, hence, $[A, A]$ is the linear span of the commutators $[a_i, a_j] = 2a_i a_j$, $1 \leq i < j \leq m$.

The product $a_{i_1} \ldots a_{i_k}$ is nonzero only if all indices $i_1, \ldots, i_k$ are different. It follows that $h_1 \ldots h_m = 0$ provided that all h_i are in $[A, A]$.

Now we consider the Lie superalgebra $H = H_0 \oplus H_1$ where $H_0 = [\Lambda]$ and H_1 (as a linear span) coincides with the Grassmann algebra generated by an infinite set of elements $\{y_1, y_2, \ldots\}$. Any commutator $[a, b]$ with arbitrary a, b in H_1 is zero. The action of H_0 on H_1 is given by

$$[x_{i_1} \ldots x_{i_m}, y_{j_1} \ldots y_{j_k}] = y_{i_1} \ldots y_{i_m} y_{j_1} \ldots y_{j_k}.$$

§4. Representability of Lie superalgebras by matrices

In other words, H_1 is isomorphic to the regular Λ-module. Obviously, an identity of the form (31) does not hold in H. For example,

$$[[x_1, x_2], \ldots, [x_{2m-1}, x_{2m}], y_{2m+1}] = 2^m y_1 \ldots y_{2m+1} \neq 0.$$

Nevertheless, we shall prove that the variety generated by H is locally representable.

It is well-known that for any algebra A (not necessary a Lie superalgebra) and an arbitrary set J of indices the free algebra of rank $|J|$ in the variety var A can be embedded in the Cartesian power A^{A^J}. For the reader's convenience we recall the construction of this embedding. If we set $I = A^J$ then A^I consists of all functions from I into A, i.e. $A^I = \{\varphi | \varphi: I \to A\}$ and $I = \{f | f: J \to A\}$. Consider the family of elements φ_j in A^I, $j \in J$, given by $\varphi_j(f) = f(j)$ for any $f \in A^J$. Then the subalgebra in A^I generated by all φ_j, $j \in J$, is a free algebra in var A. To verify this statement it is sufficient to prove that any relation among the φ_j is an identity in A.

Let $w(\varphi_{j_1}, \ldots, \varphi_{j_n}) = 0$ and $a_1, \ldots, a_n$ be arbitrary elements in A. In A^J there exists a map f such that $f(j_1) = a_1, \ldots, f(j_n) = a_n$. Therefore

$$0 = w(\varphi_{j_1}, \ldots, \varphi_{j_n})(f) = w(\varphi_{j_1}(f), \ldots, \varphi_{j_n}(f)) = w(a_1, \ldots, a_n),$$

as claimed.

Now denote by $\mathfrak{B}$ the variety of Lie superalgebras generated by H. Let $L = F_m(\mathfrak{B})$ be the free algebra in $\mathfrak{B}$ of rank m. Consider the Cartesian power $I = H^m$ of H. If we denote by H_i, for any i in I, an isomorphic copy of H then L can be embedded in the Cartesian product $\prod_{i \in I} H_i = T$. If φ_i is the projection of T on H_i then, clearly, L can be embedded in the Cartesian product $\prod_{i \in I} \varphi_i(L)$. For some fixed i in I put $B = \varphi_i(L)$. Since $\varphi_i(L) \subset H_i \cong H$, we may consider $B = B_0 \oplus B_1$ as an m-generated subalgebra in $H = H_0 \oplus H_1$. If $B = \text{alg}\{a_1 + b_1, \ldots, a_m + b_m\}$, $a_j \in H_0$, $b_j \in H_1$, $j = 1, \ldots, m$, then $B_0 \subset A = \text{alg}\{a_1, \ldots, a_m\}$. Since A is an m-generated Lie subalgebra in the Grassmann algebra $[\Lambda]$ and H_1 is a regular Λ-module, we have $[h_1, \ldots, h_m, v] = 0$ in H for any $h_1, \ldots, h_m \in [A, A]$, $v \in L_1$, as it has been shown earlier. The derived algebra of B_0 lies in the derived algebra of A. Hence B satisfies the identity

$$[[x_1, y_1], \ldots, [x_m, y_m], z] = 0,$$
$$d(x_j) = d(y_j) = 0, \qquad j = 1, \ldots, m.$$
(36)

Now the Lie superalgebra $L = F_m(\mathfrak{B})$ can be embedded in the Cartesian product of the $\varphi_i(L)$, $i \in I$, and the identity (36) holds in every $\varphi_i(L)$. It follows that (36) is an identity in L.

On the other hand, the Lie superalgebra H lies in the variety $\mathfrak{AN}_2$ since H_1 is an abelian ideal in H and H_0 is a nilpotent Lie algebra with $H_0^3 = 0$.

At the beginning of this subsection it has been shown that, for any k, $\mathfrak{AN}_k$ satisfies a family of identities of the form (30). Hence, $L = F_m(\mathfrak{B})$ is a Lie superalgebra satisfying condition c) in Theorem 4.17 and therefore $\mathfrak{B}$ is a locally representable variety of Lie superalgebras.

Comments to Chapter 6

One of the first papers concerning the representability of infinite-dimensional algebras by matrices is due to A.I. Mal'cev [Mal'cev, 1943] where the author considers associative algebras. Later the results of A.I. Mal'cev were generalized considerably by K.I. Beidar [Beidar, 1986], and we follow his paper in the proof of Theorem 4.1. The connection between representability and residual finiteness for associative algebras has been noticed by A.I. Mal'cev [Mal'cev, 1943]. For a wider class of algebras the same results have been proved in [Pchelintsev, 1989].

Local properties of varieties of algebras such as representability, residual finiteness, and the Noetherian property, were initially investigated in the associative cases only. Here the most complete exposition of these results can be found in [Anan'in, 1977].

A first article about local residual finiteness of Lie algebras is [Bahturin, 1972]. A complete classification of locally representable varieties of Lie algebras and of varieties with other local properties (residual finiteness, Noetherian and Hopf properties) was given by M.V. Zaicev in the case where the ground field is infinite [Zaicev, 1988a, b, 1989a, b]. Locally residually finite varieties of Lie superalgebras of characteristic zero are described in [Bahturin, Zaicev, 1991].

In the case of Lie algebras as well as in the case of colour Lie superalgebras the identity

$$[x, y^{(n)}, z] = \sum_{j=1}^{n} \alpha_j [y^{(j)}, x, y^{(n-j)}, z]$$

is of great importance for the study of finiteness conditions. In the Lie algebra setting this identity has appeared in several papers. For instance, Lemmas 2.10 and 2.11 have been proved by I.B. Volichenko in [Volichenko, 1980]. The method of separating variables for this identity has been worked out by S.P. Mishchenko in [Mishchenko, 1982].

Note that the residual properties of a Lie algebra L and its universal enveloping algebra $U(L)$ are closely connected. The algebras L and $U(L)$ are residually finite simultaneously [Michaelis, 1987, 1986]. Similar results for colour Lie (p-)superalgebras have been proved in [Mikhalev, 1991a].

Bibliography

Abe, E.
1980   Hopf Algebras. Cambridge University Press, 1980.
Agalakov, S.A.
1984   Notes on finite approximation of algebras with respect to entry. Omsk Univ., 1984.
1989   On the approximation of Lie algebras with one defining relation. Int. Conf. in Algebras. Abstracts in the theory of rings, algebras and modules. Novosibirsk, 1989, p. 5.
Anan'in, A.Z.
1977   Locally finitely approximable and locally representable varieties of algebras. Algebra i Logika 16(1977), 3–23 = Algebra and Logic 16(1977), 1–16.
Atiyah, M., Macdonald, I.
1969   Introduction to Commutative Algebra. Addison-Wesley, Reading, Mass., 1969.

Bahturin, Yu. A.
1972   Approximations of Lie algebras. Mat. Zametki 12(1972), 713–716 = Math. Notes 12(1972), 868–870.
1974   Identities in the universal envelopes of Lie algebras. J. Austral. Math. Soc. 18(1974), 10–21.
1978   Lectures on Lie algebras. Akademie-Verlag, Berlin, 1978.
1979   A remark on irreducible representations of Lie algebras. J. Austral. Math. Soc. 27(1979), 332–336.
1985   Identities in the universal enveloping algebra for a Lie superalgebra. Mat. Sb. 127(1985), 384–397 = Math. USSR-Sb. 55 (1986), 383–396.
1987a   Identical Relations in Lie Algebras. VNU Science Press, Utrecht, 1987.
1987b   On degrees of irreducible representations of Lie superalgebras. C.R. Math. Acad. Sci. Canada 10(1987), 19–24.
1989a   Lie superalgebras with bounded degrees of irreducible representations. Mat. Sb. 180(1989), 195–206 = Math. USSR-Sb. 66(1990), 199–209.
1989b   A note on degrees of irreducible representations of Lie superalgebras. Algebra. A collection of papers to the 80-th anniversay of A.G. Kurosh, Moscow, 1989.

Bahturin, Yu. A., Drensky, V.S.
1987   Identities of free soluble color Lie superalgebras. Algebra i Logika. 26(1987), 403–418 = Algebra and Logic 26(1987), 229–240.

Bahturin, Yu. A., Kostrikin, A.I.
1986   Second All-Union research institute "Lie algebras and their applications in mathematics and physics", dedicated to 75-th anniversary of A.I. Mal'cev. Uspekhi Mat. Nauk 41 (1986), 231–244.

Bahturin, Yu. A., Ol'shanskiĭ, A. Yu.
1988   Identical Relations. Itogi Nauki i Tekhniki. Modern Problems in Mathematics. Fundamental Branches 18(1988). 117–243.

Bahturin, Yu. A., Slin'ko, A.M., Shestakov, I.P.
1981   Nonassociative rings. VINITI, Itogi Nauki i Tekhniki, Seriya, Algebra, Topologiya, Geometriya 18(1981), 3–72 = J. Soviet Math. 18(1982), 169–211.

Bahturin, Yu. A., Zaicev, M.V.
1991   Residual finiteness of colour Lie superalgebras. Trans. Amer. Math. Soc. (to appear).

Beidar, K.I.
1986   On theorems of A.I. Mal'tsev concerning matrix representations of algebras. Uspekhi Mat. Nauk 41(1986), 161–162 = Russian Math. Surveys 41:5(1986), 127–128.

Behr, E.J.
1987   Enveloping algebras of Lie superalgebras. Pacif. J. Math. 130(1987), 9–25.

Berezin, F.A.
1983   Introduction to Algebra and Analysis with Anticommuting Variables. Izdat. MGU, Moscow, 1983.

Bergen, J., Montgomery, S., Passmann, D.S.
1987   Radicals of crossed products of enveloping algebras. Israel J. Math. 59(1987), 167–184.

Bergen, J., Passman, D.S.
1990   Delta methods in enveloping rings. J. Algebra 133(1990), 277–312.

Bergman, G.M.
1978   The diamond lemma for ring theory. Adv. Math. 29(1978). 178–218.

Berkson, A.
1964   The u-algebra of a restricted Lie algebra is Frobenius. Proc. Amer. Math. Soc. 15(1964), 14–15.

Bokut', L.A.
1976   Embeddings into simple associative algebras. Algebra i Logika 15(1976), 117–142 = Algebra and Logic 15(1976), 73–90.

Bokut', L.A., Kukin, G.P.
1987   Unsolvable algorithmic problems for semigroups, groups and

rings. VINITI. Itogi Nauki i Tekhniki, Seriya Algebra, Topologiya, Geometriya 25(1987), 3–66 = J. Soviet Math. 45(1989), 871–911.

Bokut', L.A., L'vov, I.V., Kharchenko, V.K.
1988 Noncommutative rings. VINITI. Itogi Nauki i Tekhniki. Modern Problems in Mathematics. Fundamental Branches 18(1988), 5–116.

Bourbaki, N.
1961 Algèbre commutative. Hermann, 1961
1972 Groupes et algèbres de Lie. Algèbres de Lie Libres. Hermann, 1972.

Cohn, P.M.
1951 Integral Modules, Lie Rings and Free Groups. Ph.D. Thesis, Cambridge University, 1951.
1964 Subalgebras of free associative algebras. Proc. London Math. Soc. 14(1964), 618–532.
1989 Algebra, vol. 2. Second edition, Wiley, 1989.

Computer Algebra
1983 Computer Algebra. Symbolic and Algebraic Computation. Second edition. Ed. by B. Buchberger, G.E. Collins and R. Loos in cooperation with R. Albrecht. Springer-Verlag, 1983.

Curtis, C., Reiner, I.
1962 Representation Theory of Finite Groups and Associative Algebras. Interscience, New York, 1962.

Dicks, W.
1972 On one-relator associative algebras. J. London Math. Soc. 5(1972), 249–252.

Dixmier, J.
1974 Algèbres enveloppantes. Gauthier-Villars, Paris, 1974.

Dnestrovskaya Tetrad'
1982 Dnestrovskaya Tetrad'. Unsolved Problems in Ring Theory, 3rd edition. Novosibirsk, 1982.

Doković, D.
1988 On some inner derivations of free Lie algebras over commutative rings. J. Algebra 119(1988), 233–245.

Farkas, D.R., Snider, R.L.
1974 Group algebras whose simple modules are injective. Trans. Amer. Math. Soc. 194(1974), 241–248.

Feldvoss, J.
1991 On the cohomology of restricted Lie algebras. Comm. Algebra 19(1991), 2865–2906.

Friedrichs, K.O.
1953   Mathematical aspects of the quantum theory of fields. Comm. Pure Appl. Math. 6(1953), 1–72.

Gerasimov, V.N.
1976   Distributive lattices of subspaces and the equality problem for algebras with a single relation. Algebra i Logika 15(1976), 384–435 = Algebra and Logic 15(1976), 238–274.

Golod, E.S.
1988   Standard bases and homology. Lecture Notes in Math. 1352 (1988), 88–95.

Hall, M.
1950   A basis for free Lie algebras and higher commutators in free groups. Proc. Amer. Math. Soc. 1(1950), 575–581.

Hall, P.
1933   A contribution to the theory of groups of a prime power order. Proc. London Math. Soc. 36(1933), 29–95.

Hannabus, K.C.
1987   An obstruction to generalizing superalgebras. J. Phys. A 20(1987), no. 17, L1135–L1138.

Hartley, B.
1977   Injective modules over group rings. Quart. J. Math. 28(1977), 1–29.

Hedges, M.C.
1987   The Freiheitssatz for graded algebras. J. London Math. Soc. 35(1987), 395–405.

Herstein, I.N.
1968   Noncommutative Rings. Carus Math. Monograph 15, MMA, 1968.

Higgins, P.J.
1954   Lie rings satisfying the Engel condition. Proc. Cambridge Phil. Soc. 50(1954), 8–15.

Hochschild, G.
1954   Representations of restricted Lie algebras of characteristic p. Proc. Amer. Math. Soc. 5(1954), 603–605.

Jacobson, N.
1943   The theory of rings. Amer. Math. Soc., New York, 1943.
1956   Structure of rings. Amer. Math. Soc. Coll. Publ. 37(1956).
1962   Lie Algebras. Wiley-Interscience, New York, 1962.
1975   *PI*-algebras. An Introduction. Lecture Notes in Math. 441(1975).

Kac, V.G.
1977a   Lie superalgebras. Adv. Math. 26(1977), 8–96.
1977b   A sketch of Lie superalgebra theory. Comm. Math. Phys. 53(1977), 31–64.
1977c   Classification of simple $\mathbb{Z}$-graded Lie superalgebras and simple Jordan superalgebras. Comm. Algebra 5(1977), 1375–1400.

Kantor, I.L.
1984   An analogue of E. Witt's formula for the dimensions of homogeneous components of free Lie superalgebras. Deposited VINITI, 17th April 1984, N 2384-84 Dep.

Kemer, A.R.
1984   Varieties and $\mathbb{Z}_2$-graded algebras. Izv. Akad. Nauk SSSR Ser. Mat. 48(1984), 1042–1059 = Math. USSR-Izv. 25(1985), 353–374.

Kostrikin, A.I.
1990   Around Burnside. Springer-Verlag, 1990

Kukin, G.P.
1970a   Primitive elements of free Lie algebras. Algebra i Logika 9(1970), 458–472 = Algebra and Logic 9(1970), 275–284.
1970b   On the Cartesian subalgebra of a free Lie sum of Lie algebras. Algebra i Logika 9(1970), 701–713 = Algebra and Logic 9(1970), 422–430.
1972a   Subalgebras of a free Lie sum of Lie algebras with amalgamated subalgebra. Algebra i Logika 11(1972), 59–86 = Algebra and Logic 11(1972), 33–50.
1972b   On subalgebras of free Lie p-algebras. Algebra i Logika 11(1972), 535–550 = Algebra and Logic 11(1972), 294–303.
1974   On free products of restricted Lie algebras. Mat. Sb. 95(1974), 53–83 = Math. USSR-Sb. 24(1974), 49–78.
1983   The equality problem and free products of Lie algebras and associative algebras. Sibirsk. Mat. Zh. 24(1983), 85–96 = Siberian Math. J. 24(1983), 221–231.

Kukin, G.P., Kryazhovskikh, G.V.
1989   On subrings of free rings. Sibirsk. Mat. Zh. 30(1989), 87–97 = Siberian Math. J. 30(1989), 903–914.

Kurosh, A.G.
1947   Nonassociative free algebras and free products of algebras. Mat. Sb. 20(1947), 239–262.

Latyshev, V.N.
1963   Two remarks on PI-algebras. Sibirsk. Mat. Zh. 4(1963), 1120–1121.
1988   Combinatorial Ring Theory. Standard bases, Izdat. MGU, Moscow, 1988.

Leites, D.A.
1983   Theory of Supermanifolds. Izdat. Karel. Branch AN SSSR, Petrozavodsk, 1983.
1984   Lie Superalgebras. VINITI. Itogi Nauki i Tekhniki. Modern Problems of Math. Recent Progress, 25(984), 3–50 = J. Soviet Math. 30(1985), 2481–2512.

Lemaire, J.M.
1974   Algèbres connexes et homologie des espaces de lacets. Lecture Notes in Math. 422(1974).

Lewin, J., Lewin, T.
1968   On ideals of free assocative algebras generated by a single element. J. Algebra 8(1968), 248–255.

Lichtman, A.I.
1989   PI-subrings and algebraic elements in enveloping algebras and their fields of fractions. J. Algebra 121(1989), 139–154.

Mackey, G.W.
1987   Some remarks on Lie superalgebras. Czechoslovak. J. Phys. B37 (1987), 373–386.

Magnus, W.
1930   Über discontinuierliche Gruppen mit einer definierenden Relation (Der Freiheitssatz). J. Reine Angew. Math. 163(1930), 141–165.
1935   Beziehungen zwischen Gruppen und Idealen in einem speziellen Ring. Math. Ann. 111(1935), 259–280.
1937   Über Beziehungen zwischen höheren Kommutatoren. J. Reine Angew. Math. 177(1937), 105–115.

Makar-Limanov, L.G.
1985   Algebraically closed skew fields. J. Algebra 93(1985), 117–135.

Mal'cev, A.I.
1943   On representations of infinite algebras. Mat. Sb. 13(1943), 263–285.

McConnell, J.
1982   The Nullstellensatz and Jacobson properties of rings of differential operators. J. London Math. Soc. 26(1982), 37–42.

McConnell, J., Robson, J.C.
1987   Noncommutative Noetherian Rings. Wiley & Sons, New York, 1987.

Michaelis, W.
1986   Properness of Lie algebras and enveloping algebras. II. Lie algebras and related topics. CMS Conf. Proc. 5(1986), 265–280.
1987   Properness of Lie algebras and enveloping algebras. I. Proc. Amer. Math. Soc. 101(1987), 17–23.

Michel, J.
1975 Bases des algèbres de Lie et serie de Hausdorff. Sémin. P. Dubreil Algèbres. Univ. Pierre et Marie Curie, 1973–74, 27(1975), 6/1–6/9.

Mikhalev, A.A.
1985 Subalgebras of free colored Lie superalgebras. Mat. Zametki 37(1985), 653–661 = Math. Notes 37 (1985), 356–360.
1986 Free colour Lie superalgebras. Dokl. Akad Nauk SSSR, 286(1986), 551–554 = Soviet Math. Dokl. 33(1986), 136–139.
1988 Subalgebras of free Lie p-superalgebras. Mat. Zametki 43(1988), 178–191 = Math. Notes 43(1988), 99–106.
1989 A composition lemma and the equality problem for colored Lie superalgebras. Vestnik Moskov. Univ. Ser. I Mat. Mekh. 44(1989), 88–91 = Moscow Univ. Math. Bull. 44(1989), 87–90.
The composition lemma for colour Lie superalgebras and for Lie p-superalgebras. Proc. Int. Algebraic Conf., Novosibirsk, 1989.
1990 Embedding of Lie superalgebras of countable rank in Lie superalgebras with two generators. Uspekhi Mat. Nauk 45(1990), 139–140 = Russian Math. Surveys 45(1990), 162–163.
1991a Ado-Iwasawa theorem, graded Hopf algebras and residual finiteness of colour Lie (p-)superalgebras and their universal enveloping algebras. Vestnik Moskov. Univ. Ser. 1. Mat. Mekh. 1991, No 5, 72–74.
1991b Free colour Lie super-rings. Izvestija Vysch. Uchebn. Zaved. Mat. 1991, No 10, 4–7.
1992a Images of inner derivations of free Lie algebras and Lie superalgebras. Vestnik Moskov. Univ. (to appear).
1992b On some properties of super Lie elements in the free associative algebra. Proc. Int. Algebraic Conf., Barnaul, 1991.

Mishchenko, S.P.
1982 Structure and Identities of Certain Soluble Lie algebras. Ph.D. Thesis, Moscow University, 1982.
1983 The Engel identity and its application. Mat. Sb. 121(1983), 423–430 = Math. USSR-Sb. 49(1984), 419–426.

Molev, A.I., Tsalenko, L.M.
1986 Representation of the symmetric group in the free Lie (super) algebra and in the space of harmonic polynomials. Funktsional. Anal. i Prilozhen. 20(1986), 76–77 = Functional Anal. Appl. 20(1986), 150–152.

Montgomery, S., Smith, S. Paul
1990 Skew derivations and $U_q(\text{sl}(2))$. Israel J. Math. 72(1990), 158–166.

Mosolova, M.V.
1981 Functions of non commuting operators that generate a graded Lie

algebra. Mat. Zametki 29(1981), 35–44 = Math. Notes 29(1981), 17–22.

Passman, D.S.
1972 Group rings satisfying a polynomial identity. J. Algebra 20(1972), 103–117.
1977 Algebraic Structure of Group Rings. Wiley-Interscience, 1977.
1990 Enveloping algebras satisfying a polynomial identity. J. Algebra 134 (1990), 469–490.

Pchelintsev, S.V.
1989 Locally Noetherian and locally representable varieties of alternative algebras. Sibirsk. Mat. Zh. 30(1989), 134–144 = Siberian Math. J. 30(1989), 104–112.

Petrogradsky, V.M.
1988a On restricted enveloping algebras for Lie p-algebras. Proc. 5-th Siberian Conf. on Varieties of Algebraic Systems, Barnaul, 1988, 52–55.
1988b Injective modules over restricted Lie algebras. Vestnik Moskov. Univ. Ser. I Mat. Mekh. 43(1988), 68–70 = Moscow Univ. Math. Bull. 43(1988), 66–69.
1991a Existence of identities in the restricted enveloping algebra. Mat. Zametki 49(1991), 84–93 = Math. Notes 49(1991), 60–66.
1991b Identities in the enveloping algebra for modular Lie superalgebras. J. Algebra 145(1992), 1–21.

Procesi, C., Small, L.
1968 Endomorphism rings of modules over PI-algebras. Math. Z. 106(1968), 178–180.

Razmyslov, Yu. P.
1989 Identities of Algebras and Their Representations. Nauka, Moscow, 1989.

Renault, G.
1971 Sur les anneaux de groupes. C.R. Acad. Sci. Paris, 273(1971) 84–87.

Rittenberg, V., Wyler, D.
1978a Sequences of $\mathbb{Z}_2 \oplus \mathbb{Z}_2$-graded Lie algebras and superalgebras. J. Math. Phys. 19(1978), 2193–2200.
1978b Generalized superalgebras. Nuclear Phys. B 139(1978), 89–202.

Scheunert, M.
1979a Generalized Lie algebras. J. Math. Phys. 20(1979), 712–720.
1979b The Theory of Lie Superalgebras. An Introduction. Lecture Notes in Math. 716(1979).

1982   Graded tensor calculus. Preprint Univ. Bonn, Physikalisches Inst. 1982.

Schützenberger, M.P.
1971   Sur les bases de Hall. Notes. Manuscr., 1971.

Seligman, G.
1967   Modular Lie Algebras. Ergeb. Math. Grenzgeb. 40, Berlin, Springer-Verlag, 1967.

Shirshov, A.I.
1953   Subalgebras of free Lie algebras. Mat. Sb. 33(1953), 441–452.
1956   Some embedding theorems for rings. Mat. Sb. 40(1956), 65–72.
1958   On free Lie rings. Mat. Sb. 45(1958), 113–122.
1962a   Some algorithmic problems about Lie algebras. Sibirsk. Mat. Zh. 3(1962), 292–296.
1962b   About a conjecture in the theory of Lie algebras. Sibirsk Mat. Zh. 3(1962), 297–301.
1962c   On bases of free Lie algebras. Algebra i Logika 1(1962), 14–19.
1984   Collected Works. Rings and Algebras. Nauka, Moscow, 1984.

Shtern, A.S.
1986   Free Lie superalgebras. Sibirsk. Mat. Zh. 27(1986), 170–174 = Siberian Math. J. 27(1986), 136–140.

Sweedler, M.E.
1969   Hopf Algebras. Benjamin, New York, 1969.

Ufnarovsky, V.A.
1990   Combinatorial and asymptotic methods in algebra. VINITI. Itogi Nauki i Tekhniki. Modern Problems in Mathematics. Fundamental Branches 57(1990), 5–177.

Umirbayev, U.U.
1990   On the approximation of free Lie algebras with respect to entry. Sci. Notes of Tartu Univ.-Tartu Riikl. Ül. Toimetised, 1990, No 878, 147–152.

Van Geel, J., van Oystaeyen, F.
1981   About graded fields. Proc. Kon. Nederl. Akad. Wetensch. Ser. A 84(1981), 273–286.

Vasiliev, M.A.
1985.   De Sitter supergravity with positive cosmological constant and generalized Lie superalgebras. Classical and Quantum Gravity 2(1985), 645–652.

Viennot, G.
1978   Algèbres de Lie libres et monoides libres. Lecture Notes in Math. 691(1978).

Villamayor, O.E.
1959   On weak dimensions of algebras. Pacif. J. Math. 9(1959), 941–951.

Volichenko, I.B.
1980   On the varieties of centre-by-metabelian Lie algebras. Preprint. Inst. Math. Akad. Nauk. Byelorussian SSR 19(1980).

Witt, E.
1937   Treue Darstellung Liescher Ringe. J. Reine Angew. Math. 177(1937), 152–160.
1956   Die Unterringe der freien Lieschen Ringe. Math. Z. 64(1956), 195–216.

Zaicev, M.V.
1988a   Locally residually finite varieties of Lie algebras. Mat. Zametki 44(1988), 352–361 = Math. Notes 44(1988), 674–680.
1988b   Residual finiteness and the Noetherian property of finitely generated Lie algebras. Mat. Sb. 136(1988), 500–509 = Math. USSR-Sb. 64(1989), 495–504.
1989a   Locally Hopf varieties of Lie algebras. Mat. Zametki 45(1989), 56–61 = Math. Notes 45(1989), 469–472.
1989b   Locally representable varieties of Lie algebras. Mat. Sb. 180(1989), 798–808 = Math. USSR-Sb. 67(1990), 249–259.

Zariski, O., Samuel, P.
1958   Commutative algebra, vol. 1. Princeton, N.J., 1958.
1960   Commutative algebra, vol. 2. Toronto-Sydney-New York, 1960.

Zhukov, A.I.
1950   Reduced systems of defining relations in nonassociative algebras. Mat. Sb. 27(1950), 267–280.

Author Index

Abe, E.   237
Ado, I.D.   37
Agalakov, S.A.   109, 237
Anan'in, A.Z.   236, 237
Artin, E.   212
Atiyah, M.   212

Baer, R.   161
Bahturin, Yu.A.   2, 3, 23, 25, 26, 32, 35, 38, 79, 108, 112, 122, 123, 144, 173, 178, 184, 185, 230, 236, 237, 238
Behr, E.J.   238
Beidar, K.I.   236, 238
Berezin, F.A.   13, 37, 238
Bergen, J.   145, 173, 238
Bergman, G.   82, 109, 238
Berkson, A.   174, 238
Birkhoff, G.   4, 13, 23, 27, 28, 70, 81, 85, 87, 139, 202, 217, 226
Bokut', L.A.   82, 108, 238
Bourbaki, N.   211, 212, 239
Burnside, W.   20, 21, 22

Cohn, P.M.   43, 53, 64, 78, 79, 239
Curtis, C.W.   174, 239

Dicks, W.   109, 239
Dixmier, J.   173, 239
Doković, D.Z.   53, 239
Drensky, V.S.   38, 238

Engel, F.   3, 13, 116, 184

Farkas, D.   174, 239
Feldvoss, J.   174, 239
Friedrichs, K.O.   81, 85, 240

Gerasimov, V.N.   109, 240
Golod, E.S.   240

Hall, M.   80, 240
Hall, P.   80, 240
Hannabus, K.C.   37, 240
Hartley, B.   174, 240
Hedges, M.C.   109, 240
Herstein, I.N.   112, 137, 240

Higgins, P.J.   181, 240
Hilbert, D.   90
Hochschild, G.   174, 240

Jacobson, N.   1, 21, 112, 116, 120, 129, 131, 136, 147, 155, 240

Kaplansky, I.   112, 117, 158
Kac, V.G.   11, 37, 241
Kantor, I.L.   80, 241
Kharchenko, V.K.   108, 113, 239
Kostrikin, A.I.   224, 230, 241
Kryazhovskikh, G.V.   241
Kukin, G.P.   80, 108, 145, 238, 241
Kurosh, A.G.   79, 241

Latyshev, V.N.   84, 108, 144, 241
Leites, D.A.   37, 242
Lemaire, J.M.   80, 242
Lewin, J.   109, 242
Lewin, T.   109, 242
Lichtman, A.I.   145, 242
Lie, S.   3
L'vov, I.V.   108, 239

Macdonald, I.D.   212, 237
Mackey, G.W.   242
Magnus, W.   80, 109, 242
Makar-Limanov, L.G.   109, 242
Mal'cev, A.I.   236, 242
McConnell, J.   90, 173, 242
Michaelis, W.   236, 242
Michel, J.   80, 243
Mikhalev, A.A.   37, 53, 65, 79, 80, 109, 236, 243
Mishchenko, S.P.   236, 243
Molev, A.I.   49, 80, 243
Montgomery, S.   173, 238, 243
Mosolova, L.V.   38, 243

Neumann, P.M.   123, 144

Ol'shanskiĭ, A.Yu.   238
van Oystaeyen, F.   38, 245

Passmann, D.S.   145, 173, 174, 238, 244
Paul, S.   243
Pchelintsev, S.V.   244
Petrogradsky, V.M.   144, 173, 174, 244
Poincaré, H.   4, 13, 27, 28, 70, 81, 85, 87, 139, 202, 217, 226
Posner, E.C.   137
Procesi, C.   113, 244

Razmyslov, Yu.P.   244
Rees, D.   212
Reiner, I.   174, 239
Renault, G.   174, 244
Rittenberg, V.   14, 244
Robson, J.C.   90, 242

Samuel, P.   211, 246
Scheunert, M.   11, 37, 244
Schreier, O.   53, 62, 69
Schur, I.   20, 21, 173
Schützenberger, M.P.   80, 245
Seligman, G.B.   130, 245
Shestakov, I.P.   108, 238
Shirshov, A.I.   vii, 53, 65, 80, 108, 245
Shtern, A.S.   80, 245
Slin'ko, A.M.   108, 238

Small, L.   113, 244
Smith, S.P.   243
Snider, R.L.   174, 239
Specht, W.   44, 51
Sweedler, M.E.   245

Tsalenko, L.M.   49, 80, 243

Ufnarovsky, V.A.   80, 84, 108, 245
Umirbayev, U.U.   109, 245

Van Geel, J.   38, 245
Vasiliev, M.A.   245
Viennot, G.   80, 245
Villamayor, O.E.   174, 245
Volichenko, I.B.   69, 236, 246

Wever, F.   44, 51
Witt, E.   vii, 4, 13, 27, 28, 43, 48, 70, 78, 80, 81, 85, 87, 139, 202, 217, 226, 246
Wyler, D.   14, 244

Zaicev, M.V.   236, 238, 246
Zariski, O.   211, 246
Zhukov, A.I.   246
Zorn, M.   205

Subject Index

Abelian wreath product 27
Adjoint representation 2
Algebra 1
Anticommutativity 1, 9
Antipode 92
as-regular monomial 69
Associated graded algebra 7
Associative algebra 1
— form 47
— word 39

Bilinear form ε 14
Burnside's Theorem 21

Cartesian product 12
Centralizer 20
Coalgebra 92
Colour Grassmann algebra 89
— Lie superalgebra 14
— — super-ring 78
— polynomial ring 16, 89
— tensor product 89
Commutator brackets 1
Composition 99, 100
— Lemma 101
Consequence 23
Co-product 91
Co-unit 91

Degree 5
Delta-set 124, 157
Derivation 2
Derived series 3, 11
Diagonal mapping 91
Diamond Lemma 82
Direct product 12

Elementary transformation 57, 72
— —, triangular 57, 72
Entry problem 68, 74
Equality problem 103
Equivalent identities 23
Even elements 8
— subspace 10

Filtration 7, 89
Free colour Lie superalgebra 39
— — — metabelian 27
— — — super-ring 78
— algebra 39
— product with amalgamated subalgebra 105
Freiheitssatz 104
Friedrichs Criterion 81, 93
Frobenius algebra 163

Generic Flatness Lemma 149
Graded algebra 4
— field 20
— homomorphism 5
— ideal 5
— irreducible module 152
— ring 4
— subalgebra 5
Grassmann algebra 6
— envelope 8, 17
Gröbner basis 84

Heisenberg superalgebra 112, 123, 178
Hilbert's Basis Theorem 90
Hilbert series 29
Homogeneous element 5
— identity 24
Homomorphism of L-modules 3
Hopf algebra 92
— property 175

Identical relation 23
Identity 23
— of representation 180
Inclusion ambiguity 82
Independent set 56, 70
Injective module 161
Invariant 3
— bilinear mapping 3
Irreducible element 81

Jacobi identity 1, 9

Subject Index

Leading term 49, 75
Left-normed n-fold commutator 2
Length 39, 49, 70
Lexicographical order 40
Lie algebra 1
— —, abelian 1
— —, metabelian 2, 3
— —, nilpotent 2, 3
— —, soluble 3
— superalgebra 9
— —, finitely separable 94
— —, nilpotent 11
— —, Noetherian 175
— —, residually finite 177
— —, residually in a family 201
— —, representable 177
— —, restricted 18
— —, soluble 11
— type algebra 36
Locally Wedderburn algebra 162
Lower central series 3, 11
l-homogeneous element 49, 70

Möbius function 43
Module 2
—, residually in a family 214
Monolith 205
Multidegree 39
Multihomogeneous element 49, 70
— identity 24
Multilinear identity 24
Mutual commutator 3

Nilpotent index 2
— space of transformations 181
Nil-radical 130
Non-associative monomial 39
Normal form 81
— identity 24

Odd element 8
Overlap ambiguity 82

Primitive element 92
Product variety 233
p^n-polynomial 118
p-special monomial 108
ps-regular monomial 69
p-superalgebra 18
PBW-Theorem 85, 87

Rank 23, 49, 70
Reduced set 57, 70

Reduction 81
— system 81
Regular monomial 40
— ring 161
— word 40
Representable algebra 210
— module 215
Representation 2
Residually finite module 209
Resolvable ambiguity 82
Restricted universal enveloping algebra 87
Right normed n-fold commutator 2

Schreier's Formula 62
Schur's Lemma 20
Self-injective algebra 162
Semidirect product 12
Specht-Wever Criterion 51
Special monomial 106
s-regular monomial 40
— word 40
Superalgebra 8
Superderivation 11
Supertrace 11

Triangulable module 3

universal enveloping algebra 3, 15, 85
— — —, restricted 87

Variety 23, 24
—, abelian 25
—, locally Hopf 199
—, — Noetherien 199
—, — representable 222
—, — residually finite 222
—, — soluble 199
—, metabelian 27
—, multinilpotent 33
—, nilpotent 25
—, soluble 25
—, Specht 34
Verbal ideal 23

Weak identity 220
Weight function 96
Witt's Formula 43, 48

Young diagram 26

θ-identity 122